Machines
Who
Think

Other books by the author:

Familiar Relations (novel)

Working to the End (novel)

The Fifth Generation (with Edward A. Feigenbaum)

The Universal Machine

The Rise of the Expert Company
(with Edward A. Feigenbaum and H. Penny Nii)

Aaron's Code

The Futures of Women (with Nancy Ramsey)

Machines Who Think

A Personal Inquiry into the History and Prospects of Artificial Intelligence

—

Pamela McCorduck

A K Peters, Ltd.
Natick, Massachusetts

Editorial, Sales, and Customer Service Office

A K Peters, Ltd.
63 South Avenue
Natick, MA 01760
www.akpeters.com

Library of Congress Cataloging-in-Publication Data

McCorduck, Pamela, 1940-
    Machines who think : a personal inquiry into the history and prospects of
    artificial intelligence / Pamela McCorduck.–2nd ed.
        p. cm.
    Includes bibliographical references and index.
    ISBN 1-56881-205-1
    1. Artificial intelligence–History.   I. Title.

Q335.M23 2003
006.3'09–dc21

                                                              2003051791

Printed in Canada
08 07 06 05 04   10 9 8 7 6 5 4 3 2 1

To W.J.M., whose energetic curiosity
was always a delight
and, at the last, a wonder.

"Some of the people are saying the Eight Sages took you away to teach you magic," said a little girl cousin. "They say they changed you into a bird, and you flew to them."

"Some say you went to the city and became a prostitute," another cousin giggled.

"You might tell them that I met some teachers who were willing to teach me science," I said.

The Woman Warrior: Memoirs of a Girlhood among Ghosts,
—Maxine Hong Kingston

"Everything should be made as simple as possible, but not simpler."

—Albert Einstein

Contents

Part I Beginnings

1 Brass for Brain 3

Surveys attempts before the twentieth century to create artificial intelligences, both literary and real. Argues that twentieth-century artificial intelligence is only the latest, though the most successful, instance of a long Western tradition, and shares much with its ancestors.

2 From Energy to Information 37

Delineates the early attempts of philosophers, and later psychologists, to define the mind, and the parallel attempts of mathematicians to turn human logic into rigorous mathematics. Shows that the dominant model for these thinkers was from physics, and that not until cybernetics introduced a new model—information in an open system, as opposed to energy in a closed system—could the computer successfully be used as a medium for intelligent behavior.

3 The Machinery of Wisdom 59

Focuses on the computer as a medium for intelligent behavior, with close attention to Alan Turing, a brilliant British logician who helped design one of the first computers and was responsible for much early work in artificial intelligence. Discusses the American pioneer John von Neumann, who held a view opposite to Turing's, that computers would never be able to "think."

4 Meat Machines 85

Traces the growing conviction that brains are a species of machine, the failed attempts to equate the on-off logic of the computer with the on-off logic of brain neurons. Reviews the early work of McCulloch and Pitts, as well as self-organizing systems, the Perceptron, and other attempts in the United States and Great Britain to link brains and machines.

Part II The Turning Point

5 The Dartmouth Conference 111

Reports on a conference that in 1956 gathered the threads of an assortment of projects with the same general underlying assumptions. Everyone was there, and the conference served as a rite of tribal identification that set patterns for future research and personal relations.

6 The Information-Processing Model 137

Covers the pioneering work of Newell, Shaw, and Simon, which set the tone for the next decade of research. The influence of these men on computing and cognitive psychology so pervades those fields that their discoveries and models are simply taken for granted.

7 Fun and Games 171

Describes how computers were taught to play chess and checkers, and why this development made IBM so nervous. Analyzes the scientific and social significance of computer games.

Part III Resistance

8 Us and Them 195

Discusses the critics of artificial intelligence, who give the lie to the idea that science is a disinterested enterprise. Includes some speculations as to why artificial intelligence seems to provoke people to extremes, and covers the first extremist, Mortimer Taube, author of *Computers and Common Sense*.

9 L'Affaire Dreyfus 211

Considers the bitterest critic of artificial intelligence, who examined the evidence and found it wanting. However, a computer walloped him at chess, and his demands continue to be met. (Or do they?) Dreyfus and his book *What Computers Can't Do* are introduced as fixtures in the artificial-intelligence community, and the man is presented as an example of the committed mind that finds only what it expects to find, even as scientists with the opposite commitment find what they expect to find. Relates prophecies, insults, and other scientific high jinks—a case study in the subjective side of science and philosophy.

Part IV Realizations

10 Robotics and General Intelligence 243

Investigates the robot and the difficulty of general, as opposed to special-purpose, intelligence. Presents two approaches, the General Problem Solver and the Advice Taker, and inspects some of the problems and solutions robot makers have encountered. Focuses on Shakey, the SRI robot, who rolled into the hearts and fears of millions—thanks to some scurrilous publicity.

Foreword

—

Machines Who Think has its own modest history that may be worth telling. In the early summer of 1974, John McCarthy made an emergency landing in his small plane in Alaska, at a place called (roughly translated) the Pass of Much Caribou Dung, so remote a spot he could not radio for help. Fortunately, John was rescued. It occurred to me then that before mortality claimed them, somebody ought to go around and ask these guys who'd begun this odd field called artificial intelligence, what they thought they were doing, and why.

Eventually, I appointed myself. Later that summer, John McCarthy, Ed Feigenbaum, and I had lunch at the Stanford Faculty Club, where I proposed my project. Ed, whom I'd approached first, was enthusiastic, but John was less so. It was much too early, he said. There I'd be, waking up in five years and slapping my forehead: If only I'd waited for so-and-so to come on the scene! I should think about doing something else instead (and John had a number of suggestions, most of them much too mathematical for my tastes). But when he saw I would not be deterred, he agreed to cooperate.

My husband, Joseph Traub, and I returned from that summer stay at Stanford to Carnegie Mellon in Pittsburgh, where he was then head of the computer science department, and where the faculty included Allen Newell, Herb Simon, and Raj Reddy. I made my proposal to each of them. They were encouraging, and all agreed to help—calling their colleagues elsewhere to introduce me, taking time out from extremely busy research lives to sit still for interviews, and in Herb Simon's case, opening his archives to me.

Allen Newell in particular showed his good faith by arranging for some funding for travel and transcription of the interview tapes I'd begun to make—money that long after publication I learned came from Ed Fredkin's foundation. I owe both Newell and Fredkin deep and continuing thanks, and if these public thanks are belated, that doesn't make them less heartfelt. The book would not have seen the light of day without them.

I'd certainly tried to find funding elsewhere. I approached a number of foundations and government agencies, and nearly everyone agreed that this might be an interesting project, but I, alas, was merely a writer and not a trained historian of science. Sorry, no money was available to somebody who'd only published a couple of novels. I explained that while I lacked credentials as a trained historian of science, I hoped only to tell a story—this was surely a science that had found its genesis in myth and legend, and might be something that would have an impact on everybody. Still no dice. One government agency, nameless out of courtesy, was explicit about its dedication to projects that married the humanities and the sciences, and so I presented myself to them: a member of the faculty of a university English department, ready to tackle a science I thought was significant to anybody in the humanities. That agency couldn't say no fast enough.

Amazing, when you think about it. Here was a new science, its founding fathers all active, and surely prime material for historians of science at least. As I gathered material for my own story, I expected to find those scholars in large numbers, seizing the opportunity to interview what might possibly be the Leibnizes, the Newtons, the Einsteins of a new field. That prospect was fine by me; I was aiming to tell a story to people who didn't normally read histories of science.

But this isn't how it turned out. Instead, I was alone. The scientists I interviewed were delighted to help, and I could harvest their patient answers to my questions when those responses were still fresh and spontaneous. Much, much later, when I asked a credentialed historian of science about this, she shook her head. "Nobody knew whether it would be important. Nobody wanted to take a chance."

If funders were reluctant, publishers were downright dismissive. As graduate students in Columbia's School of the Arts, we'd all memorized the depressing statistic that Joseph Heller's classic, *Catch-22*, had gone to 27 publishers before somebody took a chance on it. *Machines Who Think* was to go to something like 33. It was the mid-1970s, and publishers returned the proposal with comments such as: "We've already done a book about computers." Another favorite: "Very interesting. Too bad it's too late."

I, the author, lost heart. Doubts beset me. The rejections were painful, eventually too painful, to read. My husband offered to act as an intermediary, a shield—I let him. Gladly. He wouldn't let me quit. ("He never let me down," I say in the original introduction to this book; we still laugh that it's code for "He never let up on me.") Furthermore, though he was not a credentialed literary agent, he took matters into his own hands and soon found Peter Renz, then at W. H. Freeman, the book-publishing arm of *Scientific American*, who took a chance on the project and on me. For Joe Traub, who has always been my greatest champion, even I'm at a loss for sufficient words of thanks.

I knew when I finished *Machines Who Think* that for a while, I personally wanted to leave the general field of artificial intelligence to itself. Science moves in rhythms, in seasons, with periods of quiet, when knowledge is being assimilated, perhaps rearranged, possibly reassessed, and periods of great exuberance, when new knowledge cascades in. We can't always tell which is which. Technology changes, permitting the formerly infeasible, even unthinkable. From the perspective of a few decades, cries of triumph and cries of disappointment sound different from how they sounded at the time. Other histories of the field came out, some admirable, some not. I found other topics to write about.

Then, a few years ago, Herb Simon emailed me a copy of a note he'd written to Ed Feigenbaum "I am repeatedly frustrated by the kind of history of AI that continues to be transmitted through all kinds of channels, most of them thoroughly unreliable. I cite the [X] book as a case in point, but the history chapters or paragraphs in textbooks are only marginally better. For my money, *Machines Who Think* continues to be the most reliable source on the first couple of decades. What are the chances of getting it back into print ...?"

Soon, he emailed me directly: "Pamela: Do consider what might be done about bringing *Machines Who Think* back into print. More machines are thinking every day, and I would expect that every one of them would want to buy a copy. Soccer robots alone should account for a first printing."

Impossible to resist such flattery. So, 25 years later, I've gone back. The afterword that follows the original text of *Machines Who Think* is a summary, not an exhaustive history, of the field's work during this past quarter century, and it surely overlooks excellent accomplishments, particularly outside the United States. For this, I apologize in advance, and I hope some young writer with the energy and *Sitzfleisch* I had 25 years ago is making her way through the international field of artificial intelligence, tape recorder (or whatever) in hand, evoking from the next generation why it does what it's doing. My afterword is also much more personal, partly because I became a participant-observer in the sociology of artificial intelligence, and partly because—well, I'm not a trained historian of science, you know.

It's unlikely I'll write the history of the next 25 years in artificial intelligence research, but somebody—or even something—will. I look forward to reading that one.

Let me thank those who gave me advice, pointers, and other help this time around: first and foremost, my good friend Ed Feigenbaum; then, Duane Adams, Danny Bobrow, Bruce Buchanan, Bob Balzer, Ron Brachman, Harold Cohen, Bob Eisenstein, Jill Fineberg, Takeo Kanade, Sarah Keisler, and other members of the Nursebot Pearl research group; John Kender, Victor Lesser, Kathy McKeown, Tiger Merrin, Chuck Thorpe, and Manuela Veloso. None of these generous people is responsible in any way for my errors of commission, omission, or interpretation. How pleasant it would (and will) be to have an automated personal assistant to blame!

I owe thanks to my copy editor, Michelle Peters, who was indefatigable in dotting my I's and crossing my T's, and to Darren Wotherspoon, who cast my words into inviting typography. As this edition was in press, associate editor, Kathryn Wert struggled heroically but always with good humor to impose some neatness on my scruffy ways. Above all, I'm grateful to Alice and Klaus Peters of A K Peters, Ltd., for bringing a new edition of *Machines Who Think* into print, and nurturing the transformation of what was promised as a brief afterword into another near-book.

Pamela McCorduck
Santa Fe, New Mexico
November 2003

Preface

This book is a history of artificial intelligence, that audacious effort to duplicate in an artifact what we humans consider to be our most important, our identifying, property—our intelligence.

In two senses, this history is a personal inquiry.

First, I've not intended to be exhaustive. Instead, like the ant in Herbert Simon's *Sciences of the Artificial,* I've wandered over the landscape where I would, dwelling on what fascinated me and barely addressing what didn't. Thus for example, though considerable work in artificial intelligence has been done in theorem proving, you'll find little of it reported here. Neither my training nor my taste gives me appetite for this subject, and I rationalize such neglect by telling myself that theorems would only scare away the nonspecialist reader this book is intended for. Another example: there's a distinct American bias in my history, which may have been appropriate in earlier days, but is no longer. Vigorous research groups now exist in Europe, the Soviet Union, and Japan, and the British effort echoes throughout the history of artificial intelligence, sometimes merrily, sometimes sadly, but always a presence.

On the other hand, my training and my taste have made me endlessly fascinated by the human personalities who figure in this history, and so they've received as much attention as their work. The history is personal in this sense too.

Now, this tendency toward personalities isn't solely a matter of partiality. When young James Boswell, that biographer to whom English letters owes so much, asked his hero, Dr. Samuel Johnson, how he came to create a monumental dictionary, Johnson put the young man off, saying that "it was not the effect of particular study; it had grown up in his mind insensibly."

But Boswell, nobody's fool, wasn't satisfied with that. An examination of the text of the dictionary would tell you some important things. For example, you could see that Johnson was a city man, who didn't really know much about the sea or the countryside. He mixed up meanings of windward and leeward, and when a lady later asked him how it was possible he'd misdefined pastern as the knee of a horse, Boswell reports that "instead of making an elaborate defense, as she expected, Dr. Johnson at once answered 'Ignorance, madam, pure ignorance.'"

No, Boswell said, there's more to a history than its texts. "There is," he wrote, "perhaps in every thing of any consequence, a secret history which it would be amusing to know, could we have it authentically communicated."

And so with artificial intelligence. Its texts tell us many things of consequence, and some references to those are in this book. From the texts we see that artificial intelligence in one form or another is an idea that has pervaded Western intellectual history, a dream in urgent need of being realized. Work toward that end has been a splendid effort, the variety of its forms as wondrous as anything humans have conceived; its practitioners as lively a group of poets, dreamers, holy men, rascals, and assorted eccentrics as one could hope to find—not a dullard among them. Its visionaries have lifted our spirits and made us transcend our own species, its poets have told us things about ourselves we never suspected, and its fast talkers have set everybody's teeth on edge.

But much of this book is also the secret history, amusing to know when it is authentically communicated. The human beings who were present when this art was transformed into a science—indeed, those who caused that transformation—speak for themselves in these pages, communicating not only how it came about, but their personal hopes and dreams as well. And they speak not only in the interests of authentic history, but for still another reason.

That reason is one of several that made me write this book. It's the desire to show that science is above all a human endeavor. Science is peopled by humans, and not by a solitary, abstract Truth (though there are truths in it). Here I address my fellow humanists more than anyone else, because so many of them are convinced that science is somehow alien to the humanities. I want them to see that it is not. I want them to see a science whose genesis was in literary texts they cherish. I want them to see a science whose thinkers are humans, with human motives and goals. I want them to see a science on its way to a place where, if it arrives fully realized, it will have deep, deep consequences for the human species. Whether those consequences are good or bad for us is still an open question, and this book is also an invitation for anybody with an interest in the future of the human race to participate in the inquiry.

Busy readers can read the first and last chapters for the main points of this argument. The rest is commentary.

Two organizing principles govern the book, time and topic. Luckily for both writer and reader, topics or areas in artificial intelligence have often been taken up and developed sequentially. Some things were easy to do and got done first, other things came harder, and more things resist being done even yet. An advance in one area often revitalized another area that was languishing. Critics appeared at appropriate stages. Those who said a thing could never be done were later replaced by those who had to concede that it could, but then said it ought not to be.

I am unabashedly present, believing, as Sartre once said, that commitment, *engagement*, is simply the writer's total presence in what has been written. Readers will see that I write with optimism. This attitude may not be so much a matter of reasoned judgment as natural inclination.

These biases and lapses laid bare, I feel better about declaring my main reason for writing this book—I too share that enduring human fascination with intelligent artifacts. What they have already told us and will continue to say about human nature is deeply thrilling. It is nearly as thrilling as the promise they make of opening the universe to us in a new way, bringing us face to face with intelligences besides—even beyond—our own. At the very least, I expect that to be an exhilarating encounter.

Many busy people allowed me time to interview them for their recollections. They are Paul Armer, Alex Bernstein, W. W. Bledsoe, Ted Brain, Bruce Buchanan, Max Clowes, L. Stephen Coles, Hubert Dreyfus, Edward Feigenbaum, Julian Feldman, Edward Fredkin, Herbert Gelernter, I. J. Good, Leon Harmon, John McCarthy, Donald MacKay, Donald Michie, Marvin Minsky, Joel Moses, Allen Newell, Nils Nilsson, Seymour Papert, Bertram Raphael, Raj Reddy, Nathaniel Rochester, Charles Rosen, Arthur Samuel, Oliver Selfridge, Claude Shannon, J. C. Shaw, Herbert A. Simon, James Slagle, Ray Solomonoff, Fred Tonge, Joseph Weizenbaum, and Lotfi Zadeh. These interviews were all taped, and transcripts have been patiently produced by Mercedes Bogi. The transcripts, along with the tapes and other supporting documents, will be stored in appropriate archives and made available to other scholars.

Of those interviewed, special thanks must go to Edward Feigenbaum, John McCarthy, Marvin Minsky, Allen Newell, and Herbert Simon. Each not only sat cheerfully through multiple interviews, but also led me through the maze of an unfamiliar field when I was very uncertain, answered the questions I hadn't the wit to ask, and gave me patient bibliographic guidance. Newell and Simon allowed me access to their scientific notebooks and were ever generous with their personal libraries. J. C. Shaw gave me invaluable documents on a variety of topics that will delight future historians. Many others dug into their own files for reprints, notes, and photos, and I'm deeply grateful to each of them. I owe thanks to Ian Mitroff for many lively discussions on the sociology of science.

As the project moved along, I incurred other debts, which these acknowledgments hardly discharge. Among my creditors in this sense are Raj Reddy, Ed Fredkin, Allen Newell, and Ed Feigenbaum. Indeed, to the latter two my debt moves into that ill-defined area of friendship and moral support that is at least as essential as any travel or secretarial funds, though those were generously provided by grants from Carnegie Mellon University and Massachusetts Institute of Technology.

In the last stages, Marvin Minsky, Allen Newell, and Herbert Simon each read the entire manuscript and made detailed comments. If I've been willful enough to resist some of those suggestions, they cannot be held responsible. They did their best to keep me honest.

My final debt is to my dearest friend and companion, my husband, J. F. Traub. He never let me down.

Pamela McCorduck
June 1979

Time Line:
The Mechanization of Thinking

This time line lays out the sequence of efforts throughout Western history to mechanize thinking, beginning with the earliest mythological and literary examples, followed by philosophical tracts, mathematical formulations, automata and other kinds of devices, most importantly the digital computer, that have been proposed as ways to automate thought. A second time line, called The Evolution of Intelligence (p. 523) places these efforts into a larger intellectual context, with a new point of view that emerges thanks to recent research. This new point of view shows that while individual reasoning is fundamental to intelligent behavior, so too is collaborative thinking, and access to collections of static and dynamic knowledge.

Before the Common Era

Sixth century | Though composed earlier in oral form, Homer's poem *The Iliad* is codified, introducing into written literature assorted automata from the workshops of the Greek god Hephaestos

Sixth – fifth centuries | Hebrew Torah is canonized, including the second of the Ten Commandments, the prohibition against making graven images

Fifth century | Aristotle lays out the epistemological basis in the West for the division of knowledge into categories, with theory the most important, art least important; he also introduces syllogistic logic, the first formal deductive reasoning system

Common Era

Late First century   Heron of Alexander builds fabled automata and other mechanical marvels

Fifteenth century   Pope Sylvester II, Bishop Grosseteste, Roger Bacon and Albertus Magnus are said to have "brazen heads," simultaneously sources and proof of their owners' wisdom; Ramon Llull, Catalonian mystic and theologian, invents his "Ars Magna," a machine for discerning truth by "bringing reason to bear on all things," and based on the Arabic *zairja* he had seen

Fifteenth–sixteenth century   Mechanical clocks, the first modern measuring machines, appear in European towns; Paracelsus provides the recipe for a homunculus, an intelligent "little man"

1580   "The Golem," said to be created by Rabbi Judah ben Loew in Prague

Seventeenth century   Automata appear on European clocks or are produced to work alone as amusements for the rich

1642   Blaise Pascal invents a mechanical calculator, the Pascaline

1664   *Treatise on Man*, by René Descartes, is published posthumously and codifies the mind/body problem

1673   Gottfried Wilhem Leibniz invents the Step Reckoner, an improved mechanical calculator, and envisions a universal calculus of reasoning to decide arguments mechanically

Eighteenth century   Philosophers (Leibniz, Spinoza, Hobbes, Locke, Kant and Hume) and scientists (La Mettrie, Hartley) try to formulate laws of thought

1738	Jacques de Vaucanson presents his mechanical duck to the European public
Nineteenth century	Literary artificial intelligences proliferate, such as Hoffman's *The Sandman*, Goethe's *Faust* (part II), and Mary Shelley's *Frankenstein*; the beginning of empirical psychology
1822	Charles Babbage begins but never finishes the Difference Engine
1843	Ada, Countess Lovelace, publishes her account of Charles Babbage's Analytical Engine
1854	George Boole publishes *An Investigation of the Laws of Thought*; von Kempelen's fraudulent chess-playing machine perishes in a fire
1890	Herman Hollerith conducts the US census using machines that encode information on punch cards
1914	A. Torres y Quevedo builds electromechanical machines for chess endgames
1923	"Robot" introduced into English in a London production of Karel Capek's play, *R.U.R.*, *Rossum's Universal Robots*
1937	Alan Turing proposes an abstract universal computing machine
1938	In developing his Z1 computer in Berlin, Konrad Zuse realizes the technology will eventually become an artificial brain
1941	Automatic decryption of German intelligence messages undertaken by Turing and others at Bletchley Park, England

1943   McCulloch and Pitts publish "A Logical Calculus of the Ideas Immanent in Nervous Activity;" Rosenblueth, Wiener, and Bigelow publish "Behavior, Purpose and Teleology," introducing the term cybernetics

1944   ENIAC (Electronic Numerator, Integrator and Computer) developed by Eckert and Mauchly, comes online at the University of Pennsylvania

1945   Turing writes a pioneering, but unpublished paper, "Intelligent Machinery"

1947   Norbert Wiener publishes *Cybernetics*; Grey Walter builds his electromechanical "turtle"

1949   Mark I, the first stored-program computer, comes online at Manchester University; Turing and his colleagues attempt to program it to play chess

1950   Turing publishes "Computing Machinery and Intelligence," proposing the Turing Test; Isaac Asimov offers his "Three Rules of Robotics" in *I, Robot;* Shannon publishes a detailed analysis of chess-playing as search

1951   IAS machine, proposed by John von Neumann in 1945, comes online at the Institute for Advanced Study, Princeton

1952   Arthur Samuel begins work on a checkers-playing machine that learns, and eventually competes with human champions

1956   Dartmouth Conference, where John McCarthy proposes the term "artificial intelligence" and Newell, Shaw, and Simon demonstrate the first working AI program, the Logic Theorist

1957	Newell, Shaw and Simon demonstrate the General Problem Solver; McCarthy proposes the Advice-Taker; US President Dwight Eisenhower approves funding for the Defense Department's Advanced Research Projects Agency (ARPA)
1957	McCarthy invents LISP; H. Gelernter and N. Rochester produce a geometry theorem prover with a semantic component
1959	Jack Kilby and Robert Noyce independently apply for US patents for an integrated circuit, which leads to the technological improvements and increasing economies described by Moore's Law
1960s	John Kemeny develops the first time-shared system at Dartmouth College where all undergraduates are required to be "computer literate"
1961	T. Evans's ANALOGY program solves the same analogy problems that appear on IQ tests
1962	J. C. R. Licklider envisions in a series of memos what will eventually become the Internet, a worldwide medium for collaboration, information dissemination, broadcasting, and interaction between individuals, regardless of geographic location. DARPA begins and sustains its support, and a number of scientists, especially L. Kleinrock, L. Roberts, P. Baran, R. Kahn and V. Cerf, make crucial breakthroughs in the next few years. J. Slagle's SAINT program solves calculus problems at the college freshman level
1964	D. Bobrow's STUDENT program understands natural language well enough to solve algebra word problems

1965   Lotfi Zadeh invents fuzzy logic; Ted Nelson begins but never completes his Xanadu hypertext system; publishes his first papers about hypertext; D. Engelbart develops the computer mouse as a way of implementing his NLS (oN Line System) hypertext and collaborative workspace; J. Weizenbaum's ELIZA, interactively mimics a psychotherapist

1966   R. Quillian's PhD dissertation demonstrates the power of semantic nets

1967   *A scientific turning point in AI, where knowledge is seen to be as important as reasoning in intelligent behavior.* DENDRAL, the first successful knowledge-based program for scientific reasoning; MACSYMA, the first successful knowledge-based program in mathematics; MacHack, a knowledge-based chess-playing program, achieves a class C rating in tournament play; first version of LOGO, an interactive learning environment, appears

1969   ARPANET, the precursor to the Internet, is established; Shakey, a mobile "intelligent" robot, roams SRI's halls

1971   H. Cohen first demonstrates AARON, an autonomous art-making program

1974   First "expert system," T. Shortliffe's MYCIN program demonstrates the power of rule-based systems for knowledge representation and inference in medical diagnosis and therapy

1975   Minsky proposes frames as a representation to integrate different sources of knowledge; MetaDendral produces the first scientific discoveries by a computer to be published in a refereed journal

Late 1970s | Stanford's SUMEX-AIM Lab demonstrates the power of the ARPANET for scientific collaboration

1978 | Simon wins the Nobel Prize in Economics for his theory of bounded rationality, a keystone of AI (and human behavior) known as "satisficing;" Moravec's cart is the first computer-controlled autonomous vehicle

1981 | Commercialization of AI begins; Japanese announce the Fifth Generation project with significant AI goals

1982 | Newell et al., create SOAR, an architecture for general intelligence; US embarks on the Strategic Computing Project to achieve AI goals

1985 | R. Brooks demonstrates "Allen," the first of his autonomous reactive robots, to be followed by an explosion of this species

1987 | Minsky publishes *Society of Mind*

1988 | Berners-Lee begins work on the World Wide Web at CERN in Geneva

Late 1980s | The AI Winter

Early 1990s | Another turning point in AI: intelligent behavior is recognized to be collaborative as well as single-agent

1997 | Deep Blue defeats Garry Kasparov, world's chess champion, ending the single-agent, single-task model of intelligence as a significant AI goal; first official Robo-Cup soccer match, the new paradigm

2000 | Robot pets, smart toys, become commercially available; C. Breazeal creates Kismet, a robot that exhibits emotions

2001   Berners-Lee et al., begin work on the Semantic Web, an international effort to bring about the global exchange of commercial, scientific and cultural data on the World Wide Web, using AI techniques of logic, inference, and action

2003   DARPA initiates three major AI projects: the "LifeLog," new reasoning cognitive systems, and new real-world reasoning systems; his and hers multifunction robots offered in the 2003 Neiman Marcus Christmas catalog for $400,000 (by coincidence, the same sum that John von Neumann requested in 1945 to build his IAS machine)

Part One

Beginnings

One age cannot be completly understood if all the
others are not understood. The song of history
can only be sung as a whole.
— *José Ortega y Gasset*

You can close your eyes to reality
but not to memories.
— *Stanislaus Lec*

Chapter One

⌒

Brass for Brain

"Can a machine think?" This question is in a class with those snappy vaudeville comebacks: does a chicken have lips? And like them, it ought to end the discussion at once by its self-evident nonsense. After all, we agree, our one essential, identifying property is thinking. Don't we call ourselves *homo sapiens*; declare that we think, therefore we are; and consider ourselves lifted above the rest of the earthly beasts by our capacity for symbol making? And if we're lifted above the beasts, no use even to talk about machines. Can a machine think? Does a chicken have lips?

Yet for all its absurdity, we find the idea irresistible. Our history is full of attempts—nutty, eerie, comical, earnest, legendary, and real—to make artificial intelligences, to reproduce what is the essential us, bypassing the ordinary means. Back and forth between myth and reality, our imaginations supplying what our workshops couldn't, we have engaged for a long time in this odd form of self-reproduction.

Looked at in one way, ours is a history of self-imitation. To the point of madness we have reproduced ourselves in the flesh. Under the various banners of religion, art, or even entertainment, we have adored statuettes modeled after ourselves, and taken centuries of pleasure in representations of ourselves on cave walls, canvas, and film. We are ten times more fascinated by clockwork imitations than by real human beings performing the same task.

What stunned them in medieval Strasbourg stuns them today in Disneyland, or in any television studio where everyone watches the monitors instead of the flesh-and-blood performers. We aren't fooled for a moment, just enchanted. The Narcissus legend resonates through our life and times, as emblematic of our foolishness and poignancy as ever it was.

Much social apparatus, not surprisingly, is set up to magnify the glory of our species. We study "the humanities" with reverence, and write earnest little proposals to foundations for support in the oddest lot of activities, all with the stated aim of adding one more speck of understanding to human knowledge of human beings. If you play at being a visitor from another planet, you can't help but wonder if the whole human race isn't on some vast, endless ego trip.

Here we go again, then, about to embark on a history of artificial intelligences, those attempts to reproduce the quintessence of our humanity, our faculty for reason. We'll trace several different routes, though in the beginning it will be nearly impossible to distinguish one from another, especially since we humans haven't always been fastidious about keeping fact from fancy. The first is the route of imagination, what might be. Next is the route of philosophical inquiry, which provides the bridge between imagination and what is. The third, of course, is what is: in this case, artificial intelligence as it has been realized since the development of the digital computer. This last line of inquiry constitutes the main part of my story, but the other two are indispensable to it.

Perhaps the earliest examples of the urge to make artificial persons are the Greek gods, those wonderful superhumans who seem to behave as we would if only we had the means. It was 850 B.C. or so when *The Iliad* codified with great beauty what were surely already ancient traditions. From Homer we hear about that poor, ill-favored son of Hera, Hephaestus, the god of fire and the divine smith, who, having been cast out of Olympus by his disgusted mother and crippled as a result of his fall (or is he born crippled? accounts differ), has to fashion attendants to help him walk and assist in his forge:

These are golden, and in appearance like living young women.
There is intelligence in their hearts, and there is speech in them and
strength, and from the immortal gods they have learned how to do
things.

(Lattimore, 1951)

His forge is the delivery room for a host of other wondrous automata, including twenty tripods that will propel themselves on their golden wheels to the Olympian feasts and bring themselves home afterwards. As a present from Zeus to Europa, Hephaestus makes Talos, a man of bronze whose duty is to patrol the beaches of Crete three times a day. He thwarts invaders by hurling great rocks at them, or by heating himself red hot and squeezing trespassers in a warm embrace. But Hephaestus's most famous creation is probably Pandora, a creature commissioned by Zeus to punish mankind for accepting Prometheus's gift of fire. Pandora is sent to us with her infamous casket, which she has been forbidden to open. But curiosity—known in masculine circles as the fine urge to explore the unknown—overcomes her, and she opens it up, thus releasing all the world's evils. This story certainly raises some questions about individual responsibility; thus Zeus shares culpability too. So Prometheus and Pandora stand mythically for the two sides of human knowledge, the light and the dark, the gift and the culpability. This theme resounds through science, but never louder than in this field of man-made beings. After all, centuries later Mary Shelley subtitled her famous *Frankenstein wife Or, the Modern Prometheus.*

Misogyny accounts for Pygmalion's contribution to the history of artificial intelligences. He creates Galatea in ivory because of his disgust with flesh-and-blood women; seeing his completed work, he falls in love with it—though whether what he feels is love or vanity about his own creation is hard to say. We will see that same confusion weaving endlessly through the history of men and their self-imitations (and also its opposite, as creators recoil from what they have made). In any case. Aphrodite obliges Pygmalion by breathing life into Galatea, and the two seem to have lived happily ever after.

To step from myth to artifact for a moment brings us to Daedalus, that master craftsman, who was highly esteemed and credited with a great many of the lifelike statues that abounded in the ancient world, statues that wheezed and blinked, and in some cases, scuttled about, all to the amazement of everyone who saw them.[1] In his thorough and entertaining history of automata, *Human Robots in Myth and Science*, John Cohen (1966) quotes a description by one Pausanias that can apply to the creations of many a future worker in the field: "All the works of Daedalus are somewhat odd to look at but there is a wonderful inspiration about them."

In Hellenic Egypt, automata were certainly wonderfully inspired. Statues of gods spoke, gestured, and prophesied, and Heron of Alexandria (circa 200 B.C.) left an account of his own automatic theaters that were elaborate indeed. What with bursting flames, dancing bacchantes, and spinning deities, they sound like nothing so much as prime Busby Berkeley. These were religious shows, intended to awe and instruct. They seem to have been exceptionally clever mechanisms (detailed plans are in Max von Boehn's *Puppets and Automata* [undated] and the well-illustrated *Les Automates* by Alfred Chapuis and Edmond Droz [1949]), working sometimes by quicksilver, sometimes by hydraulics, and sometimes by a priest pulling strings. The fact that they were artifacts, patently man-made, doesn't seem to have diminished the awe that Egyptians felt in their presence. It was assumed that such statues had a sort of soul, called a *ka*, which could represent a god or a dead person; probably it was believed that the *ka* made use of

---

[1] Purists make distinctions among sorts of AI machines, which are these: automata are any self-locomoting contrivances, for example, Hephaestus's tripods and the hardy Talos. But Talos is a humanlike automaton, therefore called an *android*. Golem and homunculi are special cases of androids, made from organic matter, and are products of the single craftsman, so to speak, such as Dr. Victor Frankenstein. *Robots*, on the other hand, are also a special case of androids, but are mass-produced. The word *robot* was introduced to English by Karel Capek's influential play, *R.U.R.* (1923). It comes from the Slavic *robota*, meaning forced labor or slavery. "What distinguishes the robot from other machines is not a soul but a mind," says Robert Plank (1965)—though you may not want to make that distinction.

the priest and his apparatus as a medium. A little later the Romans too had such statues, with the priest puppeteer plainly in view, which strongly suggests the notion of human as agent but not necessarily as trickster.

We shall have to come to terms right now with the fact that the boundary between trickery and honest tries is imprecise; this fact too is part of the history of thinking machines. My own inclination is to believe, if not in a given piece of machinery, always in the astonishing and delightful imagination of its inventor. From the start, the tangle between fact and fiction has been difficult to unravel, as I've said, and if we can be confident that a little empirical rigor would have shown Paracelsus that his recipe for a homunculus didn't work, how sure can we be about the talking bronze head Albertus Magnus was said to own? Von Boehn describes it as "a lovely woman who could speak," which so offended Albertus's pupil, the young Thomas Aquinas, that he burned it upon the death of his teacher. What on earth did she say? Alas, the story loses some of its piquancy with the fact that Albertus outlived his celebrated pupil by some six years.

We're ahead of ourselves. Around 1200 B.C., somewhat before the codification that Homer represents, another codification was taking place in the desert of Sinai—the Ten Commandments. The second commandment interests us here: "Thou shalt not make unto thee any graven image or any likeness of any thing that is in heaven above or that is in the earth beneath, or that is in the water under the earth; Thou shalt not bow down thyself to them nor serve them, for I the Lord thy God am a jealous God. . . ." Only one commandment is considered more important, Jehovah's claim to his people's exclusive devotion ("Thou shalt have no other gods before me"), and the second can even be said to be simple insurance for the first. But is that all it is?

When we look at human images and idols in cultures without such prohibitions, we find them used in a variety of ways: to bring injury or death to enemies, to insure the fertility of the land and abundant crops, and even to capture the soul of a dying person. These uses all have in common one element, and that is magic. By fashioning a doll to represent the woman who has rejected you, you can transcend your own unattractive self and win her anyway. To transcend your own puny power is—what else?—to participate in the powers of the gods. You become, for a little while at least, as a god yourself. This simple-sounding idea is fraught with implications, and it isn't far-fetched to me to think that Jehovah—or the projection of the human psyche that Jehovah represents—is jealously guarding his magic as much as anything. Hermes Trismegistus, the Hellenic author of the Hermetic writings, comes right out with it. Speaking of the animated statues I've described, he says, "They have *sensus* and *spiritus*. ... By discovering the true nature of the gods, man has been able to reproduce it... unable to create souls, man invoked the souls of demons and angels and, by sacred rituals, infused them into the statues which thereby acquired the power of doing good or evil" (Cohen, 1966).

In any case, the children of Israel seem to have interpreted the second commandment with some flexibility, and even Jehovah himself is inconsistent, directing that the Ark of the Covenant be built to his specifications, which include adorning it with images of two golden cherubim, cherubim surely being a "thing that is in heaven above." The Israelites were also partial to teraphim, figurines in human shape made of clay or semiprecious stones, perhaps consulted for divination or worshipped as household deities. Isaiah, for one, is full of scorn for such things. Here he is, grumbling about the Judeans: "Their land also is full of idols; they worship the work of their own hands, that which their own fingers have made." And perhaps there's something to the view that holds it unsavory for the creator to worship his own artifact. But is it the creation, or the worship that is objectionable? We've already seen that this distinction is not easy to make. And what if the creation acquires autonomy?

So here we have the two fundamental attitudes that inform Western views on the subject of thinking machines. One, the Hellenic, says that they are useful, praiseworthy, and appealing; the other, the Hebraic, says that they are fraudulent, wicked, and even blasphemous. (Of course, these two labels, Hellenic and Hebraic, are a historian's convenience to describe two casts of mind, and don't address the evolution of living cultures or religions as they have actually come to be practiced by human beings.) The statements made by partisans and critics of artificial intelligence show that the tensions between those two attitudes exist to this day.

In the person of a remarkable medieval pope called Sylvester II, two important lines in our history momentarily converged— the line of humanlike automata, and the line of the computer, which a thousand years later would make the most plausible claims yet to having intelligence, if not quite *sensus* and *spiritus*. Gerbert, as he was known before he became pope, is credited with having made a statue with a talking head. "It spake not unless spoken to," writes a twelfth-century observer quoted by Cohen, "but then pronounced the truth, either in the affirmative or the negative. For instance, when Gerbert would say, 'Shall I be pope?' the statue would reply 'Yes.' 'Am I to die ere I sing mass at Jerusalem?' 'No.' " Perhaps more believably, Gerbert is said to have introduced the abacus to Europe after learning about it from the Arabs in Spain, and one source even says he introduced Arabic numerals too (which in fact aren't Arabic but Indian in origin). Certainly both innovations revolutionized European mathematics. Of Gerbert, B. V. Bowden writes in his historical essay about computation, which begins one of the first books devoted to the modern computer called *Faster Than Thought* (1953), "Until his time, Europeans had been concerned to investigate the properties of individual numbers rather than their combination." Sylvester earned a wide and lasting reputation for wickedness as the result of all his occult knowledge, and nearly six hundred years after his death, Montaigne professed himself scandalized by a biography of Sylvester he was allowed to see.

Not surprisingly, the flowering of Arab mathematics from around A.D. 750 to the 1400s nurtured a parallel growth in precision instruments, particularly timepieces. Despite the strict Muslim prohibition against such representations, many of these timepieces were decorated with moving figures of humans and animals. Chapuis and Droz say that in 802 the Emperor Haroun-al-Rashid (he of the *Arabian Nights*) presented the Emperor Charlemagne with an elaborate clock which, among other things, sent out a dozen cavaliers from a dozen windows each noon and returned them back again. The ingenuity behind this and many other timepieces and automata designed by the Arabs was luckily transferred, along with the mathematical knowledge that supported such craft, through such persons as Gerbert, who wandered into Spain and Sicily where Christians and Arabs mingled, and also by means of the Crusades, where this knowledge was put to work in less amusing machinery.

From a philosophical point of view, the Arab heirs to the Hellenes may have been the first to state formally that a distinction existed between natural and artificial substances. Such a distinction did not mean that the natural was superior to the artificial, only different. But not very different: one excellent means of knowing the natural was declared to be by studying the artificial (Labat, 1963). We'll hear the same assertion from modern workers in artificial intelligence.

Arab science from the eighth through the twelfth centuries had other aspects in common with modern science. Unlike the Greeks, whose science had been fixed to certain persons, who in turn were fixed to certain places, Arab science was international, and knowledge flowed with fair rapidity among scientific centers. Moreover, these men were not mere custodians of Greek thought, waiting to hand it over when the barbarians of Europe should finally wake up. Instead they were experimenters, subjecting Greek doctrines to empirical tests. As a result, they made contributions not only to mathematics, but also to medicine. They seem to have gone in for that other modern phenomenon, teamwork. In any event, a group of Arab astrologers is credited with constructing a thinking machine that they called the *zairja*, which was based on a scheme whereby the twenty-eight letters of the Arabic alphabet represented the twenty-eight classes of ideas of Arab philosophy. By combining numerical values assigned to classes and letters, some sort of insight was reached.

The *zairja* caught the imagination of an unusual missionary who had traveled from Spain to North Africa afire with religious zeal. His name was Ramon Lull, and I shall describe him in some detail, for he stands at a branching point in our story. Born in 1234 of a noble Catalonian family, he was sent to the court of James I to be the companion and tutor to the royal princes. He led a rather wild life until a series of religious visions reformed him, and he settled down to learn Arabic with the aim of converting the Muslims to Christianity. As sometimes happens with missionaries, Lull seems to have absorbed more from the Arabs than they from him; in particular he decided to design a Christian version of the *zairja*, called, more grandly, the *Ars Magna*. Cohen describes it as a logical machine that aimed "to bring reason to bear on all subjects and, in this way, arrive at truth without the trouble of thinking or fact-finding." But it is clear that the *Ars Magna* sits dead center in medieval Christian dogma. Lull held it to be a perfect and indestructible scheme, directly and literally inspired by God. Its plans call for a series of categories of knowledge and faculties, set up in segments of concentric circles, which could then be matched in different permutations in order to answer questions of theology, metaphysics, morals, and even natural science. Its apparatus consisted of discs, probably of metal and pasteboard, to be spun and matched, and Lull's description of it was impressive enough that both Leibnitz and Hegel would refer to it (Gardner, 1958).[2]

The history of thinking machines branches here. Lull was the last traveler allowed to claim that a work of pure imagination was science. Now new tests were devised by men who believed that skepticism, far from being heretical, was a moral obligation. Some of them even went so far as to say that humans themselves were knowable, that our bodies, and even our minds, would yield their secrets to the scientific method. Perhaps Descartes was the first to

---

[2] In addition to inspiring cults that sprang up to study him and his mystical writings, Ramon Lull also inspired Jonathan Swift's satirical description of the *Ars Magna* in *Gulliver's Travels*, Part III, Chapter 5. I will refer again to Swift's machine in Chapter 12.

make that branching, to tread cautiously out where no path existed. A path would exist there in years to come, a path that would widen, yield up milestones, grow rutted in some stretches and in others be quite lavishly paved. It would support temporary shelters and busy, permanent villages, and its travelers are really the subject of this book.

We will return to empiricism in Chapter 2, but first we'll make an excursion further along in the realm of fantasy. Though these two routes diverge they are not without connections. For hundreds of years, travelers on either one were within hailing distance of one another, and more than a few believed that both routes would end at the same destination in Neverneverland. I suppose they haven't yet been proved altogether wrong.

So to fantasy. The tradition of brazen heads belongs here. I've already referred to it in connection with both Sylvester and Albertus Magnus; others, such as Roger Bacon, Bishop Grosseteste, Arnold of Villanova, and Don Enrique de Villena, have also come down to us as having owned and consulted such marvels.

It's hardly surprising that the art of making clocks decorated with animated figures added much credence to the belief that learned men kept robots. To most people, there could be little difference between a human figure that nodded, bowed, marched, or struck a gong at a precise and predictable moment, and a human figure that answered knotty questions and foretold the future. By the middle of the fourteenth century, large elaborate clocks with moving figures had become public monuments—Strasbourg, Nürnberg, Lübeck, and Berne followed the Italian cities with them—and talking brass heads had become as closely associated with learned men as cats are with witches.

We might ask what it was about a brass head—an automaton presumably constructed by the sage himself, as both evidence and source of his wisdom—that allowed it to be so much wiser than its creator. It's easy to guess that the rapid proliferation of public automata would be the inspiration for such legends, but why were the legends necessary at all? What psychological purpose does the brazen head serve? We may be back where we started, dealing with the need to mediate science or any arcane knowledge as a branch of the supernatural through unnatural means. And something tells me that a kinship exists between the need to posit these supersmart machines and the very common modern view that machines can't be said to "think" unless they show superhuman skills. This latter notion is so widespread that Seymour Papert of MIT, who presently works in the field of artificial intelligence, has coined a phrase for it. He calls it "the superhuman human fallacy."

For our purposes, Paracelsus belongs in the tradition of fantasy. He was real—the grandaddy medicine showman of them all. He was born in 1493 and died in 1541, leaving the world much the richer in nonsense as well as substance. He was a physician, and was thought well enough of to be offered a chair of physic and surgery at Basel, where he was driven out less than two years later for offending his colleagues with his loud denunciations of them. He was famed for his cures—his real contribution seems to lie in pharmacology—and even more famous for his bombast and erratic behavior (Rosen, 1959). Paracelsus traveled endlessly, gathering disciples wherever he went and making enemies among the established, particularly when he twitted his fellow physicians for being more interested in cash than cures. We are interested in him because he claims to have created a homunculus, a little man, and its recipe is worth repeating. Begin with human sperm:

> If the sperm, enclosed in a hermetically sealed glass, is buried in horse manure for about forty days and properly magnetized, it begins to live and to move. After such a time it bears the form and resemblance of a human being, but it will be transparent and without a corpus. If it is now artificially fed with the arcanum of human blood until it is about forty weeks old, it will live.
>
> (Rosen, 1959)

Not only will it live, but it will have intelligence, and Paracelsus gives some detailed instructions on how it is to be educated. More to the point, he gloats, "We shall be like gods. We shall duplicate God's greatest miracle—the creation of man."

Poor Paracelsus. I'm sorry to say he died in poverty, and though he'd left instructions for his disciples on how his parts were to be preserved in horse manure for the magical forty days so he could be resurrected, it seems that one of them said the right spell at the wrong speed and ruined it all. He was the archetypal alchemist, always seeking and seldom finding—arrogant, rebellious, but withal an irresistible old smoothie to those who followed him. He inspired Ben Jonson to satire and Goethe to sublimity; and apparently he also inspired one of the more caustic modern critics of artificial intelligence to compare current work with the follies of alchemy. I'm speaking here of Hubert Dreyfus, whose book *What Computers Can't Do; A Critique of Artificial Reason* (1972) began life as a report entitled *Alchemy and Artificial Intelligence*. Yet Paracelsus, with his mad combination of swindle and science, is somehow more appealing a figure to me than many a sober scientist and philosopher who follows him. That is the way of the world.

In 1580, some forty years after the death of Paracelsus, there appeared another sort of artificial man. Its name was Joseph Golem, and it was the creation of a remarkable rabbi named Judah ben Loew, known rather irreverently in English as "The High Rabbi Loew." The High Rabbi Loew was an historical figure, the Chief Rabbi of the city of Prague, and he enjoyed friendships with such men as Tycho Brahe, the famous Danish astronomer who had come to Prague at the invitation of the Emperor Rudolf II, and Johannes Kepler, whose work is the foundation of modern astronomy. His important friendships were no protection for the Jews of Prague against the occasional pogroms that flared up, however, and so, inspired by God, the rabbi decided to fashion a spy to go among the Gentiles and report back on whatever they might be up to. The rabbi and two of his assistants went out in the dead of night to the banks of the Moldau, and from the clay of the riverbank they formed a humanlike figure. After the regulation incantations and prayers, the figure gradually assumed life, coming altogether alive when the Holy Name was implanted on his forehead. The creature was mute, which must have caused some difficulties in conveying back intelligence, but with a gift for improvisation, the rabbi seems to have managed.

When Joseph Golem wasn't spying on the Gentiles, he was used as a sort of janitor around the temple. He was the rabbi's exclusive property, though. Whenever someone else, in particular the rabbi's wife, ordered Joseph Golem to do something, it always led to mischief. For example, preparing for the Sabbath one day, the rabbi's wife asked the golem please to bring in the water from the well. Since, in modern parlance, the rabbi's wife had not specified the task precisely enough, Joseph Golem began bringing in the water from the well—all of it. The rabbi had to be called from his devotions to the rescue. Eventually the golem became too rambunctious, attacking his creator and forcing the rabbi to end its life, some say by taking away the name of God implanted on his forehead.

Several familiar themes knit together here: supernatural power (the name of God), which is necessary to give life to the inanimate; the impossibility of the uninitiated trying to use the magical contrivance; and the creation used not as a source of knowledge, but as a servant or slave, which will make it rebel and try to overcome its creator. This last theme especially will recur in future literature.

Rabbi Loew's was the most famous, but not the only golem. Cohen says that Eleazar of Worms (c. 1160–1230) records a recipe for making an artificial man by combining the letters of the Holy Name, and Elijah of Chelm is said to have made a golem in the middle of the sixteenth century. Curiously enough, several present-day researchers in artificial intelligence have told me that they grew up with a family tradition that they are descendants of Rabbi Loew, though they doubt this belief has had much influence. Among them are Marvin Minsky and Joel Moses of M.I.T. Further, Moses tells me that a number of other American scientists have considered themselves to be descendants of Rabbi Loew, including John von Neumann, the computer pioneer, and Norbert Wiener, who coined the term cybernetics and wrote a famous little book on automation called *God and Golem, Inc.*[3]

From the sixteenth century on, a population explosion of automata took place. The idea of these mechanisms gripped the imagination in tales and legends and actually came to life in wonderful toys for the very rich. I admire the impulse that says a wide distribution of bread is to be preferred to a concentration of frivolity, and I know that while craftsmen were fashioning mechanical nymphs to be chased by mechanical satyrs through the royal grottos of Saint Germain (the work of one Solomon of Caus, 1576–1626) peasants, overtaxed to pay for such things, were starving beside the palace gates. Alas, automata were only one form of self-indulgence for the rich and powerful; there were also costly expeditions to the New World, bloody wars, the overthrow of kings, the painting of pictures, the writing of plays. This is the age of Shakespeare as well as starvation.

Fantasy has its high purposes. If Descartes is willing to assert that animals, at least, can be viewed as machines, it may very well be because he is contemporaneous with this bursting forth of animated statues. We shall speak of Descartes in more detail later.

The art of mechanical statues flourished in the seventeenth and eighteenth centuries, probably reaching its zenith in the work of Jacques de Vaucanson. His most celebrated work was his duck, which appeared in 1738 and was an immediate and immense success. It could beat its wings (one observer says "it flew"), drink water, eat grain "in an incredibly natural way," which it then digested and excreted by means of elaborate tubing in its stomach. It led an odd life—Vaucanson in fact grew busier with industrial applications of automation and neglected his sideline of mechanical statues—and the duck ended up deplumed and paralyzed in Prague, where, a century after its debut, it was found and restored by one J.-B. Rechsteiner, an employee of a traveling museum of automata. He was himself a famous builder of mechanical statues, but he was astounded by the

---

[3] At the 1977 International Joint Conference on Artificial Intelligence, held in Cambridge, Massachusetts, workers from the Czechoslovakian Technical Institute in Prague reported on their robot named GOALEM, whose punning acronym stands for GOAL-oriented Electrical Manipulator.

duck's innards, and felt that their sophistication represented work worthy to be called an invention in its own right. Vaucanson had been interviewed for an encyclopedia of the sciences in 1777, and was deliberately vague about the secrets of the duck's insides. He did not intend, he said, a perfect imitation of the digestive processes, with nourishment and blood manufacture, and so forth; rather, he hoped to imitate the larger aspects, which would include intake, maceration, and an obvious chemical change prior to excretion (Chapuis and Droz, 1949). Vaucanson's aim serves as a good example of simulation, and his duck should be kept in mind when we come to the simulation of human thought processes.

Somewhere between realized automata, such as Vaucanson's, and purely imaginary kinds, such as Frankenstein's, are the frauds. Von Kempelen's chess-playing machine was the first and most famous. It was an android got up to look like a Turk. It sat at a chess table which presumably housed its mechanism, but which in fact cleverly concealed a human chess player (though whether he was a legless Pole trying to escape the Russian secret police, as one legend has it, is doubtful). In any case, the Turk won chess games all over Europe, scoring its most famous victory over Napoleon in 1809 (or over some of Napoleon's marshalls; our sources here are show biz, none too reliable). The Turk was mated once and for all in a fire in Philadelphia in 1854.[4]

By the beginning of the nineteenth century, the artificial intelligences that penetrated and dwelled in people's imaginations were

---

[4] Its twentieth-century descendant, in the form of a household robot billed as "The Ultimate Home Appliance," made a tour of the United States in 1977, gulling the public, the wire services, and the news magazines. Two Carnegie-Mellon graduate students, Mark Fox and Brian Reid, exposed the fraud in some detail, but the only response made by its promoters was that someday such a machine would be possible, even if this one didn't quite live up to its public billing. Only a handful of the newspapers that had originally been beguiled by the robot published the expose: truth was done in by a much stronger combination of the wish to believe in magic and the disagreeableness of admitting gullibility.

composed of the printed word rather than wood and metal and cloth. The reasons are not hard to see. It was the height of the machine age, and if machines hadn't quite lost their mystery, they certainly had lost their novelty; people were growing used to seeing them do all sorts of odd things. If the human senses of mystery and the supernatural were still to be served, now it was best done by words. The spread of education—and the concomitant rise in the number of books circulated—meant a greater audience for a story than ever could crowd into fairgrounds, no matter how peripatetic and hospitable a fairowner might be.

Here then was E. T. A. Hoffman's *The Sandman*, drafted about 1815 but published later, which introduced Olympia, the mechanical grandma of many a robot woman. Hoffman himself had been inspired by an obscure story of Jean-Paul Richter's called "The Death of an Angel." In his turn, Hoffman inspired Delibes to write the music for the ballet *Coppelia* in 1870, and Offenbach to write the opera *The Tales of Hoffman* in 1880, each work containing mechanical dolls that come to life. Such stories were much in vogue in the nineteenth century. Not only outright androids, but machines themselves taking on human attributes—in a rather sinister way—are found in the popular works of Jules Verne and Samuel Butler. Major and minor writers alike took up the theme of the extraordinary creation of humanlike beings. Goethe gave us Faust again (Part 2, published posthumously in 1833), modeled this time on Paracelsus and exhibiting all the paraphernalia we shall come to associate with the mad scientist, including his odd but faithful assistant, in this case Wagner. In the following passage we are in Faust's laboratory, "after the style of the middle ages; extensive unwieldy apparatus for fantastical purposes." Wagner addresses Mephistopholes, who has just blown in:

WAGNER   A man is in the making.

MEPHISTOPHOLES   A man? And what enamoured couple have
          you got locked up in your furnace?

WAGNER   God forbid! We declare all that a farce, That common
         mode of him-and-herness.
              . . .
         A beast may still find that it gives enjoyment
         But man with his great gifts must now begin
         To look for a higher, higher origin.
              . . .
         The stuff evolves! More clearly moving—
         Conviction stronger, stronger proving:
         The mystery that in nature earned one's praise
         We dare essay by rational incubation,
         And what she managed in organic ways
         We bring about by crystallization.

                                  (MacNiece, 1951)

Indeed. The homunculus is a cheerful little chap, usefully clairvoyant and more interested in finding himself a body than in doing great mischief. Wagner might well be elated. Moxon, on the other hand, an unlucky fellow who has the bad grace to best his own chess machine, gets bludgeoned to death by it in Ambrose Bierce's little chiller "Moxon's Master" (1893), repeating for us the theme of Golem, of Erewhon, and, at not too large a leap, of Lucifer himself.

Mary Shelley's *Frankenstein* (1818) is the most famous of these tales, and it is interesting not only because it is the source of many derivations (which themselves are hoary enough to have spawned parodies) but also, due to the time and place it springs from, it combines nearly all the psychological, the moral, and social elements of the history of artificial intelligence. It also has a few odd historical connections with the computer, which may or may not be significant. To make all this clear, it is necessary for me to tell a story as complicated as the plot of a Victorian novel.

In 1815, England's most famous living poet, George Gordon, Lord Byron, married an adoring woman named Annabella Millbanke in the hopes of putting an end to the rumors about his liaison with his half-sister, Augusta Leigh. The rumors in fact were true, and there was a child from that union, but she won't concern us here. What does concern us is that a scandal developed and Byron's marriage ended in separation after a year, and after the birth of his only legitimate child, his daughter

Ada. Remember Ada: she will be important in the history of the computer. Byron exiled himself to the Continent, but before he left, he had a quick affair with one Claire Clairmont. She was the stepsister of Mary Wollstonecraft Shelley, and when Mary and the young poet Percy Shelley decided to elope, Claire not only accompanied them, but eventually manipulated them to Geneva so they could all meet with Byron. It's likely that these two famous, self-exiled poets would have met anyhow, and they came to form an uneasy friendship—uneasy largely because of Shelley's sister-in-law, the aggressive Miss Clairmont—until Shelley's death a few years later. Mary, Percy, and Claire took a cottage on the lake, and not long thereafter Byron moved his party into a villa just up the hill from it. The two groups visited each other daily, the poets having long philosophical discussions and the others amusing themselves as best they could. It was a dreary summer with much rain, which kept them all in the house. They would read aloud to each other, and a set of German ghost stories, translated into French, inspired Byron with an idea. Mary later described the project it engendered:

"We will each write a ghost story," said Byron; and his proposition was acceded to.

I busied myself to *think of a story*—a story to rival those which had excited us to this task. One that would speak to the mysterious fears of our nature, and awaken thrilling horror—one to make the reader dread to look round, to curdle the blood and quicken the beatings of the heart. . . .

Many and long were the conversations between Lord Byron and Shelley, to which I was a devout but nearly silent listener. During one of these various philosophical doctrines were discussed and, among others, the nature of the principle of life, and whether there was any probability of its ever being discovered and communicated. They talked of the experiments of Dr. [Erasmus] Darwin (I speak not of what the doctor really did, or said that he did, but as more to my purpose, of what was then spoken as having been done by him), who preserved a piece of vermicelli in a glass till by some extraordinary means it began to move with voluntary motion. Not thus, after all, would life be given. Perhaps a corpse would be reanimated; galvanism had given a token of such things; perhaps the component parts of a creature might be manufactured, brought together and endued with warmth.

(Marshall, 1889)

The nineteen-year-old Mary went to bed and imagined such a creature at length during that sleepless night, and woke with the knowledge that she had her story "to speak to the mysterious fears of our nature." And speak to these fears Frankenstein did, though not, as Walter Evert points out, in the way that Gothic horror stories had (1974). Rather, the tale is frightful on a deeper, more significant level, the level that Mary refers to when she says, "Frightful must it be; for supremely frightful would be the effect of any human endeavor to mock the stupendous mechanism of the Creator of the world" (Marshall, 1889). (Stupendous mechanism. The man as machine is already a comfortable notion.) *Frankenstein* has survived, though sometimes in a rather odd form, not only because it is the story of a man who tries to make a man by unnatural means, and is thus part of a venerable tradition, but, perhaps more important, because Frankenstein's story is a paradigm for all science, of the man who yearns to know and drives madly ahead without real thought to the consequences. It isn't even clear that *all* the consequences could have been anticipated no matter how hard Victor Frankenstein had thought about his project. Our hindsight drastically prunes the possible paths events could have taken; our foresight can never be so efficient.

Since about half a dozen films have widely propagated distorted versions of the original, it is helpful to recapitulate the plot here. *Frankenstein* is a story within a story. The frame is the tale, told in letters, of a young adventurer named Walton who is attempting an expedition to the North Pole. He has to overcome serious difficulties, but eventually gets underway, and finally in the frozen wastes comes across the oddest of sights, a straggler on the ice floes. The straggler is Dr. Victor Frankenstein, and it is Walton who mediates his tale. The frame exists for several purposes, but the most important one for us is that Walton is another obsessed seeker after knowledge, a scientific altruist [says Evert] driven at whatever cost to himself, to extend the range of human knowledge for the ultimate benefit of all posterity. Dr. Frankenstein is no more the "mad scientist" of later fictions than is the young explorer Walton. Both are idealists of the possible, adventurers in a world of human perfectability that could not even have been dreamed of until modern science and technology provided the means. Unfortunately, while both were willing to suffer everything and risk everything, including the most extreme consequences of failure, neither considered the horrible possibilities of success.

(Evert, 1974)

The horrible possibilities of success are the kernel of *Frankenstein*'s horror for us; I think they are the kernel of the horror that characterizes most people's reaction to the idea of artificial intelligence itself.

The story Victor Frankenstein tells Walton is one of the ultimate research scientist. "None but those who have experienced them can conceive of the enticements of science," he rhapsodizes. "In other studies you go as far as others have gone before you, and there is nothing more to know; but in a scientific pursuit there is continual food for discovery and wonder." Frankenstein has been led to the scientific banqueting table by one Professor Waldman, who promises, in phrases nearly identical with ones in *Faust*, penetration into the recesses of nature, new and almost unlimited powers, and ascent into the heavens. In retrospect, Victor sees these as "words of fate, enounced to destroy me." But he begins with the highest of purposes. Sickened human bodies move him to pity; is there some way he can understand the human organism and alleviate pain and suffering? He studies hard, and after "days and nights of incredible labor and fatigue" discovers the secret of life. He has a few moments of hesitation, and even thinks about making a simpler organism first, but overcomes his doubts. "A new species would bless me as its creator and source; many happy and excellent natures would owe their being to me. No father could claim the gratitude of his child so completely as I should deserve theirs." All that incredible labor and fatigue seems to have sapped his common sense.

Off he goes to the hospitals and charnel houses, but not for spare parts. Evert says,

> One may wonder what substances he *did* finally use for his creation. But the fact that he decided to make his first man of an eight-foot size, because of the difficulty of working accurately with the smaller fibers of a person built to normal scale, virtually rules out the possibility of his having constructed the monster from as it were "standard parts" accumulated through multiple grave robbings. Frankenstein seems to be an honest scientist, synthesizing his materials as he goes, and not just a laboratory shoemaker stitching together the odd lots lying about the shop.

In other words, Mary Shelley intended our horror to be psychological and moral, but not the sort inspired by ghouls.

The monster is brought to life on a rainy November night. As a general rule, contented people don't make good scientists, and Victor is no exception. He describes his creation as a "catastrophe," a "wretch," I think mainly because of its watery eyes and ill-fitting skin. Victor flees in horror, spends a while pacing his bedchamber, and then falls asleep, but he wakes to find his creation standing over him. There's nothing menacing about the creature, but Victor flees again, making his repulsion obvious. This is one success that Victor won't crow about at a scholarly meeting; in fact, he tells no one about the result of his two years of feverish work, but instead falls into a grave illness, recovers after several months, and takes up paler pleasures in the study of Oriental languages.

The creature, meanwhile, is left on his own, and as he puts it later, misery makes him a fiend. First he murders Victor's little brother; the murder is blamed on a servant girl who must be hanged for it. But if Victor is smart enough to put together a living being, he's also smart enough to guess who the real murderer is, and when the two finally meet on an Alpine mountaintop, Frankenstein tries to kill the nameless monster he has brought to life. Eight-foot monsters are no pushover, and in any case, the monster wants to talk, not fight. "How dare you sport thus with life?" he says ironically to Victor. He then recounts the events that have befallen him, the misery that has made him a fiend. This is Mary Shelley's chance to show that she's the daughter of Mary Wollstonecraft and William Godwin, two of the most advanced social thinkers of their time.

The monster proposes what for Victor is morally imperative and yet impossible: that he be given a soul-mate. Persuaded by the arguments, Victor sets to work on a second creature, but just as he's about to complete it, he starts thinking about what might go wrong. He looks up, sees the monster waiting anxiously outside the window for its mate, and renounces his new work on the spot; he has learned some sort of lesson. But circumstances have also changed. The doctor's broken promise leads the monster to commit mayhem—the accumulated horrors include the murder of Frankenstein's best friend and, indirectly, the death of his father.

Finding no remedy at law, Frankenstein now pursues his monster with the same obsession for its death that he once had for its life, and it's in the midst of this chase that Walton finds him on the ice floe. Just after hearing Frankenstein's tale, Walton is visited by a deputation of his own crew who, though they despair of ever getting off the ice alive, want his promise that if they do escape they will be allowed to go home instead of pursuing the enlargement of natural science. Through his fever, the old Victor speaks out:

> What do you mean? What do you demand of your captain? Are you so easily turned from your design? Did you not call this a glorious expedition? And wherefore was it glorious? Not because the way was smooth and placid as a southern sea, but because it was full of dangers and terror, because at every new incident your fortitude was to be called forth and your courage exhibited, because danger and death surrounded it, and these you were to brave and overcome.

If Walton accedes, Frankenstein will not: he owes it to his own species to persevere. Yet his ambivalence remains to the end. The monster presents himself, not to gloat but to beg forgiveness, acutely aware of his own ambivalences.

This is no ghost story. If it were, it would now be an unread curiosity, like the only other published product of that rainy summer, the vampire story written by Byron's physician. With its intellectual breadth and psychological depth, *Frankenstein* lives in its own right, and to my mind would make a fitting graduation present for any fledgling scientist.

For here is the moral dilemma of science presented in concrete and implacable terms. Good can beget evil, and this outcome is only sometimes predictable. The Waltons turn back, but someone must go forward. Who will it be? And under what safeguards and restraints? The scientists I have spoken with in the course of writing this book talk hopefully of "the net good," a clear sign that they understand there is evil to be weighed along with the good inherent in their work. Still, nearly every one of them has found more good than bad. Is their calculation disinterested? If not, whose will be? Professional moral philosophers, like economists, always seem to be better guides to the past than the future, fair at prescribing what we should have done, but no better than most at describing what we must do next. What will we do with them, once we have these artificial intelligences? Or, as some would have it, what will they do with us?

Writers continue to speculate. In 1923, Karel Capek brought his play *R.U.R.* to London and audiences lived through the man-versus-machine dilemma once more; by 1950 Isaac Asimov had formulated his "Three Rules of Robotics," for which he has said he will probably be best remembered (Asimov, 1950). They are

1. A robot may not injure a human being, or through inaction allow a human being to come to harm.
2. A robot must obey the orders given it by human beings except where such orders would conflict with the First Law.
3. A robot must protect its own existence as long as such protection does not conflict with the First or Second Law.

These laws are immutably wired into the "positronic" brains of Asimov's race of robots, and indeed the rules have proved so comforting that many other science fiction writers have adopted them as naturally as gravity—laws that are given, and broken only under the most unusual circumstances. The successive representations of the robot in Asimov's stories are worth a study in themselves: they go from childhood companion and toy to menace-in-spite-of-themselves, and finally to the ultimate artificial intelligence, a sort of deity computer that controls the world in the best interests of humankind, which include keeping that secret from us. But this reference to Asimov brings us well into a time when artificial intelligence had begun to be a serious field of study.

Six or seven years before Mary Shelley published Frankenstein — around 1812—a brilliant young mathematician named Charles Babbage "was musing over a table of logarithms at Cambridge. A friend came into the room and called out, "Well, Babbage, what are you dreaming about?" Babbage pointed to the logarithm tables. "I am thinking that all these tables might be calculated by machinery." These were the words that set Charles Babbage out on a project that would consume a good part of his life and the better part of his private fortune (Morrison and Morrison, 1961).

The idea of calculating machines was not new. I've already mentioned the abacus, which is a calculator of a kind. In 1642 Blaise Pascal built the first simple digital calculating machine (digital here means a machine with a finite set of states of being, as opposed to analog machines, which have a potentially infinite set of states). Leibnitz too invented a calculating machine, writing that "it is unworthy of excellent men to lose hours like slaves in the labor of calculation which could safely be relegated to anyone else if machines were used." Since lucid descriptions of the history of computers already exist, in particular Herman Goldstine's The *Computer from Pascal to von Neumann* (1972) and B. V. Bowden's wonderfully droll *Faster Than Thought* (1953), there's no use recapitulating them here. But I can't resist a few words about Charles Babbage, who raised personal eccentricity to heights that have not since been exceeded in the field of computing, though there have been some splendid tries.

Charles Babbage was born in 1792, the pampered son of a wealthy banker and his wife. "During my boyhood," he writes in his autobiography, "my mother took me to several exhibitions of machinery. I well remember one of them in Hanover Square, by a man who called himself Merlin. I was so greatly interested in it, that the Exhibitor remarked the circumstances, and after explaining some of the objects to which the public had access, proposed to my mother to take me up to his workshop, where I should see still more wonderful automata" (Morrison and Morrison, 1961). Babbage then describes two "uncovered female figures of silver," each about twelve inches high. One walked, used an eye-glass occasionally, and bowed frequently; her motions were singularly graceful, Babbage wrote. The other was a dancer, full of imagination and irresistible. Young Babbage was enchanted with them, but knew better than to ask, as he normally did with his toys, that they be pulled apart so he could see how they worked. "These silver figures were the chef d'oeuvres of the artist; they had cost him years of unwearied labor, and were not even then finished," he says. And since Babbage was writing in his seventies, when his own chef d'oeuvre could have been described in precisely the same terms, the parallels between him and Merlin are surely on his mind. Years after he first saw the silver figures he was able to acquire one of them, and once it was modestly draped it had a place of honor in his parlor for the amusement of visitors.

He arrived at Cambridge to discover that he'd educated himself well beyond the knowledge of his mathematics tutor. In any case, mathematics at Cambridge was moribund, still under the influence of Newton, who had lived two hundred years earlier, and unaffected by exciting things that were happening on the Continent. Together with two of his good friends, John Herschel (son of the astronomer and later an astronomer in his own right) and George Peacock, later the Dean of Ely, Babbage formed the Analytical Society, which, says Bowden, "gave the first impulse to a revival of the study of mathematics in this country after half a century of neglect" (1953). In addition to such lofty matters, Babbage adored whist and chess, and he was an ardent sailor. Once he left the university he was equally sociable, and entertained most of the major scientific figures of his day at his delightful dinners—Darwin mentions them, for example—or traveled to the Continent to seek them out.

But the automatic calculation of tables possessed him. He'd heard of the French efforts in 1784, begun by G. F. Prony, director of the École des Ponts et Chaussées, who was himself inspired by Adam Smith's *Wealth of Nations,* which describes the divisions of labor essential for manufacturing. Prony drafted skilled mathematicians for the most difficult tasks of table calculation, eight "well-trained computers" for the next level of tasks, and between sixty and eighty unskilled computers who did the simple addition and subtraction necessary for the final form of the tables. Babbage was certain that the last stage—and the typesetting, a costly source of error—could be done by machine.

By 1822 he had constructed a small working model of his automatic table calculator, which he called the Difference Engine, and the British government had been persuaded to finance a larger model that would work to twenty decimal places and sixth-order differences in the computation of tables. Such tables were essential to navigation and ballistics, and the faulty tables that had been produced by hand, that is, by human calculators working with ink and paper, were not only costly but horrendously prone to error—sometimes fatally so, when ships ran aground because of navigational calculations based on wrong numbers.

So the work was undertaken. But several things conspired against its success. Partly the project was too ambitious to be realized in the time Babbage had projected, given the primitive state of the art of machining. Hand-fitting a single timepiece was one sort of craft; the production of thousands of precision parts was altogether another—and in Babbage's time unheard of.

Also, Babbage had other interests to distract him. For example, he made many trips abroad, and from one wrote a book called

Economy of Manufactures and Machinery, which was the first study
of what we would now call operations research. He consulted for
the railways and the post office, introducing safety devices to one
and the penny post to the other. He once nearly drowned himself
with a contraption he invented for walking on water, and another
time shut himself up in a sculptor's oven to experience first hand
the effects of 265°F heat on the human body. He had himself
lowered by rope to the bottom of the crater of Vesuvius ("I was
much exhausted by the heat, though I suffered still greater incon-
venience from the vapours"), and I suppose he can be considered
the father of light shows, since it was he, bored to distraction at
the opera, who conceived of using colored lights in a theater, and
put together a "rainbow dance" to demonstrate the idea's possibilities. He mounted breathtaking verbal assaults on the Royal
Society and the Royal Observatory, but he was probably most
notorious in his own lifetime for his campaign against street noises.
Here is Bowden:

> Babbage was intensely annoyed by the cries of street musicians, who,
> so he said, made it impossible for him to concentrate on his work.
> Instead of following the example of a fellow sufferer—Thomas
> Carlyle—who retreated to a soundproof room, Babbage embarked
> on a life-long vendetta against them, and tried to have them pros-
> ecuted. This public-spirited action so enraged his contemporaries that
> jeering children followed him through the streets; drum and fife bands
> came miles out of their way to serenade him, and indignant citizens
> who had an hour or two to spare made a point of having a drink at
> some local hostelry, and then blowing bugles and other instruments
> under his windows at all hours of the day and night.
>
> (Bowden, 1953)

Some subtitles of the chapter in Babbage's autobiography headed
"Street Nuisances" will give the flavor of his argument: "Instruments of Torture—Encourages; Servants, Beer-shops, Children,
Ladies of Elastic Virtue—Invalids distracted—Horses run away—
A cab-stand placed in the Author's street attracts Organs—Mobs
shouting out his Name—Threats to burn his House—Abusive Plac-
ards—An Association for the Prevention of Street Music Proposed."

Babbage felt obliged to dash off a letter to Alfred, Lord Tennyson, about this couplet in "The Vision of Sin":

> Every minute dies a man,
> Every minute one is born.

I need hardly tell you [Babbage wrote] that this calculation would tend to keep the sum total of the world's population in a state of perpetual equipoise, whereas it is a well-known fact that the said sum total is constantly on the increase. I would therefore take the liberty of suggesting that in the next edition of your excellent poem the erroneous calculation to which I refer should be corrected as follows:

> Every moment dies a man,
> And one and a sixteenth is born.

I may add that the exact figures are 1.167, but something must, of course, be conceded to the laws of metre.

(Morrison and Morrison, 1961)

Several biographers have taken him seriously here; I cannot. He was too full of fun during that part of his life, however crusty and embittered he may have been later. I like to think that the Babbage who penned this letter is the same who devotes an entire chapter in his autobiography to "wit," and who cozens the French outrageously about the eating habits of English gentlemen during a dinner with the great Laplace. In any case, Philip and Emily Morrison point out that it is a fact that the couplet in all editions up to and including that of 1850 reads, "Every minute dies a man, / Every minute one is born," while all later editions read, "Every moment dies a man, / Every moment one is born."

Babbage helped found the Astronomical Society (1820), the British Association for the Advancement of Science (1831), and the Statistical Society of London (1834).

Meanwhile—I take a breath to say it—he was at work on the machine that would doom his Difference Engine,[5] namely, his

---

[5] Lady Lovelace points out that work on the Difference Engine had been suspended for some time (a year, actually) when Babbage conceived the Analytical Engine, and that therefore the latter cannot be said to have doomed the former. But practically speaking,

Analytical Engine, a grander, bigger, all-purpose calculating machine. It would not only be capable of arithmetical calculations, but it would also be capable of analysis and of tabulating any function whatever. It was to have an enormous storage, twenty times bigger than the EDSAC, an electronic machine built by the British more than a century later, and its mill, or what we would today call its central processing unit, was to be controlled by the same sort of punched cards that Babbage had seen used in the Jacquard loom. "It would weave algebraic patterns the way the Jacquard loom weaved patterns in textiles," Lady Lovelace put it (Morrison and Morrison, 1961).

But the production problems were gargantuan and the government had long ago withdrawn its support. Babbage was forced "to try to solve by himself and with his own resources a series of problems which in the end taxed the united efforts of two generations of engineers," says Bowden (1953).

The resources weren't entirely his own. You will remember that we left Lady Byron with a baby daughter when Lord Byron ran off to the Continent to try to forget his half-sister—and dally with Claire Clairmont (among others), and inspire Mary Shelley to write *Frankenstein*. This baby daughter grew up to be a precocious student of mathematics, tutored by the famous mathematician Augustus De Morgan. She was also an attractive and charming flirt, an accomplished musician, and a passionate believer in physical exercise. She combined these last two interests by practicing her violin as she marched around the family billiard table for exercise. Her name was Ada, and she was brought to Charles Babbage's workshop during her first London season to take a look at Mr. Babbage's odd machine.

"While the rest of the party," writes Mrs. DeMorgan in her memoirs, "gazed at this beautiful instrument with the same sort

---

Babbage's heart was in his new invention, and it was left to others to build the Difference Engine (Morrison and Morrison, 1961). The British government may not have got its Difference Engine from Babbage, despite the huge investment , but it has been estimated that the work of Babbage and his associates so transformed machining that the government more than recovered its investment. U.S. scientists like to point this out to Congress members.

of expression and feeling that some savages are said to have shown on first seeing a looking glass or hearing a gun, Miss Byron, young as she was, understood its working and saw the great beauty of the invention" (Bowden, 1953). A few years after this first encounter, Ada, now married and known as the Countess Lovelace, undertook to translate a set of notes that had been taken by one L. F. Menabrea during a lecture Babbage had delivered in Italy on the subject of his Analytical Engine. This translation (in Morrison and Morrison, 1961), together with her extensive notes (which are more than twice as long as Menabrea's original text) still stands as the most lucid contemporary report of Babbage's work, one which he immediately recognized to be better than anything he had written himself. It is in large part thanks to Lord Byron's daughter that Babbage's place in the history of computing machines is recognized, for in fact the Analytical Engine was never completed, and Babbage grew ever more irascible and embittered until he died in a dreadful solitude. Perhaps hopelessly, scholars are presently trying to determine how much of an intellectual contribution the countess made to Babbage's work. Scientific partnerships deserve more study.

During his life, Babbage never ceased trying to raise funds for constructing the Analytical Engine, and until her death at age thirty-seven, Lady Lovelace was his accomplice. At one point they entertained a scheme to build and exhibit a tic-tac-toe—playing machine to raise capital; some time later, Babbage realized that his Analytical Engine could, in principle, play chess, and he considered building a quick and dirty version for raising money that way. But their final effort together was the development of an "infallible" system for betting on horseraces, a system that plunged the countess deeply into debt and threatened her name with scandal. Twice she had to pawn the Lovelace family jewels behind her husband's back, and twice they had to be secretly redeemed by her mother, Lady Byron. When Ada called Babbage to her deathbed, it was to give him instructions for paying off a particularly obnoxious London bookmaker (Turney, 1972).[6]

One statement of Lady Lovelace's has often been quoted: "The Analytical Engine has no pretensions whatever to originate anything. It can do whatever we know how to order it to perform." And this statement has been adduced as evidence that machines cannot, in any way, be said to think. It's a true statement, but a misleading one, and will bear some looking into later on. Certainly Babbage and the countess did not propose that the Analytical Engine would "think," though the countess prudently says that the actual existence of the machine and experience with its practical results would be the only way to answer questions about the machine's intelligence with any finality. But where later researchers (John von Neumann, for example) would be willing to use human terminology for at least some parts of the computer, such as memory or judgment, Babbage chose the eminently commercial terms "storehouse" and "mill."

What drove Babbage—and his predecessors, and his successors—were practical problems. In his case it was the desperate need of a seafaring mercantile society for accurate astronomical tables essential to navigation, tables that up to his time had been computed, as Bowden puts it, "by elderly Cornish clergymen, who lived on seven figure logarithms, did all their work by hand, and were only too apt to make mistakes." Later in the century, the same sorts of considerations would inspire Lord Kelvin to build his "tidal harmonic analyser," an analog rather than a digi-

---

[6] Without coming to a sensible conclusion, I've pondered the odd coincidence that Lord Byron and his daughter were involved in two quite different aspects of the history of artificial intelligence. Mary Shelley's stepsister also had a daughter by Byron, who was Ada's half-sister, though they never met. I cannot discover if, when Mary Shelley returned to England, widowed and with a young son, she and Ada ever met. I think they would have liked each other, both being highly intelligent, generous-hearted, and unconventional women. Ada was deeply fond of her other half-sister, Medora Leigh, despite the fact that she represented the illicit love affair that had driven their father out of England. Ada and her father never saw each other after Ada was a month old, though a tenderness persisted between them, and Ada asked to be buried next to him at her death. One biographer has speculated that Charles Babbage played the father's role for Ada that Lord Byron could not.

tal computer, whose object was "to substitute brass for brain in the great mechanical labor of calculating the elementary constituents of the whole tidal rise and fall." Brass for brain. It's a fine phrase, but sounds distant indeed from Homer's golden young women with intelligence in their hearts. Even the immediate forebear of the computer as we know it today, an electronic digital computer developed at the Moore School of the University of Pennsylvania during World War II, and called ENIAC, was intended solely for the calculation of bombing tables. Its general-purpose possibilities were only discovered later.[7]

In other words, practical and pressing problems have often driven research in computing, and except in Lord Kelvin's metaphorical sense, no serious person seems to have publicly admitted that these machines were anything more than a relief from drudgery. That a science might grow up around the descendants of these machines, dedicated to studying their phenomena in the same way that physicists study matter, would have floored the nineteenth-century pioneers, and there's still controversy over whether computer science is a justifiable term. That dispute pales in comparison with the one as to whether artificial intelligence can justifiably be called a science.

But it's fascinating to me that unlike computing generally, artificial intelligence—whatever its stated purposes now—did not originate in the search for solutions to practical problems, though even its severest critics agree that it has made many useful contributions to the art and practice of computing. Such contributions might have come in the course of other pursuits, but they didn't.

I like to think of artificial intelligence as the scientific apotheosis of a venerable cultural tradition, the proper successor to golden girls and brazen heads, disreputable but visionary geniuses and crackpots, and fantastical laboratories during stormy Novem-

---

[7] Or so say the history books. In a talk at the Los Alamos History of Computing Conference in the summer of 1976, John Mauchly recollected that he and his colleagues were indeed aware of the general-purpose possibilities of their machine, but were constrained by wartime restrictions to keep that knowledge to themselves and to concentrate on the immediate military applications.

ber nights. Its heritage is singularly rich and varied, with legacies from myth and literature; philosophy and art; mathematics, science, and engineering; warfare, commerce, and even quackery. I've spoken of roads or routes, but in fact it is all more like a web, the woven connectedness of all human enterprise.

Sometimes we forget that most sciences began with ideas that seem a bit loony to us now but were sound enough in their own time. If we detect lunacy among the ancestors of artificial intelligence, we'd better admit that it's our very own, and probably here to stay. We harbor that mysterious but ancient urge to reproduce ourselves in some essential but extraordinary way. Artificial intelligence comes blessed with one of the richest and most diverting histories because it addresses itself to something profound and pervasive in the human spirit.

For some, an excess of high spirits is found here, a playfulness that somehow ought to be forbidden in the sober halls of science. Luckily, hardly anybody in the field shares that view. The urge to excess and to play is as strong as ever, with contemporary Paracelsuses tweaking the dewlaps of an outraged establishment— and contemporary Frankensteins and Babbages being tormented by their own inventions. True to its speculative origins, artificial intelligence poses a set of grave moral questions while, true to its claims to be a science, it promises answers to puzzles about the nature of intelligence.

Ours. Or anyone—or anything—else's.

Chapter Two

~

From Energy to Information

A history of psychology is nothing less than a history of the effort to explain what many people still regard as ultimately inexplicable, the workings of the human mind. As philosophers, and later psychologists, illuminated bits and pieces of the mechanism, led here by speculative argument and prompted there by empirical evidence, the notion of mind slowly gave way to the notion of brain function. This latter idea suggests that the brain is an organ, fully as explainable as the pancreas or the kidneys, though surely more complex than even those awesomely complicated structures.

We left Ramon Lull writing his mystical poetry and musing over his Ars Magna, his thinking machine.[1] Though it sat square within a mystical and Christian theological tradition, the Ars Magna was remarkable because, along with its Arab forerunner the *zairja*, it was based on the assumption that human thought could be mechanized. This assumption was implicit in the work (and probably unconscious). Lull seems not to have perceived that if you consider thought to be a mechanical process, you can view human beings as machines, though of a very special kind.

Now human thought, at least until recently, has been the exclusive business of philosophers. Their world sometimes reminds

---

[1] Indeed, Martin Gardner (1958) says the Ars Magna "amounted virtually to a satire of scholasticism, a sort of hilarious caricature of medieval argumentation."

me of the floor of the Chicago Board of Trade: all that heat and hollering, from vendors whose only customers are each other. The rest of us muddle along, largely indifferent (and helpless if we cared) to the futures of the hogbellies and soybeans of philosophy—let's say phenomenology or positivism—while the vendors shout themselves purple at each other. It has always been something of a specialist's market, but in the past it was less so: nearly anyone bright enough and so inclined could set himself up to trade. Philosophy was also taken more seriously than it is now; at least, the wrong philosophy could prove lethal to its luckless promoters. Commodities market traders have a technique called hedging, by which they hope to limit their losses, and one of the greatest hedgers in philosophy was the celebrated René Descartes (1596-1650). He was by no means the first to think of human intelligence in terms of mechanisms. As in so many other things, the Greeks offer evidence of being first at that, in particular the philosophers of the Epicurean school and, from another point of view, the physicians of the Hippocratic school.[2] But for our purposes, it's the father of modern rationalism who makes a grand, if elaborately hedged, move toward conceiving thought to be the result of a mechanism.

The first instrument for studying mind was mind itself, and we have to judge it admirable, for it raised questions about itself that have yet to be conclusively answered. In the course of examining his own astonishing mental faculties, Rene Descartes concluded that mind and body are two quite different things. Accordingly, he divided human acts up into two distinct kinds, mechanical and rational. Obviously, mechanical acts were those that could be imitated by automata: walking, eating, playing the flute. The rational, however, could not be imitated: judgment, will, choice. Descartes argued that in the pineal gland, buried

---

[2] These philosophies were unrelated to the technical achievements of Heron of Alexandria, described in Chapter 1. Mechanical arts were considered too base by Greek thinkers to have any philosophical applications. Of course, the equivalence between a human and a machine would have come as no surprise at all to slaves and women, who formed the bulk of the Greek population.

deep within the brain, soul, the director of rational behavior, met and interacted with the material body.

Though what could or could not be imitated has changed, this scheme bears some surprising structural similarities to descriptions of human cognition offered by late-twentieth-century workers. And as the modern researchers have been deeply influenced by the computer model, so was Descartes influenced by the automata that were proliferating through Europe in chateau gardens and municipal clock towers. Some brave anatomists were beginning to probe the human body, and the parallels with the animated statues—and of course clocks—were irresistible. The body had a heart that "pumped" blood, or acted as a "spring"; the blood circulated through "tubes" and "valves," and so on.

Descartes, struggling with the nature of mind and body, eventually declared that animals were wonderful machines. Human beings were too, *except* that they possessed a mind. That enormous reservation would be the source of endless debate about whether animals could be said to possess souls; Descartes, acutely aware of the persecution of Galileo, hedged his bet. In his *Treatise on Man*, he seems to be at the edge of declaring at last that humans too are machines, wonderful machines, but machines nonetheless. He resisted and Cartesian dualism—the mind-body distinction—became a fact of Western intellectual life. The battle began and was taken to extremes by Descartes' enemies and advocates alike. A group of theologians published objections to his work; they proposed that the beast-automaton notion can only lead its supporters to conclude that there is a continuity in intelligence between animals and human beings. Just so, replied the Cartesians.

Descartes himself was actually more subtle. He argued that animals might lack the ability to think and the language to communicate to one another, but that they were not necessarily without consciousness, memory, emotion, or perception. Perhaps it's fairer to Descartes to say that if some model besides human beings had occurred or been presented to him where a physical system could clearly be seen to embody a symbolic one, as the computer was to do three hundred years later, he would have risked the wrath of church and state and declared mind and body one. But no such model existed, and his intellectual honesty compelled him toward what must have been an unsatisfying dichotomy. In any event, the barrier set up by Cartesian dualism that separated mind and body had to be chipped away in pieces over a long period of time with the chisels of increased scientific knowledge. It still exists in everyday speech— for example, "The spirit is willing but the flesh is weak."[3]

Descartes also held that there were two kinds of ideas. Derived ideas came directly from sensory experience—a smell, a sound, the perception of a stone bench in a garden. More important were innate ideas, developing out of the mind of consciousness and independent of sensory experience. These might correspond to what we now call "wired-in" thinking, which may or may not include the ability to acquire language. Then again, he might have meant something similar to what Michael Polanyi means by his term "tacit knowledge," things humans know but cannot (yet) phrase.

Though this précis hardly does it justice, the Cartesian system was so rich and suggestive that it was to inspire a dialectic which, in its sum, constituted a major part of Western philosophy from that time on. (It isn't quite fair to say, as Norbert Wiener does, that all the schemes that followed were concerned solely with mental content and not with mental process, but there certainly was such an emphasis.) In Amsterdam, for instance, Baruch Spinoza (1632-1677) studied the new Cartesian system in detail and found that he had to reject the dualism of mind and body. These, he believed, were merely aspects of the same thing—indeed two attributes of God.

---

[3] The mind-body separation is of course fundamental to the Western patriarchal tradition, essential not only to theology, but to politics, economics, and all other areas where women and workers have been systematically subordinated. Judging from their writings, many men have viewed these two groups, especially women, as irritating and constant reminders that the human mind and body are aspects of the same indivisible system.

Among Spinoza's visitors to Amsterdam was Gottfried Wilhelm Leibnitz (1646-1716), a man of astounding mental power who had already begun to invent the integral and the differential calculus and had perfected Pascal's calculating machine. He was on his way to the court of the Duke of Hanover, where he had just been appointed librarian. The two philosophers surely discussed the French rationalist's system, each with misgivings about it. Leibnitz went on to the Hanover court, where his activities were exhaustive, ranging from the practical to the most abstract. (I find him very modern in the way he adroitly enlarged a commission to do a genealogy of the House of Hanover from a modest family history to something more to his taste, a history of the world.) Leibnitz suggested that mind and body were indeed separate, but exactly matched, giving meaning to each other in a system of corresponding monads, clocks wound up to keep time together for eternity.

In addition, Leibnitz's travels and conversations with scientists all over Europe made him yearn for a common language among scientists, so they could not only disseminate ideas, but also discuss them clearly and rationally. Thus, he dreamed of reducing reasoning to an algebra of thought, a *calculus ratiocinator*. This idea emerged again in the nineteenth century with the mathematics of George Boole, and over a half century later in the *Principia Mathematica* of Whitehead and Russell. It was to beguile more than one researcher in artificial intelligence later still, when efforts were made to find a universal grammar, common to humans and computers alike.

With these early thinkers, then, one sees the enormous force of mind examining mind. The theories are consistent with the systems they spring from, but there's no way to show that one is better than another, that one comes closer than another to describing what really goes on in the human head. You must simply trust personal preference. Yet a rational approach to mind must have been a seductive idea in the late seventeenth and early eighteenth centuries: nearly every thinker of note had a fling at it.

One reason was the inspiration of Newtonian mechanics. If the physical world could be described—no, explained—with such elegant coherence as the Newtonian synthesis permitted, why not everything else? Of course, as J. D. Bernal (1974) points out, Newton's theory of inertia, say, came from prevailing religious ideas, and his general model corresponds with the new social and economic order, where individual enterprise was replacing the fixed feudal hierarchy. Invention, discovery, and inspiration are complex, and are never accomplished in a vacuum. Nevertheless, Newton's contribution was stunning, an appeal to natural law to explain what had heretofore been explained by dogma. Newton himself wrote in the preface to the *Principia*, "I wish we could derive the rest of the phenomena of Nature by the same kind of reasoning from mechanical principles. ..."

Thomas Hobbes (1588-1679) took him at his word. Impelled by political events that shocked him, and by the new mechanics, which inspired him, Hobbes stated his theories of the mind in terms drawn from both. Every aspect of human behavior is simply evidence of internal motion, he declared, inspired not by gravity, but by fear and self-interest. Hobbes was also one of the first to observe the associative aspects of mind, that thoughts are linked to one another in ways that cannot necessarily be called logical, but are rather associated in odd, contingent ways. Thus he distinguished between free association and controlled or purposeful thinking. Hobbes was explicit in his debt to Newtonian mechanics, but the same model informed psychological thinking until the 1960s. Perhaps its most famous instance is Sigmund Freud's psychoanalytic theory of mind, which deals with pressures and discharges and drives in terms that would make a hydraulic engineer feel right at home.

John Locke (1632-1704), more optimistic than Hobbes, believed in the rationality of man; in particular, he supposed ideas to come both from experience and from inner reflection upon sense data. He suggested that complex ideas, however abstract they might seem, were built up from simple sensory ones, or from reflecting on those simple ideas. What did he mean by reflection? Nothing very clear. Gardner Murphy (1972) writes, "Locke's greatest contribution to psychology thus lay in making explicit the possibilities of a theory of association which should start with the data of experience and work out the laws governing the interconnections and sequences among experiences."

It was David Hume (1711-1776) who proposed such laws. In 1739 he published *A Treatise on Human Nature*, reemphasizing Locke's theory of complex ideas as compounds of simple ones, and going on to assert that mind is nothing more than the flow of ideas, sensations, memories, and reasoning. Impressions are what we today would call sensation or perception, while an idea is the mental experience we have in the absence of any stimulating object. Hume stressed that though complex ideas are compounded from simple ones, they need not then resemble any simple ideas, since in the course of becoming complex, novel combinations appear. His laws of association proposed the means by which this transformation takes place: resemblance or similarity, and contiguity of time or place.

Hume's work is in the mainstream of British empiricism, and the fates of two contemporary physicians probably best illustrate the division that was growing between the empiricists in Britain and the rationalists on the Continent. David Hartley (1708-1757), a British physician, had a quiet medical practice that allowed him plenty of leisure to speculate on the genesis of thought. He was acquainted with the associational theories, but believed that a physiological basis was needed to account for them. Inspired by what he knew of Newton's work with the pendulum, he suggested that nerve fibers were set in motion in an order corresponding to experience. A human being is born with the capacity for sensory experience accumulating in increasingly complex ways. Hartley eventually published a work that brought together the basic ideas of a dominant school of psychology called the associationists. For his synthesis, he was honored as the founder of this important school.

In Paris, however, in 1747, a French physician by the name of Julien Offray de la Mettrie, a ladies' man and a clown, stood the philosophical—and theological—world on its ear with the publication of his book *L'Homme Machine*. He had read Descartes, he had read the rest of them, and they all seemed to be talking rubbish, words without substance, speculation without knowledge. Practitioners tend to feel that way about theorists. La Mettrie was a physician and had seen human bodies in every state, and he knew that one's mental processes were profoundly connected to one's physical state. Indeed, the idea seems to have come to him after he suffered an attack of cholera on the battlefield, when he knew himself to be fevered and irrational. Suddenly he could gather evidence from every direction that physical substances affected thinking: diet, pregnancy, drugs, fatigue—he gave instance after instance from his own observations and experience. He too used the language of machines, but with the deliberate, conscious intention to shock: "The human body is a machine that winds up its own springs; it is a living image of the perpetual motion" (Vartanian, 1960).

At last we are away from the desk and into the laboratory, or at any rate, the surgery. Here is empirical evidence offered in behalf of a comprehensive psychological theory. But the construct is exactly that—a theory, a model, not a statement of the absolute truth of things. The notion of human as a machine is a heuristic—a rule of thumb or point of view, from the same Greek root as *eureka*—to aid in an understanding of how human beings work. To call us machines does not define us, and it will not "encompass the human essence." For La Mettrie, thinking is essentially symbolic in nature, says Vartanian, "and, as the system of coding grows in complexity and precision, thought becomes clearer and more comprehensive, or simply truer. ..."

The reception of La Mettrie and his ideas was a shabby, and from this distance, slightly comical episode in intellectual history. He had already been hounded out of Paris when he published *L'Homme Machine* anonymously in Holland; when he was discovered to be its author he had to make another dash for it, this time to the liberal court of Frederick the Great in Berlin. Here he joined such intellectual luminaries as Voltaire, but his clowning seems to have offended everyone save his patron. He died in exile, pining for France, and his fellow philosophers, thoroughly annoyed with him, set about as quickly as they could to suppress his work and make of him what we today would call a nonperson.

La Mettrie's spiritual godchild was Diderot (1713-1784), the encyclopedist who explored the relation of humans and machines —and humans as machines—in an even richer way, bringing his enormous store of knowledge and his graceful pen together to suggest that not only is the human being a machine, but that technology must be humanized. By the end of the eighteenth century, human-as-machine was a commonplace, and no one was especially surprised when Cabanis declared the brain to be an organ that somehow digests sense impressions and secretes thought.

The power here is in the metaphor. It becomes richer and more flexible, from the automata of Descartes to the brain simply as organ, something to be studied and eventually understood. There was much further to go than anyone guessed. There often is. La Mettrie and Diderot were making assertions on faith, writes Vartanian (1960), concerning the ultimate fecundity of the mechanistic method in bridging the gap between the living and the nonliving, and between the conscious and unconscious aspects of a presumably unitary nature.

The ultimate corruptibility of the mechanistic method was left for Mary Shelley to suggest.

It was Immanuel Kant (1724-1804) who embodied the Continental, antimechanistic view. In some ways he reminds me of his intellectual forebear Leibnitz, most especially in his versatility. He made substantial contributions to a theory of knowledge and to ethics and aesthetics; he lectured on logic, mathematics, physical geography, and even on fireworks. But we are interested in what he had to say about the way the mind knows. In *The Critique of Pure Reason*, Kant suggested that the mind has a priori principles which make things outside conform to those principles. In other words, the shape of the world is a function of our minds, and not of the world itself. Modern brain research seems to be confirming some aspects of Kant's beliefs, though not in precisely the way he intended; the question of how detailed and mutable these a priori principles are—in language, to give an example—is still in question. But as Seymour Papert writes,

> Kant's doctrine could neither be translated into material terms nor adequately developed on a clear, logical basis without concepts of representation and computation that would not come into being until our own period. ... A split eventually had to come between psychology, which was based on mechanism but unable to reach the complex properties of thought, and philosophy, which took the properties of thought seriously but could be satisfied with no conceivable mechanism.
>
> (McCulloch, 1965)

Thus the nineteenth century saw neo-Kantians attempting to construct a science that would combine what was being discovered in physiology, psychology, ethics, epistemology, and anthropology; it was to be a systematic, scientific approach to the problem of knowing. The German desire to unify all phenomena was in sharp contrast to the British and French practices of dividing them up and declaring some amenable to exact study and others not. It's germane to add that Kant also articulated a belief that was fundamental not only to psychology but to science itself. Quantitative descriptions, he said, though they might not get at the ultimate nature of things, are the only means we humans have of exchanging information that will be orderly, coherent, and relatively resistant to distortion. *Though they might not get at the ultimate nature of things*: it's a caveat easily forgotten.

Late-nineteenth-century investigations into learning and memory were true to that part of the Kantian spirit. An American, E. L. Thorndike, was apparently the first to use, or in any case to write about, his experiments with a "puzzle box," a box from which an animal had to figure out how to escape. Learning and intelligence became matters of how quickly the animal could solve that puzzle initially and on successive occasions.

Thus, intelligence was thought to demonstrate itself in problem solving, and that behavior could be quantified. These were the underlying assumptions of the Frenchman Alfred Binet (1857–

1911) and his followers, who devised tests to measure the intelligence of schoolchildren. In these tests, children were asked to name objects, compare lengths of lines, complete sentences, and answer questions.[4] Though these tests were revised continuously by Binet until his death, and by his followers thereafter, neither the objectives nor the methods of the tests changed. Such tests are familiar to any reader of these lines. Binet regarded intelligence as a combination of faculties—which is what he tested for—including the ability to understand directions, maintain a mental set, and correct one's own errors. And IQ, a concept that would bedevil schoolchildren and their parents alike for the next sixty years at least, was presented as the quotient of the mental age of the subject, as demonstrated by his or her ability to solve these problems, divided by the chronological age and multiplied by 100.

While Binet was confident that he knew what intelligence was, and moreover that he could measure it, endless debate ensued over how humans accomplished what everyone "knew" was intelligent behavior. The atomistic theories of psychology had given way to a consideration of organic wholes, led by the German psychologist Max Wertheimer (1880-1943), the founder of Gestalt psychology. Gestalt psychology held that the primary data of perception are not elements but significantly structured forms. The Gestaltists wished to apply the concept of Gestalt, or shape, "far beyond the limits of sensory experience," wrote Wolfgang Kohler, one of Wertheimer's colleagues. "According to the most general functional definition of the term, the processes of learning, of recall, of striving, of emotional attitude, of thinking, acting and so forth, may have to be included." Thus the Gestaltists saw the primary brain process as a dynamic system, a continuous organizing and patterning that takes place as sensory experience comes pouring in. The process is spontaneous; it does not have to be learned. The Gestaltists

---

[4] I'm especially taken by one question that any normal French schoolchild over six at the turn of the twentieth century could answer. It could not be answered with certainty during the last quarter of the century by several college professors of my acquaintance. The question is, "——— dragged the body of——— around the wall of ___." Fill in the blanks. Nearly everybody gets Troy.

attacked as artificial the kinds of experiments Thorndike had done with his animals in the puzzle box, since there the solution to the problem was hidden and the animal prevented from seeing and relating all of the elements to form a sense of the whole. Alas, the animals did solve the puzzles, artificial or not, and the Gestaltists were left open to counterattacks that they were too vague, lacking in scientific rigor, and bereft of empirical data to support elaborate theories (Murphy, 1972). In a sense, it's the Kantians versus the Humeans again, elaborate theory versus tractable, measurable, testable evidence. But the Gestaltists were on to something: in attending to process as well as content, they were beginning to respond to a new spirit.

In the United States, write Newell and Simon in their historical addendum to *Human Problem Solving* (1973), a great gap existed in research on complex human cognitive processes from the time of William James almost to World War II. They state,

> Although the gap was not complete, it is fair to say that American cognitive psychology during this period was dominated by behaviorism, the nonsense syllable, and the rat. Hull's doctoral thesis (1920) on concept formation was a notable exception, but his desertion of this problem for others more compatible with the Zeitgeist is typical of the period. Not only that, it was widely interpreted as showing the futility of a direct approach to higher mental processes.

But approaches were possible. That news came from outside traditional psychological research. And of all the odd places, it came from a fancy kind of engineering called cybernetics. Before considering cybernetics, however, I'd like to return to Leibnitz who, says Norbert Wiener, is the best candidate for the post of patron saint of cybernetics.

When I spoke earlier of Leibnitz, I mentioned his *calculus ratiocinator*, a calculus of reasoning. Leibnitz longed for a universal scientific language in which workers could exchange scientific ideas,[5] and he understood that to manipulate this universal language the

---

[5] A part of the dream has come to pass, though the language is English: at a scientific conference I'm fascinated to watch a Japanese and a French scientist speak—however haltingly—in my mother tongue about ideas that are altogether alien to me.

calculus was needed. The calculus would "facilitate the process of logical analysis and synthesis by the substitution of compact and appropriate ideograms for the phonograms of ordinary language," write Lewis and Langford in their survey of symbolic logic (1956).

So Leibnitz recognized the necessity for expressing symbolically not only the propositions of logic, but also the relations between them. For a variety of reasons he couldn't accomplish his goals, and to him belongs credit for prophecy rather than accomplished fact. It's a seductive idea, this invention of a system of symbols that would allow a compact, precise representation of the terms of a syllogism. I'm speaking here of what Aristotle considered the *sine qua non* of thinking, presented in the familiar form, "All men are mortal; Socrates is a man; therefore ..."

The important work was done by English mathematicians between 1825 and 1850. Among them were Sir William Hamilton and Augustus De Morgan (the latter we have already met as Lady Lovelace's tutor). But the eponymous hero of the story is George Boole, who taught at Queens College, Cork, and who laid the foundations of modern symbolic logic in *An Investigation of the Laws of Thought on Which Are Founded the Mathematical Theories of Logic and Probabilities*, published in 1854, and commonly referred to as Boole's *Laws of Thought*. Boole wrote in his preface, "The laws we have to examine are the laws of one of the most important of our mental faculties. The mathematics we have to construct are the mathematics of the human intellect." Indeed. Symbolic and traditional logic are concerned with general principles of reasoning, but where traditional logic uses words, symbolic logic uses ideographs, which minimize the ambiguity of natural language. Boole's system had "elective symbols," meaning arbitrary designations for classes of existing things, and "laws of thought," the rules of operation on these elective systems, rules which, most significantly, would hold in an algebra of the numbers 0 and 1.

Boole's algebra was subsequently refined by others who followed him, culminating in Whitehead and Russell's *Principia Mathematica*, which aimed to show the nature of mathematics and its relation to logic. Lewis and Langford write, "It is difficult or impossible to convey briefly the significance of this achievement. It will be increasingly appreciated, as time goes on, that the publication of *Principia Mathematica* is a landmark in the history of mathematics and of philosophy, if not of human thought in general" (1956).

Boole figured elsewhere, too. With breathtaking insight, a young engineering student at Massachusetts Institute of Technology earned his master's degree in 1937 by using Boolean algebra to describe the behavior of relay and switching circuits. He was Claude E. Shannon, and he figures prominently in the history of science in the twentieth century. We'll say more about Shannon's role further along. Now it is only necessary to say that it wasn't farfetched to ask if, since the laws of thought could express the behavior of electronic circuits, electronic circuits could express thought.

All these things—intelligence tests, logic, algebra—seemed potent. But hardly anyone raised one essential question, which is this: How much do logic and consistency really have to do with human thinking? Are these qualities, as Aristotle believed, the *sine qua non*? Or is human thinking more various, encompassing the rigor of logic among several—maybe many—other modes of thought? Is it basically irrational to exclude the irrational as a component of thinking? Are there several kinds of irrational, that is, nondeductive, nonlogical ways of thinking which form an essential part of human cognition? The evidence from brain physiology suggests that Aristotle had something. Those most abstract functions of humans, such as reasoning and language, do seem to be more or less localized in the neocortex, the part of the tripartite human brain that evolved most recently, and, by evolutionary time scales, with a burst. I'm hedging myself here because the neural connections among the parts of the brain are much more intimate than a tripartite model might suggest; what's more, the other functions of mind, such as emotion and altruism and ritual, localized in the older parts of the brain, are not to be denied as part of the human cognitive process.

But whether human thinking comprised more logic than lust, was it appropriate or even possible to capture its processes in mathematical terms? Lots of people thought so, for reasons grand and small.

If you set out to write a best-selling potboiler, there's no use writing it in, say, Sanskrit. Even if you reach every Sanskrit reader presently alive, you won't sell very many copies. Better to write it in English or Chinese, for which the apparatus for printing and disseminating is relatively efficient, and the audience is large. In the same way, the scientist who could explain a certain aspect of human thought with a mathematical expression had not only a ready-made language, but a ready-made audience. Using them has considerably more appeal than inventing your own language and teaching it to the reluctant unwashed. The grand reasons were equally compelling. A mathematical expression lends a nigh irresistible sense of universality to the phenomena it means to describe. Surely the general rules of thought, the great principles of intelligence, were universal. If an agent behaved intelligently in one task, it was only a matter of adapting those same rules to another task. By expressing such rules mathematically, one would climb to appropriately high scientific altitudes of abstraction.

In 1958, John McCarthy proposed that all human knowledge be given a formal, homogeneous representation, the first-order predicate calculus. Let us, he said, construct theorem provers that will piece together symbolic expressions to reason about the world's knowledge. And in the early 1960s, a logician named Alan Robinson published a paper on what he called the Resolution Method, a highly machine-oriented and efficient means of proving theorems in the predicate calculus. Many saw this method as "an engine which would finally realize McCarthy's dream," as one researcher put it, not to mention the dreams of all those who had gone before. And that effort was to occupy a great deal of research time in the mid-1960s. I will take up the Resolution Method further in Chapter 10 in the discussion of robotics and the problem of general intelligence.

But let us return to what is really an account of the development of information theory. Two editions were published of the seminal book *Cybernetics* by Norbert Wiener. The first came out in 1948,[6] and it was in that first edition, Wiener later wrote, that the ideas seemed odd, or even shocking. In the second edition, published in 1961, those odd ideas had become so familiar and widely used over the intervening decade that he now worried the book might be considered trite or commonplace.

Cybernetics recorded the switch from one dominant model, or set of explanations for phenomena, to another. *Energy*—the notion central to Newtonian mechanics—was now replaced by *information*. The ideas of information theory, such as coding, storage, noise, and so on,[7] provided a better explanation for a whole host of events, from the behavior of electronic circuits to the behavior of a replicating cell. One reason for this is that the old Newtonian mechanics had dealt with closed, conservative systems, while the information-theory model could deal with open systems, that is, systems coupled to the outside world both for the reception of impressions and for the performance of actions, and where energy is simply not the central issue. It's no wonder that mainstream psychology, still enthralled by concepts drawn from Newtonian mechanics, was first, not alert to, and then resistant to so drastic a change in paradigm.[8]

Norbert Wiener (1894-1964) had been brought up in Cambridge, Massachusetts, where his father was a professor of phi-

---

[6] The galleys of the first edition were proofread by Wiener's young assistant, Oliver Selfridge, who would become a key figure in early artificial-intelligence research.

[7] These terms mean pretty nearly what you'd think. *Coding* refers to "a system of signals used to represent letters or numbers in transmitting messages" (Wiener, 1961); storing means holding these signals until they're needed. *Noise* is a disturbance that obscures or affects the quality of a signal (or message) during transmission.

[8] Herbert A. Simon has brought to my attention the presidential address before the Eastern Psychological Association by Edwin G. Boring in 1946. While Boring has no notion of the information-processing theory of modeling the mechanisms of mind, he reports a discussion with Wiener who "defied me to describe a capacity of the human brain which he could not duplicate with electronic devices. I could not at once name him any, and I confess that I myself thought it would be salutary to show that all human mental functions can have their electronic analogues. I lacked, however, an inventory of the functions and thus could not be sure that there was not some psycho-

lology at Harvard. Wiener had been a prodigy—the first volume of his autobiography is called, with a touch of melancholy, *Ex-Prodigy*. He received his Ph.D. from Harvard at the age of eighteen. From there he went first to study logic with Bertrand Russell and then to Germany to study with David Hilbert, whom he admired for his combination of "abstract power with a down-to-earth sense of physical reality." The admiration wasn't precisely reciprocated. Wiener must have been an abrasive young man (Reid, 1970).

His life for the next few years was a wistful search for a niche, and around 1920 he settled down in exasperation at Massachusetts Institute of Technology's mathematics department, a place of no particular distinction at the time. Yet it was a fortunate choice: he and his department and his institution grew slowly in prestige over the next few decades until all three achieved worldwide eminence.

Wiener loved to work in what he called the boundary regions of science, the no-man's-land between the various established fields. Specialization, as necessary as it was, seemed to him only one way of doing science. The partnership of a team of scientists, "each a specialist in his own field but each possessing a thoroughly sound and trained acquaintance with the fields of his neighbors" and familiar with their intellectual customs—this was the proper means of attacking the most interesting problems. With his friend the Mexican physiologist Arturo Rosenblueth, he dreamed of an interdisciplinary institute of independent scientists "working together in one of those backwoods of science, not as subordinates of some great executive officer, but joined by the desire, indeed by the spiritual neces-

---

logical function left over, one which a nervous system could perform and an electronic system could not." Here Boring also refers to the notion of search as an intelligent procedure, a vague notion of what Simon now calls the chunking process (ways associations are made), a sense of an information-processing view of the nature of symbols, and a proposal for a Turing-type Test four years before Turing is to suggest it (see Chapter 3). A study of robotics, he says, will force psychologists to stop using vague terminology now incapable of rigorous definition. Boring and one or two other psychologists were in the minority: the dominant paradigm was stimulus-response. (Boring, 1946).

sity, to understand the region as a whole, and to lend one another the strength of that understanding" (Wiener, 1961).

One issue that fascinated them both was the kinds of analogies that could be made between electronic devices and biological devices. Before they had a chance to explore this problem in any detail, however, World War II intervened. It forced Wiener to turn his attention to defense issues. Immediately prior to this, Wiener claims to have submitted to Vannevar Bush a set of specifications for what would later be incorporated into general-purpose digital computers, namely, a numerical central processor whose mechanism would be electronic and not mechanical, based on a binary rather than a decimal system; a machine with built-in abilities to make logical decisions, and an apparatus for easy storage and manipulation of data. If a document containing this material existed, it has not been found; in any case, Wiener reports that at that stage of preparation for the war, the machine did not seem important enough to make immediate work on it worthwhile. His specifications were all ideas, however, "which are of interest in connection with the study of the nervous system," Wiener wrote hopefully.

Meanwhile, he and his colleague Julian Bigelow were at work on another project, a means of improving antiaircraft artillery. They began to study how feedback operates, and called on Rosenblueth to advise them on the problems of excessive feedback in the human body. With what Rosenblueth was able to add to what Wiener and Bigelow already knew about servomechanisms—devices that rely on feedback to operate a mechanism, such as the thermostat that operates a furnace—the three of them devised a model of the central nervous system that explained some of its most characteristic activities as circular processes, emerging from the nervous system into the muscles, and reentering the nervous system through the sense organs. This model appeared in a well-known paper called "Behavior, Purpose and Teleology," which was published in *Philosophy of Science* in 1943. It proposed a way of embodying purpose, and Rosenblueth and Wiener regarded it as the statement of a program for future experimental work in their interdisciplinary institute, which they still hoped to found after the war.

At this time, when such a systems approach to biology, and in particular, mental function, is common in biology textbooks, it's difficult for us to imagine the surprise of any biologist who happened to pick up that issue of *Philosophy of Science.* It certainly astounded the regular readers.

With Bigelow, Wiener had also turned his attention to other sorts of problems, and they found themselves making a science out of what had until then been the art of engineering design. What impressed them was "the essential unity of the problems centering about communication, control, and statistical mechanics, whether in the machine or in living tissue," Wiener wrote. But there was a depressing lack of unity in the research on these problems, and the existing terminology was awkwardly repetitive, biased toward whatever field it happened to come from. Thus they decided to call the entire field of communication and control theory, whether in the machine or the animal, by a name they coined, *cybernetics,* from the Greek word for steersman. Son of the philologist that he was, Wiener also meant by this term to honor the nineteenth-century physicist Clerk Maxwell, who had written the first significant paper on feedback mechanisms, an article on governors, which word itself is derived from a Latin corruption of the Greek *kybernetes,* or steersman.

In 1942, at a meeting sponsored by the Josiah Macy Foundation in New York, Arturo Rosenblueth presented the ideas that would appear the following year in the *Philosophy of Science* paper. In the audience was Dr. Warren McCulloch of the University of Illinois Medical School, who had long been interested in the organization of the cortex of the brain. McCulloch, a neurophysiologist, had already begun working with Walter Pitts, a mathematician, on a mathematical description of certain neural behavior. In 1943, the same year that the Rosenblueth-Wiener-Bigelow paper would appear, McCulloch and Pitts would publish "A Logical Calculus of the Ideas Immanent in Nervous Activity" in the *Bulletin of Mathematical Biophysics.*

The opening sentence of its abstract resounds with the ideas of Leibnitz, of Whitehead and Russell, and of the logician Rudolf Carnap: "Because of the 'all-or-none' character of nervous activity, neural events and the relations among them can be treated by means of propositional logic." The paper goes on to describe a logical calculus and principles for constructing a class of computing machines that would permit the embodiment of any theory of mind or behavior provided only that it satisfied some very general principles of finitude and causality, says Seymour Papert of MIT, and referring to the Rosenblueth-Wiener-Bigelow paper and the McCulloch-Pitts paper, goes on to add, "these two papers introduce so clearly the new frame of thought that their publication could well be taken as the birth of explicit cybernetics." In fact, the term itself only came into use in 1947 (McCulloch, 1965).

Thus McCulloch and Pitts invented and drew upon the information-theory ideas that were astir in the times, being formulated independently and almost simultaneously by Wiener and Claude Shannon and, before them, implicit in the work of Alan Turing, which will receive attention in the next chapter. The cybernetics theory of thought, with feedback as a central notion, seemed to promise the possibilities of imitating human cognition by modeling systems of neurons. This modeling was not based on detailed biological knowledge of the natural cell, which we didn't have (nor for such purposes do we yet have). Instead, it seemed certain that the correspondence between the on-off behavior of the neuron and the on-off behavior of the electronic switch would be sufficient to allow significant modeling of neural systems, and then intelligent behavior. Its basic assumption was that brain cells were on the whole general-purpose, organized for specific functions because of external stimuli. McCulloch was convinced of it, and his forceful personality inspired a flock of young researchers around him at MIT, where he eventually moved, to work at developing such systems. Sadly for him and for them, the approach was to prove sterile.

Digital computers would indeed come to simulate human cognitive processes, but the approach would be quite different from that of biology or cybernetics. It would be called the infor-mation-*processing* level of modeling, quite distinct from informa-tion theory, and its central idea would be the *manipulation of symbols*, as opposed to mere feedback, or on-off technology. In-formation processing was an intermediate level of modeling, which admitted the use of both mathematical and nonmathematical expressions, allowing the formulation of many more hypotheses of brain behavior while at the same time making understanding at the cell level unnecessary. The method was admirably suited to—indeed, inspired by—testing and experimentation on the digital computer, for its singular and most powerful feature was to view the computer as a species of information processor, sym-bol manipulator, a view which could also be applied to human beings. And, sure enough, the information-processing model would come to dominate cognitive psychology thirty years later, while the attempt to imitate cell behavior produced only trivial results, then withered and died.

Chapter Three

The Machinery of Wisdom

"May not one be admitted to inspect the machinery of Wisdom? I feel curious to know how thoughts—real thoughts are born. Not that I hope to win the secret," Lady Blandish writes to that ardent fan of the scientific method, Sir Austin Feverel, in George Meredith's *The Ordeal of Richard Feverel*. Had Meredith heard of Charles Babbage's remarkable Analytical Engine, or of the Countess Lovelace's description, published in 1843, which addressed the question of whether the machine can think, and cautiously left it open? Had he come across George Boole's *Laws of Thought*, first presented in 1847? I do not know. Meredith may simply have absorbed the machine imagery that saturated the mid-nineteenth century, and perhaps Lady Blandish's skepticism is his own. In any event, thought has been associated with computing machines from their beginnings, for, after all, they were conceived as a means of evading the drudgery of thinking.

Consider an intellectual descendent of Babbage's, Leonardo Torres y Quevedo, an early twentieth-century member of the Royal Academy of Sciences in Madrid. Torres devoted himself to automatic devices of various kinds, especially calculators, and he also built two electromechanical chess automata for the endgame of king and rook against king. He declined to claim that his automata were actually thinking, but in 1915 he wrote,

"The inventor claims that the limits within which thought is really necessary need to be better defined, and that the automaton can do many things that are popularly classed with thought" (Randell, 1973). The year before, he had published a paper that appeared both in France and Spain in which he took up the subject of what he called Automatics, "another type of automaton of great interest: those that imitate, not the simple gestures, but the thoughtful actions of a man, and which can sometimes replace him." Descartes would have been fascinated.

Torres drew a distinction between the simpler sort of automaton, which has invariable mechanical relationships (the self-propelled torpedo was his example here; he had developed and demonstrated such a device) and the more complicated, interesting kind, whose relationships between operating parts alter "suddenly when necessary circumstances arise." Such an automaton must have sense organs, that is, "thermometers, magnetic compasses, dynamometers, manometers," and limbs, as Torres called them, mechanisms capable of executing the instructions that would come from the sense organs. The automaton postulated by Torres would be able to make decisions so long as "the rules the automaton must follow are known precisely." But he reminded his readers of the difficulties that Babbage could not overcome, the problems of mechanical engineering that were simply intractable at the time. Torres was not only gloomy about the progress mechanical engineering had made since Babbage's time, but he didn't hold much hope for electromechanical devices either, since their reliability was so low, and one small failure could make nonsense out of the most complicated procedure. He was right to be gloomy: reliability was a problem that would plague computers until the invention of the transistor. For his part, Torres thought that the answer might lie in telekinetics, that is, machines that would execute orders given them by radio.

Servomechanisms, which is what we would call Torres's machines, appeared in great numbers in the early twentieth century, and were the immediate inspiration for the field of cybernetics. But modern digital computers, general-purpose, program-controlled machines, were not operational until World War II.[1]

The honor for the first general-purpose program-controlled digital computer up and running (and we're talking about matters of months here) doesn't seem to belong to the Americans, who have long claimed it, or even to the British, who have recently been declassifying evidence that they were ahead of the Americans. Instead it goes to Konrad Zuse, who was then a young German engineer. Nearly all of his machines were destroyed during the war, but patent applications seem to support Zuse's claim.

Impatient with the labor necessary to do certain kinds of calculating, Zuse had put together a machine that took up a good part of the parlor in his parents' Berlin apartment. Its possibilities as a thinking machine were clear in his mind: by 1943 he was wondering whether it could play a master in chess. In 1945, he had developed a programming language called the *Plankalkül*, which, he felt certain, could be used for solving not only mathematical problems but also many other symbolic problems, such as chess moves, though he believed that real artificial intelligence was one or two generations away. Isolated by Germany's defeat and postwar prohibitions against electronic development, Zuse was unable to exploit his pioneering work. When news came to him in the mid-1950s of the AI work underway in the United

---

[1] The first edition of *Cybernetics*, even though it was published in 1948, has only one chapter on computing machines, and while some attempt is made to compare the all-or-none aspects of the neuron with the all-or-none aspects of the computer, and the logic of the computer with the logic of the brain, the discussion is colored by worries about machine reliability. There are other references to computers in that first edition, including a final note on the possibilities of constructing a chess-playing machine to play chess "not so manifestly bad as to be ridiculous"—an amusing statement from a man who, Marvin Minsky recalls in an interview, played ridiculously bad chess himself. "Norbert Wiener used to sit in the Faculty Club at MIT and play chess with other MIT people, and he'd usually lose. He wasn't very good at it and he complained that the trouble with chess was that you mustn't make any mistakes. He had much better plans, he thought, for how to take all his opponent's pieces away, and get his king, but Wiener would always make a mistake and lose his queen or something early in the game, so he couldn't carry out these plans. Chess players don't get credit for these ambitious schemes because if there's the slightest flaw in them, then the whole thing collapses."

States, he was appalled, he remembers, at what seemed to be their lack of concern with the consequences of what they were doing. "They were like children playing with matches," he said to me in 1976. "I was shocked. I see no reason why machines cannot think, and it wouldn't surprise me if they someday out-think humans. But. ..."

Of course, to ask whose computer was first does violence to the reality of the situation. Roughly parallel, independent efforts to build such machines were taking place in Germany, Great Britain, and the United States. The connection between computing machines and thinking was explicit in all the major computer efforts. Nowhere was it more so than in the work of a remarkable British mathematician and logician named Alan Turing (1912-1954). I find Turing to be one of the most appealing figures in this history, though people who knew him are divided. Some remember him as a delightful if eccentric acquaintance, while others were put off by his abrasiveness.

Turing was the second son of British parents in the Indian Colonial Service, though he was born and brought up in England. According to his mother, Sara Turing, who wrote an affectionate and gentle memoir of her son after his death, he showed his remarkable gifts early (S. Turing, 1959). He studied mathematics at Cambridge, but despite his insight and genius he obtained no better than second-place honors when he took his honors examination in mathematics. Like many gifted people, he found it hard to put his mind to things that didn't immediately take his interest. At about the same time as these examinations, his first publication came out and, much to his honor at age twenty-two, he was elected into a Fellowship at King's College (to the pleasure of the provost, Sir John Stoppard, since Turing was so cheerful in contrast, thought the provost, to other mathematicians). Turing's undergraduate days echo Babbage's (whose work he knew and admired)—a lively social life, including rowing, running, and playing the violin. As to the latter, he mainly taught himself, and never got beyond a very elementary stage, as his mother puts it, though he continued scratching away at it the rest of his life.

Not long after he graduated, Turing published a paper called "On Computable Numbers with an Application to the *Entscheidungsproblem*" in the *Proceedings of the London Mathematical Society* in 1937. Had he done nothing else, this paper would have earned him a permanent place in the annals of mathematical logic. In it he proved that certain classes of mathematical problems cannot be solved by any fixed and definite process. By definite process, he meant something that could be done by the automatic machine he proposed, an abstract universal computing machine that has come to be known as a Turing machine. What the Turing machine can do is of more interest.

This machine, it must be imagined, has passing through it an infinitely long tape divided into squares, each of which either has one of a finite number of symbols on it or is blank. The machine can scan only one square at a time, and is capable of moving the tape, one square at a time, backwards or forwards. It can erase a symbol or print one. With just these primitive operations, Turing showed that his universal machine was capable of performing any number of programs expressed in the binary code of zeros and ones (remember Boole's *Laws of Thought*, expressible in an algebra of zeros and ones). In other words, if we can express precisely the steps needed to accomplish a task, the task itself can be programmed and carried out by the machine in this astoundingly abstract way. Turing's universal machine can in theory carry out any computing task that any special-purpose automaton can do.

This discovery had profound implications for mathematics and computing. It meant that the algorithm (a procedure or list of instructions for solving a problem), which had not until now been precisely defined, could at last be precisely defined in terms of a Turing machine. This is why some classes of problems could finally be shown to be unsolvable algorithmically.

Curiously, nothing like this machine really existed when Turing wrote his paper. The imagined device had some similarities to Babbage's Analytical Engine—I've already said that Turing knew and admired Babbage's work—and its logical design contains ideas that would later be incorporated into all digital computers. Without the technology to produce real computers, Turing perfected a conceptual model capable of embracing all known computers. Like Babbage, Turing was eager to see his ideas realized. In reporting this, his obituary in the London *Times* later observed, "The description that he then gave of a 'universal' computing machine was entirely theoretical in purpose, but Turing's strong interest in all kinds of practical experiment made him even then interested in the possibility of actually constructing a machine on these lines."

A strong interest in all kinds of practical experiment. I like the story I first heard from Seymour Papert, which Sara Turing also reports. Turing once had a bicycle whose chain would fall off periodically as he was riding it. He observed and soon discovered that the chain disengaged after a certain number of revolutions of the pedals. So he would keep count and leap from his bike just before the chain was due to come off, and fix it back on. After a while it was tedious to count, so he rigged up a counter on the handlebars. The problem, as it happened, was a faulty link in the chain that met a bent spoke; a bicycle mechanic could have fixed it in five minutes. But Turing relished doing it *his* way. Another time, en route to the United States in the fall of 1936, where he was to spend two years at Princeton doing graduate work in mathematics, he struggled aboard with a cumbersome old sextant so he could read his position all the way across the Atlantic.

At Princeton Turing worked with the logician Alonzo Church, and was eventually offered a position as John von Neumann's assistant, which he turned down to return to King's College, Cambridge in 1938. But by 1939 it was clear that Britain would soon be at war, and Turing left the university to go to work for the British government on what was one of the most strategically significant, and at the same time intellectually interesting, projects of the war. It was called "Ultra," and more than thirty years after the end of the war, much of its work is still classified. However, we know the outlines.

In September 1939, a center was set up in Bletchley Park, a country house some fifty miles north of London, to take advantage of the astonishing fact that British intelligence had somehow acquired a German cipher machine, called "Enigma." The purpose of the machine was to scramble letters in the words of messages in such a way that they could only be deciphered by a receiver with a key to set his own receiving machine accordingly. If the British could find that key, they would be able to receive any and all messages that the German High Command sent to its military via this system. Along with a small group of other mathematicians, Turing was invited to Bletchley to see if the key could be found.

It could and was. And it was done by a computer in whose design Turing had played a leading part. "I won't say that what Turing did made us win the war," says I. J. Good, now a professor at Virginia Polytechnic Institute, who was then Turing's statistical clerk, "but I daresay we might have lost it without him." Donald Michie, who was also at Bletchley, describes the machine as incorporating two synchronized photoelectric paper tape readers, capable of reading two thousand characters per second. "Two loops of 5-hole tape, typically more than 1000 characters in length, would be mounted on these readers. One tape would be classed as data, and would be stepped systematically relative to the other tape, which carried some fixed pattern, by differing in length from it by, for example, one character. ... The machine, and all its successors, were entirely automatic in operation, once started, and incorporated an on-line output teleprinter or typewriter" (Randell, 1973).

Here's the description of an earlier electromagnetic machine, the immediate predecessor of the machine described above, by F. W. Winterbotham, who was in charge of security for Enigma, and who transmitted the decoded messages directly to Winston Churchill: "I am not of the computer age nor do I attempt to understand them, but early in 1940 I was ushered with great solemnity into the shrine where stood a bronze-coloured column surmounted by a larger circular bronze-coloured face, like some Eastern Goddess who was destined to become the oracle of Bletchley, at least when she felt like it. She was an awesome piece of magic" (Winterbotham, 1974). By the end of February 1940, the Luftwaffe began to put practice messages on the air and the bronze goddess deciphered them successfully. Winterbotham goes on to claim that the entire Bletchley operation was essential to the British, and then the Allied, war effort against the Germans. If he is correct, given their state of military unreadiness, it's hard to see how the Allies could have won the war without Ultra. (On the other side, young Konrad Zuse had been rescued from the front lines, but nobody took him and his odd machines seriously enough to pay much attention to him.)

In 1942, Turing made the risky trip across the Atlantic to the United States, where he remained from November to the following March. Here he seems to have talked to John von Neumann, whom we'll consider later in this chapter, but there's no direct evidence that he learned of other work close to his own, such as the McCulloch-Pitts paper, or the Rosenblueth-Wiener-Bigelow paper. But with von Neumann, there was evidently a fruitful exchange of ideas.

After the war, Turing went to the National Physical Laboratory in Teddington, and began work on designing the ACE (Automatic Computing Engine—its name Turing's homage to Babbage's Analytical Engine). There were pleasant times at Teddington. Here he took up running seriously for the first time, coming in fifth in the Amateur Athletic Association Marathon Championship in 1947, which encouraged him to train for the Olympic trials. Unfortunately, he displaced something in his hip, which forced him to give up the idea of competitive running, though he continued to take ten and fifteen mile recreational runs for the rest of his life. Sara Turing reports:

> During his time at Teddington he occasionally stayed near Dorking with Professor Champernowne's mother, who remembers his appearing in the drawing room one evening with a pair of white socks which he proceeded to darn. The work finished, he said he had found it very soothing. Actually he darned extremely neatly and put a very neat darn in his dark-coloured trousers—but unfortunately in white or some other light colour. It was in Mrs. Champernowne's garden that

he used to play a complicated form of chess with Professor Champernowne. In this game, each player had to run round the garden after his move, and if one arrived back at the board before his opponent had moved, he was allowed to have an extra move. Fleetness of foot probably helped to counterbalance Alan's not being a very good chess player. The purpose of this game was to throw light on the physiological effects of violent exercise on the functioning of the brain.

(S. Turing, 1959)

Nevertheless, Turing began to grow restless with the slow pace of progress on the ACE. He felt he was wasting his time, especially since regulations forbade him to go to work on the engineering side. He asked permission to take a year's sabbatical at King's, and there, in September 1947, he wrote a paper which has only recently been published, called "Intelligent Machinery" (Turing, 1969).

In the spring of that year, Norbert Wiener visited him at Teddington where, according to Wiener, they talked over the fundamental ideas of cybernetics. Talk with Wiener surely stimulated Turing, but the thoughtfulness of the 1947 paper seems to show that Turing had already been speculating about the possibilities of intelligent machinery before their meeting. On another visit two years later, Wiener would pronounce Turing's work "an original combination of the modern stream of mathematical logic with the theory and practice of control and communication apparatus" (Wiener, 1961).

Turing's abstract for "Intelligent Machinery" is bold:

> The possible ways in which machinery might be made to show intelligent behavior are discussed. The analogy with the human brain is used as a guiding principle. It is pointed out that the potentialities of the human intelligence can only be realized if suitable education is provided. The investigation mainly centres round an analogous teaching process applied to machines. The idea of an unorganized machine is defined and it is suggested that the infant human cortex is of this nature.

There follows a lucid paper that first outlines the arguments against intelligent machines—that humans will not admit the possibility of rivals in intellectual power; that such a machine, if possible, would be irreverent; that machinery and humans are vastly different; that Gödel's Theorem[2] shows that machines are inherently incapable of solving problems which humans can overcome; and that, anyway, machine intelligence is no more than a reflection of the intelligence of its creator—and then refutes them.

Turing soon comes to the central issue, man as machine. He points out various machines that already imitate parts of man— the television camera as eye, the microphone as extended ear, and servomechanistic robots that imitate actions of the limbs, and so forth.

> Here [he writes] we are chiefly interested in the nervous system. We could produce fairly accurate electrical models to copy the behavior of nerves, but there seems very little point in doing so. It would be rather like putting a lot of work into cars which walked on legs instead of continuing to use wheels. The electrical circuits which are used in electronic computing machinery seem to have the essential property of nerves. They are able to transmit information from place to place, and also to store it.

He fancifully suggests that one way to build such a "thinking machine" is to build an analog man, with cameras, microphones, loudspeakers, and so on, an idea that probably appealed to Turing's somewhat fey sense of humor (even in this sober paper, he pictures the analog careening around the countryside, a serious danger to the ordinary citizen), but he dismisses this possibility as too slow and impracticable. Instead, he suggests that a "brain" be built that might apply itself to any of the following tasks: games such as chess, tic-tac-toe, bridge, and poker; the learning of languages; the translation of languages; cryptography; and mathematics.

This list turns out to be the main part of the program that occupied artificial-intelligence researchers for the next two decades.

---

[2] That in any sufficiently powerful logical system, statements can be formulated which are neither provable nor unprovable within that system, unless the system is logically inconsistent.

Based, as we now guess, on his success during the war, Turing wrote,

> The field of cryptography will perhaps be the most rewarding. There is a remarkably close parallel between the problems of the physicist and those of the cryptographer. The system on which a message is enciphered corresponds to the laws of the universe, the intercepted messages to the evidence available, the keys for a given day or a given message to important constants which have to be determined. The correspondence is very close, but the subject matter of cryptography is very easily dealt with by discrete machinery, physics not so easily.[3]

He then compares the newborn human's brain to an unorganized machine, waiting to have impressed upon it the routines, rules of thumb, and general knowledge that will fit it as a citizen of the world. The brain-machine comparison is a fairly detailed scheme, with some vague ideas about "teaching" the machine in some of the ways we teach children. At first he speaks in terms of discipline, a system of rewards for suitable behavior and punishments for unsuitable, but he also speculates about how one might copy the notion of initiative, and suggests ways of inculcating it—directly, along with the reward and punishment, and indirectly, gradually allowing the machine to make choices or decisions about what it will do. Ironically, twenty-five years later, researchers at MIT undertook to teach children using methods they had used successfully with computers, the unexpected reverse of Turing's proposal.

What is most striking in the paper is Turing's discussion of intellectual searches. He suggests that nearly all intellectual problems can be phrased in the form "find a number n such that . . .," and that search is the proper mode in which to consider problem-solving behavior. These are precisely the terms that Allen Newell and Herbert A. Simon used eight years later when they tried to express problem solving in a formal way. Even more remarkable,

---

[3] In the late 1970s, cryptography was undergoing profound changes thanks to an ingenious application of an idea from theoretical computer science.

Turing suggests a good beginning might be to provide this machine with a program that

> corresponds to building in a logical system (like Russell's *Principia Mathematica*). This would not determine the behavior of the machine completely: at various stages more than one choice as to the next step would be possible. We might arrange, however, to take all possible arrangement of choices in order, and go on until the machine proved a theorem, which, by its form, could be verified to give a solution of the problem. This may be seen to be a conversion of the original problem into another of the same form. Instead of searching through values of the original variable n one searches through values of something else. In practice when solving problems of the above kind one will probably apply some very complex "transformation" of the original problem, involving searching through various variables, some more analogous to the original one, some more like a "search through all proofs." Further research into intelligence of machinery will probably be very greatly concerned with "searches" of this kind,

And indeed it was. Newell and Simon even used Whitehead and Russell's *Principia* as a task for their first working intelligent program. Turing's paper was unpublished until nearly thirty years after he wrote it and unknown to Newell and Simon, who arrived at the same ideas independently in their own work during the fall of 1955.

A paper of Turing's that was published just after it was written, and that received wide attention, came out in October 1950. It was called "Computing Machinery and Intelligence," and it addressed the question of whether machines could think. Here Turing proposed what came to be known as Turing's Test, where an interrogator is separated from the person (or machine) being interrogated, and can communicate only by a teletype. Turing suggests that if the interrogator cannot tell for certain whether he or she is communicating with the person or with the machine, then a machine can indeed be said to think.

As usual, Turing took on the doubters. One of them this time was Professor Sir Geoffrey Jefferson, F.R.S., who delivered the Lister Oration for 1949. Turing quotes him:

Not until a machine can write a sonnet or compose a concerto because of thoughts and emotions felt, and not by the chance fall of symbols, could we agree that machine equals brain—that is, not only write it but know that it had written it. No mechanism could feel (and not merely artificially signal, an easy contrivance) pleasure at its successes, grief when its valves fuse, be warmed by flattery, be made miserable by its mistakes, be charmed by sex, be angry or depressed when it cannot get what it wants.

(Turing, 1950)

As Turing argues, this is the solipsist point of view. We cannot know that anyone or anything thinks, according to this position, except by the signals we receive from them. "Instead of arguing continually over this point, it is usual to have the polite convention that everyone thinks," Turing observed drily, and then proposed a variation of his imitation game, a *viva voce* or oral examination to discover whether a student has learned something or merely imitates it parrot-fashion. Here's Turing's dialogue:

INTERROGATOR   In the first line of your sonnet which reads "Shall I compare thee to a summer's day" would not "a spring day" do as well or better?

WITNESS   It wouldn't scan.

INTERROGATOR   How about a winter's day? That would scan all right.

WITNESS   Yes, but nobody wants to be compared to a winter's day.

INTERROGATOR   Would you say Mr. Pickwick reminded you of Christmas?

WITNESS   In a way.

INTERROGATOR   Yet Christmas is a winter's day, and I do not think Mr. Pickwick would mind the comparison.

WITNESS   I don't think you're serious. By winter's day one means a typical winter's day, rather than a special one like Christmas.

(Turing, 1950)

And Turing concludes that Professor Jefferson, faced with such a sustained dialogue, could not call it "an easy contrivance." But Professor Jefferson could have argued that no such machine then existed, nor was it on the horizon, and he would have been quite right.

By now Turing had moved up to Manchester University where he was assistant director of "Madam," the Manchester Automatic Digital Machine. After coming back from his sabbatical at King's, he'd again been disappointed by the progress on the ACE, but whether he was dismissed from the National Physical Laboratory or resigned is unclear; Sara Turing seems to imply both. His instincts appear to have been right, for the ACE was not dedicated until 1958, nearly ten years later.

Officially, Turing was working on the design of Madam at Manchester, but at the same time he'd become fascinated by biology. He was invited to join the Ratio Club (named by Albert M. Uttley, who consulted an etymological dictionary and liked the way it combined reasoning, relations, and number), an informal group of physiologists, physicists, mathematicians, and engineers who met periodically in London to discuss problems of mutual interest. He submitted two papers to the group, evidence of his belief that a mathematical inquiry into some biological problems might yield interesting results.

Professor Donald MacKay, a physicist who had worked on radar during the war and who was a founding member of the Ratio Club, says Turing was a stimulating associate, and in a conversation with me recalled one of Turing's presentations. "He started off: 'A nerve cell is something that grows in a way which is more like the growth of a, a, a tree than a, a, a horse.' He had a bit of a stammer, and we all burst out laughing. It was such an unexpected sort of contrast."

In the correspondence between Turing and the biologist J. Z. Young, it is clear that Turing by then knew of Warren McCulloch's work in the United States and was likewise convinced that a mathematical approach was more fruitful than an anatomical one to the problem of brain function. "I am afraid," he wrote to Young on February 8, 1951, "I am very far from the stage where I feel inclined to start asking any anatomical questions. According to my notions of how to set about it, that will not occur until quite a late stage when I have a fairly definite theory about how things are done" (S. Turing, 1959).

He was not working on that problem at all, he went on to tell Young, but rather on a mathematical theory of embryology, which he saw as connected to brain structures, for it had to be achieved by the genetical embryological mechanism. "What you tell me about this growth of neurons under stimulation is very interesting in this connection," Turing wrote. "It suggests means by which the neurons might be made to grow so as to form a particular circuit, rather than to reach a particular place" (S. Turing, 1959).

Turing contributed to B. V. Bowden's *Faster than Thought* (1953), published to celebrate the development of the digital computer. Both Turing and Bowden collaborated with Audrey Bates and Christopher Strachey on the chapter dealing with game-playing machines, which, along with the chapter by Bowden that follows it, gives a good idea of the state of research in the field of intelligent machinery in the late 1940s and early 1950s in Britain. There they describe a program of sorts for a chess-playing machine that bears some resemblance to the chess machine Shannon proposed in 1949. This one is Turing's, based on work done in 1947 and 1948, and he concludes, "If I were to sum up the weakness of the above system in a few words, I would describe it as a caricature of my own play. It was in fact based on an introspective analysis of my thought processes when playing, with considerable simplifications. It makes oversights which are very similar to those which I make myself, and which may in both cases be ascribed to the considerable moves being inappropriately chosen." Even worse, the machine couldn't run around Mrs. Champernowne's garden either.

This scheme was later adapted to the Manchester University machine to make it the first machine capable playing a complete, though very slow game of chess. In the words of Newell, Shaw, and Simon (1958) it was "not a very good chess player, but it reached the bottom rung of the human ladder." The Manchester group also took up checkers, a game of intermediate difficulty between the easily understood and programmed games, such as tic-tac-toe, and the most complex ones, for instance chess. It's unclear whether they were aware of Arthur Samuel's work in the United States, which had got underway in 1948, but in any event, they produced a program that could play "a tolerable game until it reaches the endgame."

The essay in *Faster than Thought* dealing with thinking and machines is more speculative than concrete, full of sentences that begin, "The machine might be made to. . . ." The recurring theme is that the differences between the operations of the machine and those of a human being are of a degree rather than kind. The notion of information theory is present here, as are early ideas about the physiology of learning. Bowden draws back at the conclusion, willing to grant that machines might be said to think in some limited way, but "no machine is ever likely to undertake the work of those few extraordinary men whose dreams and whose efforts are responsible for the growth and the flowering of our civilization." But how many humans undertake those dreams and efforts? Even Bowden calls them few and extraordinary. Are the rest of us nonthinkers?

If Turing exceeded his colleague in optimism, his foresight was not exhaustive. Seymour Papert points out that in some ways, Turing's vision was limited by the primitive state of computers in his day, and so he did not suspect the time would come when very high-powered languages (many a direct product of artificial-intelligence research) would be available to allow us to program computers with relatively little effort. Hence, he worried about the time it would take to accomplish such tasks. ("At my present rate of working I produce about a thousand digits of programme a day, so that about sixty workers, working steadily through fifty years might accomplish the job, if nothing went into the wastepaper basket. Some more expeditious method seems desirable" (Turing, 1969). Nevertheless, Turing refused to be misled by the obvious differences he could see between the digital computer and human neural physiology. He understood that at another level, these days called the information-processing level, the brain and the computer had much in common, and that insight into the organization of one would surely give insight into the organization of the other.

Turing died in 1954, just before his forty-second birthday. There is dispute over the circumstances. There was in Turing what his mother called a childish streak, which showed itself in a number of ways, including his affectionate friendships with children throughout his life, and in his devotion to the BBC's "Children's Hour," a favorite being something called "The Toytown Programmes" (he even called his mother long-distance to alert her to especially good ones coming up). In this he was no different from many another creative scientist, which is what makes living with them alternately a joy and an exasperation.

One more manifestation of this childish streak was his desert island game, where he tried to see how many chemicals he could produce from household substances using only home-made apparatus. One of the chemicals he produced that way was the highly poisonous potassium cyanide, and a spoon coated with it was found in his home laboratory. Sara Turing reminds us of his carelessness and his childishness, and suggests that, since his work was going very well and he had no money worries, and since according to his immediate friends and neighbors all his behavior just prior to his death was normal, the death must have been accidental.

But the verdict of the inquest was that the poison was self-administered, "while the balance of his mind was disturbed."

Friends who knew him suggest a sadder story. It was Turing's grave misfortune to be a homosexual in a time and place where extremely harsh criminal and social sanctions existed against homosexuality. Perhaps we can hardly conceive of the appalling disgrace exposure would have meant in the Britain of the early 1950s. All we can do is imagine what it must have been like for a man as sensitive as Turing to be threatened (and it seems he was) with the loss of everything that mattered to him, his science, livelihood, the esteem of friends. And we must muse on the savage price exacted by a society that without him might not have survived to demand it.

Another intellectual giant associated with computers did not share Turing's optimism about whether they would eventually think. He was John von Neumann, the extraordinary mathematician who was born in Budapest, and had come to the United States in 1930 from his post as a *privatdozent* at the University of Berlin to lecture in mathematical physics at Princeton University. The history of von Neumann's involvement with and contributions to the digital computer are detailed in Herman Goldstine's *The Computer from Pascal to von Neumann* (1972). Von Neumann is credited with conceiving the idea of the stored program, that is, controlling the computer by means of a program stored in the computer's internal memory. (Von Neumann seems not to have made this claim for himself, and crediting him with the idea is a constant irritant to those, such as John Mauchly and Prosper Eckert, who believe the idea of the stored program was a group development that came out of the Moore School of Electrical Engineering at the University of Pennsylvania, led, unsurprisingly, by Mauchly and Eckert.)

Brian Randell writes that the advent of the stored program meant that "for the first time it became a practical and attractive proposition to use a computer to assist with the preparation of its own programs, thus opening the way to the development of programming aids such as assemblers, compilers, operating systems, etc." (Randell, 1973).

Von Neumann had become interested in computing in 1943 when he visited the United Kingdom (perhaps at Turing's invitation; he had known Turing at Princeton and seen him during Turing's visit to the United States from late November 1942 to March 1943). Von Neumann's work at the Manhattan Project convinced him that high-speed computing was surely to be essential in both science and mathematics. He had been associated with the group at the Moore School at the University of Pennsylvania, which built the ENIAC, and he was asked to design a bigger, faster machine, whose specifications he wrote in 1945. It is

John Von Neumann

interesting to read this early paper and see that von Neumann makes explicit comparisons between the parts of the computer he is proposing and the human nervous system. Not only does he use such terms as *memory* and *control* organs to designate certain functions of his new computer. He also says, "The three specific parts . . . correspond to the *associative* neurons in the human nervous system. It remains to discuss the equivalents of the *sensory* or *afferent* and the *motor* or *efferent* neurons. These are the input and the output organs of the device, and we shall now consider them briefly" (Goldstine, 1972).[4] Further on, he cites the work of Pitts and McCulloch whose 1943 paper on a logical calculus to describe nervous activity had clearly impressed him. Along with them, he ignored some of the messy complexities of neuron functioning and showed that a simplified notion of it could "be imitated by telegraph relays or by vacuum tubes whose speed and

---

[4] So far as I can discover, von Neumann was the first to do computer anthropomorphizing on such a scale (an oddity, considering his lifelong skepticism about the connection between human brains and computing). Edsger Dijkstra, the well-known Dutch computer scientist, considers this habit pernicious, and continuing indulgence of it a big obstacle to the maturation of the entire field of computer science. "I should have guessed it was Saint Johnny," he grouched to me once. Others do not share his view.

reliability are essential to the functioning of the computer" (Goldstine, 1972).

In October 1945 a document was written to persuade the trustees of the Institute for Advanced Studies at Princeton where von Neumann had returned after the war, that they should support the building of a successor to the Moore School's ENIAC and EDVAC (an unusual departure from the pure research that had always taken place at the institute). In it, Frank Aydellote, then director, wrote, "Curiously enough, the plan of such a machine is partly based on what we know about the operation of the central nervous system in the human body." The institute machine was to be all-purpose, unlike its predecessors; these had been built for specific tasks, such as the calculation of nautical and bombing tables. Von Neumann was excited about the possible tasks such a machine might be put to: the complicated problems of shock waves, quantum theory, and stellar astronomy. More important, he wrote, "such a machine, if intelligently used, will completely revolutionize our computing techniques. ... The projected machine will change the possibilities, the difficulties, the emphases, and the whole internal economy of computing so radically, and shift all procedural options and equilibria so completely, that the old methods will be much less efficient than new ones which have to be developed. These new methods will have to be based on entirely new criteria of what is mathematically simple or complicated, elegant or clumsy" (Goldstine, 1972). Some of these developments could already be visualized. However, the main work would have to be done when the machine was completed and available, by using the machine itself as an experimental tool. We see this theme again and again in computer science, and most certainly in artificial intelligence. The concrete systems themselves are so complicated that their nature can only be discovered empirically. Von Neumann's machine, known variously as the Princeton machine and the IAS machine, was funded and built, and served as the prototype for many machines to come.

By now the term "giant brains" had captured the public imagination. That computers could do low-level sorts of intellectual tasks, such as arithmetic and collation, seemed to suggest to the overly enthusiastic that high-level intellectual behavior was just around the corner. In a December 1950 issue of the *Saturday Evening Post* an advertisement for Shell Oil bragged that the company's industrial lubricants were being used by the "Oracle on 57th Street," depicted as an enormous, berobed woman meditating on a lengthy scroll of scripture, the base of her pedestal the showrooms of IBM at the corner of 57th and Madison Avenue in New York.

Von Neumann was obviously fascinated with the idea, but he simply saw no ultimate way that the connection between human thinking and machine performance could be made. In a paper written in 1951, he took pains to point out the differences between the human nervous system and a computer. There are, first of all, the physical differences, and these aren't incidental. The components of computers are large, awkward, and unreliable, compared with the miniature and reliable cells of the brain. The human nervous system also shows clear signs of both discrete and continuous behavior, whereas computers must be either discrete (digital) or continuous (analog). But the major reason for his despair over ever getting a computer to think was the lack of a logical theory of automata. The lack of such a theory, based in formal, logical terms, prohibited machine builders from ever building machines with much more complexity than was possible in 1951, von Neumann argued, and such complexity was absolutely essential to the production of anything like intelligent behavior.

He is full of praise for the work of McCulloch and Pitts:

> It has often been claimed that the activities and functions of the human nervous system are so complicated that no ordinary mechanism could possibly perform them. It has also been attempted to name specific functions which by their nature exhibit this limitation. It has been attempted to show that such specific functions, logically, completely described, are per se unable of mechanical neural realization. The McCulloch-Pitts result puts an end to this. It proves that anything that can be exhaustively and unambiguously described, anything that can be completely and unambiguously put into words is ipso facto realizable by a suitable finite neural network.
>
> (von Neumann, 1951)

So much for theory. In the real world, even for the simplest act, such altogether unambiguous specification might fill up the universe.

Von Neumann was invited to give the Silliman Lectures at Yale in 1956, but before he could complete his preparations, he was stricken with cancer. The incomplete notes for the lectures were published posthumously, called *The Computer and the Brain* (1958). Here he is trying, he says, to approach an understanding of the brain from a mathematician's point of view.

But again, the comparison is between the two sets of hardware,[5] rather than between the two sets of functions. It's as if we were considering two modes of transportation, say, private automobiles and jet planes, and were flummoxed by the fact that one vehicle moves on wheels along the ground, powered by an internal combustion engine and restricted by roads and legal speed limits, while the other flies through the air at something like ten times the speed and has a jet engine. These are essential differences to their respective designers and mechanics, but viewed another way, cars and planes are two instances of people movers, each capable of that function and each most appropriate in a given situation.

As intrigued as von Neumann was by the idea of trying to express human neural behavior in rigorous, mathematical terms, he seems to have been convinced to the end of his life that such expression was a hopeless task. That the terms might be rigorous though not necessarily mathematical, and that the matching might be done at the functional instead of the hardware level, seems not to have occurred to him.

For convenience, I've traced the influences upon artificial intelligence through separate fields: computer design and construction, cybernetics, mathematical psychology and physiology, and formal logic. John von Neumann is the perfect example of why such divisions are ultimately nonsense. He was first a mathemati-

---

[5] *Hardware* comprises the things you can put your hands on in a computer; *software* is programs.

cian, a student at the Swiss Federal Institute of Technology in Zurich of George Polya (who would later also teach Allen Newell at Stanford); he was grounded in formal logic, which was expressed profoundly and elegantly in his designs for computers. He had joined in December 1944 with Howard Aiken, the Harvard-based co-inventor of the Mark I computer, and with Norbert Wiener the cyberneticist, Walter Pitts the logician, and Warren McCulloch the neurophysiologist, to form something called The Teleological Society, to discuss "communication engineering, the engineering of control devices, the mathematics of time series in statistics, and the communication and control aspects of the nervous system." The work of Pitts and McCulloch, who were then at the University of Illinois, influenced his thinking about the design of a computer just as it was to influence many people's designs for an explicit means of imitating human thinking.

Personal recollections of von Neumann are a staple in any history of computer science, but the truth seems to be that von Neumann was an overwhelming personality. Professor Leon Harmon, now a theoretical neurophysiologist at Case-Western University, has recollections that typify the stories told about von Neumann.[6]

Von Neumann was a true genius, the only one I've ever known. I've met Einstein and Oppenheimer and Teller and—who's the mad genius from MIT? I don't mean McCulloch, but a mathematician. Anyway, a whole bunch of those other guys. Von Neumann was the only genius I ever met. The others were supersmart .... And great prima donnas. But von Neumann's mind was all-encompassing. He could solve problems in any domain. . . . And his mind was always working, always restless. He walked into my living room one night and a half dozen people were already having cocktails, and he disappeared into a corner and stood with his back to us, hands behind him, and after about two minutes turned to me and said, "About two thirds of a liter a week, Leon."And I had to think about it for three or four

---

[6] Quotations or extracts not followed by citations or presented in the discussion of a particular work are from personal taped interviews conducted by the author.

minutes, and finally I said, "Yeah, Johnny, that's just about right."
He'd walked up to the nine-gallon tropical fish aquarium that stood
on a table in the corner, had noted the temperature of the water, had
made an estimate of the surface area, had seen the gap that existed
between the overhead light and the glass to keep the fish from jump-
ing out, made an estimate of the particular escape velocity of the wa-
ter molecules, integrated and found out how much added water was
needed each week for that aquarium. And he was right within a few
percent. That's the kind of thing he did all the time. Another thing
that he isn't known well for was his sense of humor. He really enjoyed
dirty limericks. And though we never said anything to each other
deliberately, it sort of evolved that whenever we came together, whether
it was an hour or a month later, the name of the game was to see who
could rush up the fastest and unload the largest number of new lim-
ericks. It turned out to be a delightful game. He had oodles of them; I
was hard put to keep up with him. His memory was just beyond
conception, a photograph for everything he ever learned or saw. Light-
ning calculator and head screwed on to boot—he put all of those
together with a huge creative talent.

Perhaps it isn't entirely surprising that neither von Neumann
nor his friends were persuaded to view the two systems they were
confronted with, namely human beings and computers, as two
instances of information processors, and to examine a grosser level
of functioning than cells and diodes. But a younger group of scientists, in many cases the students of these same men, did perceive those functional similarities, and at least in the case of von
Neumann and Wiener, confronted their mentors with such a view.
"Write it up," von Neumann told a young mathematics student
named John McCarthy at Cal Tech, after McCarthy had heard
von Neumann talk at the Hixson Conference on Cerebral Mechanisms and Behavior in 1949. McCarthy did set to thinking about
his ideas, but didn't write them up just then, because he felt he
couldn't relax his mathematical standards—a scruple that would
cause him some future grief.

They were all connected: Wiener, Shannon, McCulloch, Turing, von Neumann. They were connected by friendship, by prox-

imity, by their fascination with the dawning possibility that tools were finally at hand for understanding at least some aspects of human thought, which up to then had frankly eluded anything but proof by assertion. I don't mean that we can declare a direct lineal descendancy from one generation of researchers to the next. Turing, with his appreciation of the rich possibilities of organizational and functional comparisons between human brains and the computer—as opposed to comparisons at the cell level—should have been the intellectual forebear of artificial intelligence, but he seems to have had little or no influence on the American researchers. They took their cues—sometimes inspiration and sometimes untempered rejection —elsewhere. And his influence on artificial intelligence in Britain was almost nil. Max Clowes, professor of artificial intelligence at Sussex University, says that when he was at the National Physical Laboratory where Turing had worked, Turing's legacy seemed to be hardware—the ACE computer—but not a sense of continuity with work that was going on at the information-processing level. The information-processing model that was dominant in the U.S. somehow failed to become a major influence in Great Britain in any big way until the mid-1960s.

That's a curiosity, for it might seem now that the information-processing model for formalizing intelligent behavior had been anticipated by workers in the 1940s, and was just waiting to be seized by anyone with open eyes. Indeed, how was this approach to be avoided? Those concerned were open-eyed people—the Ratio Club in London, the Teleological Society, and the Macy meetings in the United States—all devoted to the mathematical analysis of the nervous system (and as Wiener notes, you cannot study the nervous system without studying the mind), and the digital computer was at hand, a medium for realizing all those formalisms shimmering with promise. Surely psychologists and physiologists sang hosannas to celebrate the possibility of a scientific solution to the mind-body problem and, metaphorically speaking, hoisted to their shoulders these heroic pioneers.

No. Science is a human institution, and things don't work that way. While some did leap at the new ideas and wanted to apply them to everything under the sun, the evidence of the scientific journals is that the new thinking was a long time being adopted, and that in fact cybernetics had all but disappeared as a field by the time its contributions were coming to be widely applied.

I point this out only to give some perspective. I should also add that none of these researchers was starving in a garret. Still, with the exception of Wiener and perhaps von Neumann, none has had the renown one might expect in proportion to his effect on our lives now and in the future.

Chapter Four

Meat Machines

"The brain," MIT's Marvin Minsky declared a few years ago, "happens to be a meat machine." The phrase seems to have shocked a lot of people—and given Minsky's delight in provoking anyone within earshot, it was probably so intended. People who are scandalized by such a statement take it as one more instance of the generally irreverent, even misanthropic, attitudes that they are convinced pervade artificial-intelligence work.

They might just as well raise such an objection to all brain research, for though the neurophysiologists don't put it in quite the colorful terms Minsky does, their assumptions are the same. The brain is an electrical and chemical mechanism, whose organization is enormously complex, whose evolution is barely, understood, and which produces complex behavior in response to an even more complex environment.

The problem, I suppose, is our own associations with the notion of mechanism, or machine. *Machine* conjures up steam engines, drill presses, things that clank. Brains clank softly, the AI researchers like to joke, and by that they mean several things. First, the behavior of the brain is accessible to understanding at several different levels. Thus, at the physiological level, it will someday be explicable in detailed electrochemical terms. But there are other ways of understanding mental behavior, other levels of modeling. A metallurgist, a mechanical engineer, and a transportation specialist each regard an automobile from a point of view appropriate to their specialty, quite differently from one another, but within their separate contexts the work of each is perfectly sound. Some illumination will occur among disciplines—the metallurgist may be able to suggest ideas to the mechanical engineer, the mechanical engineer to the transportation specialist. In the same way multiple levels of modeling, more or less related, exist for human mental behavior. A moral philosopher will address human ethics, proposing a model of good behavior in the human situation. A psychoanalyst explains mental behavior in terms of stress and tensions, conflicts and drives. The information-processing model, with which artificial-intelligence research concerns itself, explains thinking in terms—unsurprisingly—of the processing of information in the human brain, representing it in symbolic form, storing it, the means of recalling it, controlling it, and so forth. This model often casts the act of thinking into terms of a problem to be solved, and examines the techniques by which thinkers, whether man-made or begotten, solve them.

Presently no complete, coherent model exists that explains all aspects of mental behavior, but most researchers are agreed: there's no ghost in the machine. Everything from symphonies to simultaneous equations to situation ethics is finally produced by those electro-chemical processes. This view can be considered mechanistic.

> The fact is [says Minsky] that throughout history the people who I think considered themselves to be mechanists tended to be something else. I don't know if there's a word for them. There should be—let's say simplists. Striking examples are people like Pavlov and Watson and the whole family of people who believed in conditioning as a basis for learning, the mechanical associationists. Although on the surface they could be considered mechanists because they seem to talk more openly about the mind being a machine, their real trouble is that their image of the machine is precomputational.

So the response of those scandalized by the notion of the brain as a meat machine is two centuries late. The concept was highly arguable when it was fresh; as noted, it drove Julien Offray de la Mettrie to run for his life after he declared that humans are indeed machines, and, in particular, that our mental behavior is finally the result of the machine in our heads.

These assumptions needn't rob us of our awe as we confront the human brain. On the contrary, the brain seems to me all the more awesome without a magic explanation. For my own taste, the elegant structure of DNA, say, is more satisfying in every way than any proposal of an *élan vital*. Nor need these assumptions limit the views we can take of human mental behavior to the "merely mechanistic." Some scientists are at great pains to stress that the biological, which is to say the mechanistic, account of the development of the brain and its functions is perfectly compatible with divine origin. Several of them I talked to during the course of preparing this history are deeply religious. They include both Christians and Jews, and all of them believe that a scientific approach to the mind—whether through physiology or through psychology—can coexist with a religious reverence. *In particular, declaring the brain to be a meat machine does not absolve the person whose brain it is from ethical obligations* any more than it exempts humans from love and grief and ambition and disappointment and satisfaction. The information-processing model does suggest a way of describing these highly symbolic phenomena, and even of achieving them in a physical system, such as a human body. But no one I talked to dreams of denying their existence.

In the twentieth century studies of mind have largely been undertaken by either physiologists and psychologists, though philosophers continue to fret about the nature of mind without paying too much attention to the results obtained by scientists. This involvement by several disciplines isn't necessarily a bad thing. The human mind seems to me complex enough to deserve a variety of approaches. Perhaps it's a matter of taste whether you prefer one set of symbols to another. By and by, one model does come to predominate, and while its dominance has something to do with its ability to account for a wide variety of the phenomena under scrutiny, it also has to do with a certain fitness with the times. This book will only now and then touch on what philosophers and physiologists have to say about mind, but this is such a moment, for one of the most influential persons in the germination of modern artificial intelligence was the neurophysiologist Warren McCulloch.

As a freshman at Haverford College in 1918, McCulloch remembers being questioned by the Quaker Rufus Jones:

> "Warren," said he, "what is thee going to be?" And I said, "I don't know." "And what is thee going to do?" And again I said, "I have no idea, but there is one question I would like to answer: What is a number, that a man may know it, and a man that he may know a number?" He smiled and said, "Friend, thee will be busy as long as thee lives."
>
> (McCulloch, 1965)

And so he was. McCulloch left Haverford and went to Yale, where he majored in philosophy and minored in psychology, took an M.A. in psychology at Columbia, then transferred to the medical school there. He interned and did his residency at Bellevue Hospital, then undertook research in epilepsy and head injuries. After a couple of years at Rockland State Hospital, he went back to Yale to study the activity of the central nervous system, and then moved on to the University of Illinois to direct the Laboratory for Basic Research in the Department of Psychiatry, where he remained until 1952. He then moved once again, to become a member of the staff at the Research Laboratory of Electronics at Massachusetts Institute of Technology, where he remained until his death in 1969.

What is a number, that a man may know it, and a man that he may know a number? McCulloch credits Bertrand Russell with answering the first part of his question. A number, Russell said, is the class of all those classes that can be put into a one-to-one correspondence with it. What is a man—at least, how can he be defined so that he can know a number? That was the part of the question that preoccupied McCulloch all his life. In 1943 he published a paper with a young mathematician he had met at Illinois named Walter Pitts,[1] and the aim of that paper was to define a man so logically—as a net of interconnected neurons—that how he might know a number could be shown.

The paper, as noted earlier, was called "A Logical Calculus of the Ideas Immanent in Nervous Activity," and it appeared in the *Bulletin of Mathematical Biophysics* in 1943, "where, so far as biology is concerned," McCulloch wrote much later, and with some exaggerated modesty, "it might have remained unknown; but John von Neumann picked it up and used it in teaching the theory of computing machines." Nineteen forty-three was the same year that Arturo Rosenblueth, Norbert Wiener, and Julian Bigelow published their celebrated paper "Behavior, Purpose and Teleology," also referred to earlier. Each of these papers proposed an information-processing model that would allow greater flexibility in hypothesizing about mind.

In an introduction to some collected essays of McCulloch's (1965), Seymour Papert writes,

> Between the class of trivial combinational functions computable by simple Boolean logic and the too general class of functions computable by Turing machines, there are intermediate classes of computability determined by the most universal and natural mathematical feature of the net—its finiteness. This is pure mathematics. The theo-

---

[1] Manuel Blum, now a professor at the University of California in Berkeley, heard this story when he was a student from Warren McCulloch: Walter Pitts was forced to drop out of high school by his father, who wanted him to go to work and earn money. Rather than do this, young Pitts ran away from home and ended up in Chicago, penniless. The fifteen-year-old boy spent a lot of time in the park, where he met and began to have conversations with an older man he knew only as Bert. When Bert detected the boy's interests, he suggested that young Pitts read a book that had just been published by a professor at the University of Chicago by the name of Rudolf Carnap. Pitts did, and showed up at Carnap's office. "Sir," he said, "there's something on this page which just isn't clear." Carnap was amused, because when he said something wasn't clear, what he meant was that it was nonsense. So he opened up his newly published book to where young Pitts was pointing, and sure enough, it wasn't clear; it was nonsense. Bert turned out to be Bertrand Russell. It was three years later, when Pitts was about eighteen, that he and McCulloch published their famous paper.

retical assertion of the paper is that the behavior of any brain must be characterized by the computation of functions of one of these classes.

McCulloch and Pitts had not proved the proposition that whatever can be completely described can be realized by a net of neurons. What they did was to provide a new definition of computing machine, which allowed the brain to be construed as a machine in a more precise way than before.

They proposed that their neural net—a mathematical model of various phenomena associated with nerve behavior—be compared to a Turing machine. That is, every net, if provided with tape, scanners, and so on, could compute only such numbers as could a Turing machine (though not all, amends Papert later), and could, in the manner of a Turing machine, imitate or model itself. The components of their net, synapses connecting all-or-none neurons, would prove to be considerably—and some researchers would say misleadingly—more simple than organic neurons and synapses. But that was in the future, and for 1943, "A Logical Calculus" was startling enough, stating as it did that the laws governing mind should be sought among the laws governing information rather than energy or matter. It was no wonder that the computer began to be called a thinking machine. No artifact in the long history of artifacts produced for that purpose had ever shown more promise of imitating, and perhaps illuminating, the mind of human beings.

Not only did McCulloch and Pitts enable researchers to think of the brain as a computing machine in a more precise way than ever before, but their paper also focused on where the remaining problem would lie: that knowledge is complex, the neurons of the brain relatively simple (though not so simple as the state of neurophysiology then led them to believe), and the interactions between those two are what remain to be described in far more detail. Present-day neurophysiologists might rewrite that statement to say that knowledge is complex, the neuron looks much more complicated than we ever dreamed, and understanding the mapping between the two might well be a hopelessly insoluble problem.

Even in the late 1950s, when it began to become clearer—based on the evidence acquired by the microelectrode, which could be implanted in a single nerve cell—that McCulloch's model of the neuron was misleadingly simple, he had a band of loyal followers who were busy through the mid-1960s building nets based on his early work (and even simplifications of it), nets which failed to yield any but the most trivial behavior. For this McCulloch cannot be blamed. He said before his death, "It is in the interest of science to expose the formation of hypotheses to the criticism of fellow scientists in the hope of experimental contradiction. Facts have often compelled me to change my mind. . . ." (McCulloch, 1965).

McCulloch's all-or-none neuron had been based on the neurophysiological research of his time, and the sudden avalanche, as one researcher put it, of information about the natural neuron that came in the 1940s, 1950s, and 1960s, showed it to be "a complex system of electrical and chemical mechanisms which interact nonlinearly and which combine both discrete and continuously variable processes" (Harmon, 1962). Oversimplification of the neuron was also a result of the lack of mathematical tools to treat anything more complex, a situation that was slowly to be rectified over the next few years. No one was more aware of this limitation than McCulloch. He was aiming, not arriving. "Don't bite my finger," Papert remembers him saying habitually, "look where I am pointing."

It's fun to read McCulloch. He writes in a by-gone style, full of allusion to nearly everyone who ever put pen to paper, from Saint Bonaventura to Shakespeare, as the titles of his papers indicate: "Where is Fancy Bred?", "Why the Mind is in the Head," "Through the Den of the Metaphysician."[2] From the cover of a collection of his papers, he stares out with the intensity of a playwright of the Absurd. The impression is altogether misleading,

---

[2] On the relation between thoughts and the molecular motions of the brain, Clerk Maxwell despaired: "but does not the way to it lie through the very den of the metaphysician, strewn with the bones of former explorers and abhorred by every man of science?" McCulloch responds: "Let us peacefully answer the first half of his question 'Yes, ' the second half 'No, ' and then proceed serenely."

for while McCulloch was of a most whimsical nature, his aim was the very opposite of the Absurd. He spent his life attempting to show that rationality could be applied to the most mysterious phenomena of the human mind. What he hoped to find was nothing less than a physiological theory of knowledge. He would have been perfectly comfortable with Marvin Minsky's description of the brain as a meat machine.

McCulloch was convinced that problems of philosophy, whether ethics or epistemology, could be stated and solved only in terms of the anatomy and the physiology of the nervous system:

> In these terms, we are inquiring into the a priori forms and limitations of knowing and willing determined by the structure of the nervous system and by the mode of action of its elements. We ask two kinds of questions. Of universals, or ideas, we would know how nervous activity can propose anything concerning the world and how the structure embodies this or that idea. Of values, or purposes, we should know how nervous activity can mediate the quest of ends and how the structure of the system embodies the possibility of choice.
> (McCulloch, 1965)

He believed it was reasonable to assume that humans inherited only a few fixed universals, such as the qualities of sensation and sense of position and motion, "and those reflexes and appetites without which we and our kind would perish. For all else, we must begin with *random nets*."

Now these random nets are reminiscent of the *tabula rasa*, the clean slate of early philosophers of mind, but they are not the same thing. The nets are composed of neurons with the ability to be connected in any number of ways. Given a stimulus, the net arranges its connections in one way; another kind of stimulus would cause another kind of arrangement, and subsequent stimuli, if they are strong and persistent enough, will cause the net, even though it is no longer random, to change its configuration once again. This reforming in response to stimuli is learning, McCulloch speculated. And a circular net could embody not only learning, but memory, prediction, and purpose. This idea was appealing, and it led to a series of research efforts over the next few years, including Marvin Minsky's doctoral dissertation at Princeton. Far more important than McCulloch's ideas was his certainty that mind could be known and described in scientific terms. He was an enormous influence on Minsky, much greater than, say, Turing, whose work Minsky knew somewhat, but regarded as somehow off to one side. It was McCulloch's certainty that convinced Minsky and many another that machine intelligence was possible—and to get going on the problem.

In 1959 McCulloch addressed the Chicago Literary Club and read some of the poems he had written over his lifetime. I like this one, written in his third year of college:

Appointments
November 16, 1919
(His Birthday)

Yesterday:

Christ thought for me in the morning
Nietzsche in the afternoon

Today:

Their appointments are at the same hour.

Tomorrow:

I shall think for myself all day long.
That is why I am rubbing my hands.

And he finished his address with this one, written not long before he read it aloud:

farewell sweet morrows hopes deferred and all
crisp years fat earnest in defect of youth
Indian summers quicken to keen fall
as brisk October blazons time no ruth

i cry no quarter of my age and call
on coming wits to prove the truth
of my stark venture into fates cold hall
where thoughts at hazard cast the die for sooth

from me great days are gone and after none
array the ardour that i scarce compress
in temperance terrible charged i abide
the desperate victor of my last race run
wanting bold challenge to lifes dread excess
to fire that frenzy i must else wise hide

McCulloch, scarcely compressing his ardour in poetry or life, was by no means the only neurophysiologist convinced that physiology, if we only understood it well enough, would explain thought and knowledge. He expressed an idea that was just beginning to take hold, but it was an idea fitting with the times. He would meet and talk with others, in particular with the cyberneticists, and together with Wiener and von Neumann, he formed that informal group of scientists interested in just these matters, the Teleological Society. More formal were the Josiah Macy meetings throughout the 1940s where these ideas were explored. That the shift in paradigm from energy to information had begun is also suggested by the fact that a similar group had grown up in England. Each group originated its own line of inquiry, but both the American and British groups read each other's papers and to some extent were influenced by one another. McCulloch visited the British, argued with and delighted them, and they returned the favor.

It was in Great Britain in 1949 that a group of physicists, engineers, and physiologists met at Grey Walter's laboratory in Bristol to discover what common grounds they might have for thinking about the brain. Among them were John Bates, a neurosurgeon at National Hospital, and physicist Donald MacKay. These two rode back together from Bristol on the train, and agreed that the meeting had been so stimulating that they should have another—that in fact, perhaps regular meetings should be arranged. It happened that Warren McCulloch would be visiting shortly, and on September 14, 1949, the first meeting of the Ratio Club took place in London, to coincide with McCulloch's visit. McCulloch, having had some experience with the Teleological Society in the United States, was full of practical suggestions for how to run a free-floating association of this kind. For example, MacKay recalls, one of the Ratio Club's rules was that any member who reached the rank of full professor, where he would have potential power or control over other members, automatically ceased to be a member. It was the only way the free exchange of ideas could be assured.

Scientists came from all over England—Cambridge, Bristol, Manchester. The meetings were initially packed with short talks until everyone got a good idea of everyone else's interests, and then the meetings settled down to a normal talk or two per time. Central to the Ratio Club was the notion of processing information in animals and machines—cybernetics, in a word, except that that particular word was quickly becoming sensational and imprecise. The variety of approaches members of the Ratio Club took describing mind—or "minding" as Donald MacKay prefers to call it—was due to their diverse backgrounds. Except for Albert M. Uttley and Alan Turing, none of them had much connection with digital computers, and so the computer didn't necessarily suggest itself as a useful model for the brain. Indeed, for many of them who were physiologists, the all-or-none logic that had so enchanted some researchers seemed, in MacKay's words, perverse, because it didn't provide for the kind of graded signals that the brain's chemicals seemed to carry.

Being an analog man myself [stated MacKay], I wasn't going to take easily to the suggestion that the brain is a digital computer. I started thinking what sort of more general mechanisms one could conceive of, artificial mechanisms, which would handle information in a more general sense than a digital computer does. For instance, a digital computer is unable to represent the concept of in between.

He tried to imagine a machine with the virtues of digital machines, which have great accuracy but, he felt, no way of representing gradations, and analog machines, which can represent gradations but are limited as to accuracy:

I was accustomed to getting a bit of flak from the computer-based chaps on the ground that it was all very well to talk in these terms but you couldn't formalize it. And there was this great feeling that unless you could formalize something mathematically, you weren't really under-standing it. I've resisted this all my life because I think it's absolutely not true. There is so much of understanding which is essentially qualita-tive—the sort of aha! you get by seeing that somebody's pulled the plug out of the bath and that's why the water's going down, and all sorts of analogous situations where formalization is just a waste of time.

Formalization is needed, MacKay believes, when interactions are so complicated that we are unable to understand—get an in-tuition—for what's going on otherwise.

When MacKay came to the United States on a Rockefeller Fellowship in 1951, he cheerfully stuck out his neck with such sentiments and found more than a few axe wielders among the digital computer people who were annoyed by such wooly think-ing about information processing. MacKay understands, he says, that it is often trivial to set up a model on a digital computer of a given analog process:

There is no pattern of behavior which, in principle, is beyond the power of the digital computer to simulate. But that's rather like say-ing that if you want to draw some lines on the board, there is no line you can draw which is beyond the power of a man with a fine pen point to reproduce, dot by dot. Sure. But what's the merit in using a fine pen point to reproduce, dot by dot, when with a coarse pen point you can take one sweep and there it is. I know the practical answer, of course, is that it would take you a long time to use a soldering iron and actually hook together the stuff you need for the other, whereas you've got a digital, high-power, general-purpose machine there, and all you've got to do is try it. But I think we're coming more and more to the frontier where what artificial intelligence is waiting for is the mechanization in hardware of the sorts of parallel interaction which will allow an order of magnitude increase in intelligent behavior with a given outlay of cost. I think that parallel interaction to allow a multidimensional graded approach is the crucial thing.

MacKay's work these days is on modeling the human brain, but in the late 1940s and early 1950s, the days of the Ratio Club, he was wondering whether there was such a thing as the measurement of information, and in particular whether limits existed, in the way that the Heisenberg Principle limits certain kinds of physical measurements. Though he was acquainted with Claude Shannon's work in information theory, he didn't see that it bore directly on his own work, which was using a different sort of model for describing the phenomena surrounding information.

In 1949 MacKay privately circulated a document that addressed the problem of combining digital and analog techniques in the design of what he called Analytical Engines. Here he envisaged an "autonomous artefact" which, if it met his specifications, among them being the ability to respond and adjust its behavior to a new set of information in order to achieve equilibrium (in other words, to learn), could be considered genuinely autonomous. That is, it would be able to pursue a disciplined existence independent of human intervention, an existence with at least an abstract sense of purpose (to seek equilibrium). MacKay clearly saw that the behavior of such a machine need not be altogether predetermined by its designer; that in fact it would develop purposes of its own, congruent with an overriding design principle of, say, seeking equilibrium.

MacKay believed that the data used in human thinking were not exact, could not ultimately be reduced to yes-or-no propositions, and he wanted to introduce instead the idea of partial truth or probability. (This idea would come up again in the future for the same reasons MacKay raised it. Herbert Simon, for example, would introduce the concept of "satisficing," of finding a good enough if not perfect solution to a problem; and in the 1960s, Lotfi Zadeh would invent the concept of fuzzy logic.) In a 1951 paper, MacKay proposed a statistical mechanism that would display many of the attributes of human cognition, including prejudices, preferences, originality of a kind, and learning. He chose to call such behavior mindlike, rather than thinking, but in the last part of the paper, he did address the question that has teased philosophers and psychologists since at least the time of Descartes— the problem of how mind arises from matter. It isn't the descriptions of mind or matter that are exclusive, he concluded, but the logical backgrounds in terms of which they have meaning. Like the dual descriptions of light as waves and as particles, each description is valid in its own right. This seems to me a safe flight over but not a trip through the den of the metaphysician.

In 1955 MacKay again looked at these problems, this time at the invitation of Claude Shannon and John McCarthy, who were compiling their landmark volume, *Automata Studies*. His focus this time was more philosophic—the paper's title is "The Epistemological Problem for Automata." Here again he proposed a statistical model that could, in some sense, learn, and was in principle capable of developing its own symbols for abstract concepts.

The well-known "turtle" was the brainchild of Grey Walter, the brain physiologist at whose laboratory the Ratio Club was conceived. He had made significant discoveries about the electrical activity of the brain, and in 1948, to study some of these further, he built the turtle, a dome-shaped electromechanical device that rolled its way around obstacles, and retired to its hutch when its batteries needed recharging. Walter intended it to show that complex behavior—what an observer might see as purpose, independence, and spontaneity—was in fact the result of rich interconnections of a relatively small number of original elements. This theme would be taken up again by Herbert Simon, using the metaphor of an ant crawling along a pebbled beach, in the Karl Compton Lectures he gave at MIT in 1968 (Simon, 1969).

In front of me as I write is a picture of Grey Walter stepping through a doorway, a briefcase in one hand and the turtle's carrying case in the other. He looks for all the world like a country gentleman hurrying off to the Westminster Kennel Show, with what has to be a champion terrier in that traveling case.

The turtle spawned offspring everywhere, it seemed—the idea of negative feedback was fascinating, and its execution provided enormous fun in the name of science. To this day, remote descendants of Grey Walter's turtles amuse and instruct the schoolchildren of Cambridge, Massachusetts.

W. Ross Ashby, a psychiatrist and another member of the Ratio Club, had long been certain that much human mental behavior could be accounted for mechanistically. In 1948 he published a celebrated paper which he later expanded into a book; both are called *Design for a Brain* (1952). Here he proposed a means of imitating the brain's ability to produce adaptive behavior, or what we might call learning. "Many other workers have proposed theories on the subject, but they have usually left open the question whether some different theory might not fit the facts equally well," Ashby wrote in his preface. "I have attempted to deduce what is necessary, what properties the nervous system must have if it is to behave at once mechanistically and adaptively."

Ashby's brain model is presented in two forms, first in plain language and then in mathematical terms.

> Having experienced the confusion that tends to arise whenever we try to relate cerebral mechanisms to psychological phenomena, I made it my aim to accept nothing that could not be stated in mathematical form, for only in this language can one be sure, during one's progress, that one is not unconsciously changing the meanings of terms or adding assumptions, or otherwise drifting toward confusion. The aim proved achievable.

Design for a Brain was an inspiring book. For some it was richly suggestive, chockablock with ideas to explore further. For others it was inspiring in a slightly different way. "It was maddeningly full of holes," Herbert Simon recalls, "holes I wanted to fill." "It must have had a very good effect on a lot of people," says Marvin Minsky, "because it was beautifully written, and had great logical clarity; it was easy to understand; and instead of talking about advanced ideas of heuristic search, it talked about very simple ideas of nonheuristic search. Lots of people could understand that. McCulloch said even doctors could understand it, so it was an enormous achievement! And not only could they understand it, but they could see that it wasn't good enough to explain thought, and perhaps think of the next step, so it would turn on people who might do more work."

Ashby's model is reminiscent of the cybernetic principles Wiener enunciated, though not identical with them. "The free living organism and its environment, taken together, form an absolute system. . . . The two parts act and re-act on one another" (Ashby, 1952). This notion is not new, not with Ashby or even with Wiener, for Ashby quotes scientists as early as 1906 who made the same observations. But Ashby refines it, introducing other concepts such as stability (what MacKay called equilibrium), a mode of survival in the organism. Or in the intelligent system of any description. A key passage focuses this idea: "A determinate 'machine' changes from a form that produces chaotic, unadapted behavior to a form in which the parts are so coordinated that the whole is stable, acting to maintain certain variables within certain limits—how can this happen?" The answer is that the machine is a self-organizing system, a system that responds to stimuli, changing its behavior, and in some sense its shape, in order to achieve stability—what Ashby chose to call ultrastability.

Such a machine had indeed been built by Ashby: he called it the homeostat. It's a cluster of four units, each unit able to emit direct current output to the others and to receive theirs in turn. Since definite values were assigned to various governing devices in the units, the homeostat would begin to exhibit definite patterns of behavior relative to the settings of those governing devices, always seeking to stabilize itself. Ashby extended this principle to living organisms, suggesting that their adaptive, learned behavior could be expressed as a system that organizes itself to seek stability.

He pointed out that his own aim and the aim of a person who designs "a new giant calculating machine" might both be described as trying to design a mechanical brain. But the latter wants a specific task performed, preferably better than a human can do it and not necessarily by methods humans might use, while Ashby's aim "is simply to copy the living brain. In particular, if the living brain fails in certain characteristic ways, then I want my artificial brain to fail too, for such a failure would be valid evidence that the model was a true copy" (1952).

Here Ashby articulated the distinction that would subsequently define two major branches of artificial intelligence: one aimed at producing intelligent behavior regardless of how it was accomplished, and the other aimed at modeling intelligent processes found in nature, particularly human ones. That division was to turn out to be less distinct than researchers in the early 1950s imagined.

The homeostat adapted in various ways to a segment of the universe, or a linked series of segments (as indeed do all organisms: even that most adaptable creature *homo sapiens* cannot live without artificial aid under water, or beyond the boundaries of certain temperatures), and itself linked the adaptations it made to each of those several environments. It seemed very brainlike.

Marvin Minsky, whose provocative statement about human brains as meat machines I've already quoted, was himself deeply concerned with trying to understand how the brain works, and trying to understand it in terms of neurons, using as a model the on-off cells of the digital computer. He recalls,

Most young people in this field had been strongly influenced by the writing of McCulloch and Pitts in the 1940s. It took me a long time to switch from trying to understand how the brain works to understanding what it does; in particular to try and make up theories of how any kind of thinking machine might work.

Prodded by his colleague Ray Solomonoff, who complained that there must be a more direct way of working on ideas of intelligence, such as trying to define what the behavior was instead of the parts it comprised, Minsky moved away from the neural-net idea.

Just as in computing, we now know enough not to worry about what the flipflops and resistors and wires do in the computer, but instead we try to describe what the procedures should do, and get at what's essential to those procedures. Newell and Simon started out with that idea for some good reason that I don't know, but those of us under the influence of McCulloch and Pitts kept worrying about how, not what.

Minsky had grown up in New York City—he was a Bronx High School of Science alumnus—and had done his undergraduate work at Harvard, where he was nominally a physics major:

I wandered around the university and walked into people's laboratories and asked them what they did. I didn't know anything about the social life of the undergraduates, but I knew when the department teas were, and I'd go and eat cookies and ask the scientists what they did. And they'd tell me.

He might have set a record: he had three laboratories of his own. In the biology laboratory he was given an abandoned darkroom, where he took apart crayfish, connected their nerves to an electrical source, and manipulated their claws to pick up a pencil and put it in a jar, by coincidence foreshadowing the robot arms that would attempt the same sorts of tasks in his laboratory nearly two decades later. He also had a laboratory in the physics department where, together with a friend from Princeton, he assembled three or four hundred tubes to build a learning machine based on stochastic reinforcement (a popular mathematical theory of learning at the time). It simulated four rats running around in a maze, bumping into and learning to avoid each other.

It was really spectacular [Minsky says], but by the end of the summer, it was clear that we had a bad theory, and that left the laboratory with the problem of what to do with half a roomful of stuff. It sat there for quite a few years, and then some Dartmouth student read the memo describing it, was greatly taken with it, and the machine went up to Dartmouth, where they worked on it for a year or so. But I don't know what happened to it then.

It's uncommon for an undergraduate to have his own laboratory for anything, even at Harvard. For Minsky to have not one, not two, but three, was simply unheard of. Yet these three laboratories—in biology, physics, and psychology—stand as an elegant illustration of the kinds of interests early workers in artificial intelligence had; they also stand for the ardor such researchers brought to their work, a yearning to draw anything germane to a theory of thinking from any discipline possible.

For Minsky it was all a little premature. His hope of reproducing mindlike behavior by simulating the physiology of the brain came to naught, except to earn him a Ph.D.:

I wasn't stuck in the sense that there wasn't anything to do. There were many ways to push that theory and make it more complicated and see how those machines might learn more things, but none of them seemed very exciting. The whole thing was pretty tedious at that point.

In storybook science, a good scientist would simply abandon a fruitless line of research and go on with something else, and indeed, Marvin Minsky did that. But that kind of wholesale abandonment is a storybook ideal; it seldom happens in real life, and Minsky's move away from the physiological to the information-processing model was unusual. Though he acknowledges the facts of human behavior in such situations, he explains his unusual departure with simplicity: "It's very boring to be doing something that doesn't work. It's really a question of why people keep doing the same thing when they don't like it."

Thinking of politicians, surgeons, professors, real estate brokers, married persons, and countless others who find themselves bored with their initial commitments but unable to change, I suggested that ego might have something to do with it.

It's more complicated than that [Minsky replied]. I think most people don't like learning something new. It's unpleasant. And when you're very good at something, if your main goals don't succeed, there are always lots of other things you can do with those skills you've developed.

You either have to like learning new things, or have some kind of masochistic streak so that you enjoy suffering through a change of view. I don't know quite which of those I have.

Were there other times in his life when he'd seen he was going down the wrong track and made a change?

Sure, yeah. In fact, I always sort of look forward to it expectantly. It's very nice, see, because you don't have to finish that thing. Once, I started making a piece of machinery, and learned all about machining and precision grinding and polishing. I spent about six months making a steel bar very flat, but after a while I thought there must be a better way to make optics work than just be very accurate. I enjoyed quitting that. Another time, I read about a new kind of microscope and tried to build the instrument very precisely, but then I learned about making precision instruments without doing any work, and I remember saying, "Oh, gee, the whole thing I've been doing is quite wrong. What fun! Now I can do something else."

The neural-net approach took as its basic assumption that organic brain cells were largely undifferentiated, and only organized themselves into purposeful behavior because of experience and perceptions, the hypothesis Warren McCulloch had advanced in the late 1940s. Thus, for example, when certain kinds of damage occur to some parts of the brain, other parts of the brain can assume responsibility for the functions the damaged part used to do, particularly if the damage occurs early in the brain's existence. In other words, there seems to be a good deal of built-in redundancy in an organic brain, although it is eventually limited. Moreover, the way those connections between neurons is made seems more or less random; thus electrical impulses can detour when one route is blocked or damaged.

This model represented an overwhelming temptation for researchers using new technology to build systems that would likewise organize themselves (so-called self-organizing systems), that would do so randomly, and that would learn and adapt in the way that organic brains seemed to.

One of the largest such efforts was a system called the Perceptron, which was the work of a group of researchers at Cornell led by

Frank Rosenblatt, who had been a classmate of Minsky's at Bronx Science. It originally had three levels. The first was a grid of photocells corresponding to the retina of the eye, which reacted to light stimulus. Below this level, associator units collected the impulses transmitted from the photocells, to which they'd been randomly wired, and those in turn signaled to response units. Because we know that animals, including humans, are born knowing some things, Rosenblatt and his coworkers modified the Perceptron in such a way that not all of it was randomly wired, which improved its performance in recognizing and "learning," say, a letter.

As time went on, the Perceptron began to acquire a certain amount of notoriety. Besides its simplicity, there was another reason for its growing fame, and that was Frank Rosenblatt himself. Present-day researchers remember that Rosenblatt was given to steady and extravagant statements about the performance of his machine. "He was a press agent's dream," one scientist says, "a real medicine man. To hear him tell it, the Perceptron was capable of fantastic things. And maybe it was. But you couldn't prove it by the work Frank did."

I. J. Good, who had been Turing's assistant during the war, and who is now a professor at Virginia Polytechnic Institute, says, "In fact, it was fairly clear that Rosenblatt's original work did not contain real proofs of results that he claimed. It was later work by H. D. Block and also by Seymour Papert that showed there were certain theorems that could be proved."

Case-Western's Leon Harmon, who worked on the von Neumann machine at the Institute for Advanced Study at Princeton, and who describes himself as perhaps the first computer operator, still seethes about walking into the Smithsonian and discovering that beside the von Neumann machine, which well deserved to be there, stood a Perceptron, sharing floor space as if it were equally important. Harmon doubts that we'll ever learn much about brain operation from studying electronic hardware, and believes that the really interesting and potent things the computer in our head does are inscrutable. If he was once enamored of the work Grey Walter and W. Ross Ashby were doing, it's an infatuation he's outgrown. Rosenblatt only irritated him.

"He *did* irritate a lot of people," says W. W. Bledsoe of the University of Texas, speaking of Rosenblatt, "but he also charmed at least as many, and I count myself among them. Just when you were thinking that Frank didn't have another trick up his sleeve, along he'd come, and he'd be so darn convincing, you know, he just had to be right." Rosenblatt also influenced a number of people at the Stanford Research Institute who still speak respectfully and affectionately of him. But his Perceptron ran into both practical and theoretical difficulties, and Rosenblatt's accidental death seemed to rob the effort of its energy to continue.

Another who was irritated by Rosenblatt was Marvin Minsky, perhaps because Rosenblatt's Perceptron was not unlike the neural-net approach Minsky was alternately intrigued and frustrated by. Many in computing remember as great spectator sport the quarrels Minsky and Rosenblatt had on the platforms of scientific conferences during the late 1950s and early 1960s. While he was trying out other approaches to the problem of making thinking machines, Minsky kept playing around with the Perceptron idea. In 1961 he attended a conference in England to present some work on a Perceptron-like machine, and there he met another young researcher who had proved pretty much the same theorem about Perceptrons, using quite different methods. This was Seymour Papert, who was later brought to MIT by Warren McCulloch, and who was to become Minsky's research partner in a number of efforts, including the last word (so far) on the subject, a book called *Perceptrons* (Minsky and Papert, 1968).

One problem with a Perceptron was that it could classify stimuli it received but it lacked an internal representation of that act. Therefore it couldn't refer in some symbolic way to the act of perception, but had to recapitulate the act itself exactly, which put it in the position of being no better off when repeating the act again. Symbolic representation, by which machines (and humans) deal with a host of phenomena without having to reiterate them in their totality, is central to intelligent behavior, to memory, and to consciousness.

Since Minsky and Papert had found the whole theory of Perceptrons to be very muddy, they decided to turn it into a crisp theory. But an odd thing happened. After working on the problem of Perceptrons for some three years, and coming to understand them at least partially, and proving some theorems about them, Minsky and Papert laid out their book. In the process of writing, loose ends appeared, and the two scientists kept working, tying up the loose ends and delaying publication. By the time the book was finished, it left only four or five difficult unsolved problems. An enthusiastic review by Allen Newell expressed the hope that the work Minsky and Papert had done would now be extended, but in fact nothing happened. Minsky says,

We feel that we made a mistake in solving all the problems that looked easy; there's nothing left for beginners to do. It's not that we understand the thing nearly as well as we want, but with three years' practice at solving all the problems that still looked easy to us, we solved all the ones that might tempt beginners. There isn't anything near the surface left to do. That's advice to anyone who writes a book about a theory: don't tell all you know, or you'll never find out anything else.

And so much for the hope of making a machine think by trying as literally as possible to imitate the brain, the meat machine, at the cellular level. It didn't work, as John von Neumann had said it wouldn't. We didn't then—nor do we now—know enough about how the brain works at the cell level to make such a model, and there's presently speculation that in any event, a serial machine simply won't be able to imitate what is very likely a series of parallel processes.[3]

But researchers in artificial intelligence had ideas that there were other equally useful and effective ways of modeling or simulating thought, and it was these alternative models that were to preoccupy them in the future.

---

[3] Most computers presently in operation are serial machines. They have one processor which executes instructions consecutively, although at such high speeds—millions of instructions per second—that these machines appear to naive human observers to be doing many things at once. The single-processor machine is very fast but its capacity is limited, and an innovation has appeared in the form of parallel machines. These have a number of processors, and their virtue is an increase in speed over the single-processor serial machines.

Part
Two

—

The Turning Point

When you have eliminated the impossible, whatever remains, however improbable, must be the truth.
— *Sir Arthur Conan Doyle*

Anyone who is practically acquainted with scientific work is aware that those who refuse to go beyond fact rarely get as far as fact.
— *Thomas Henry Huxley*

Chapter Five

~

The Dartmouth Conference

On the leafy campus of Dartmouth College in the summer of 1956, a handful of scientists met to talk about the work they were doing toward making machines behave intelligently. Although they came from different backgrounds—there were men trained as mathematicians, as psychologists, as electrical engineers—and although some worked for industry and others were at universities, they had in common a belief (more like a faith at that point) that what we call thinking could indeed take place outside the human cranium, that it could be understood in a formal and scientific way, and that the best nonhuman instrument for doing it was the digital computer.

Four of them had collaborated on a proposal to the Rockefeller Foundation which said, "We propose that a two-month, ten-man study of artificial intelligence be carried out during the summer of 1956 at Dartmouth College in Hanover, New Hampshire. The study is to proceed on the basis of the conjecture that every aspect of learning or any other feature of intelligence can in principle be so precisely described that a machine can be made to simulate it."

The four were John McCarthy, a young assistant professor of mathematics at Dartmouth, Marvin Minsky, then a Harvard Junior Fellow in mathematics and neurology, Nathaniel Rochester, manager of information research at IBM's research center in Poughkeepsie, New York, and Claude Shannon, a mathematician at Bell Telephone Laboratories who was already well known for his statistical theory of information.

McCarthy, who was really the main organizer and mover behind the conference says, "At the time I believed if only we could get everyone who was interested in the subject together to devote time to it and avoid distractions, we could make real progress."

He thought of Marvin Minsky, whom he'd known when they were graduate students together in mathematics at Princeton. They'd had some talks about artificial intelligence then. "Not very productive ones, other than to establish that we were in fact allies on the subject and that we agreed on a number of things—I don't think Minsky and I changed each other's ideas much in the areas that we didn't agree, but in the areas in which we did agree, we reinforced each other." In 1952, they had both been hired to work for the summer with Shannon at Bell Labs (though Minsky got married that summer, and for all practical purposes disappeared), and thus McCarthy knew that Shannon was sympathetic to these ideas, and he too was invited to join in the proposal. In connection with the gift of a computer IBM was making to MIT, McCarthy had run across Nathaniel Rochester and had discovered his interests in intelligent machines.

Rockefeller provided some $7500, and the initial four invited others who shared their belief that "every aspect of learning or any other feature of intelligence" could be simulated. Among them were Trenchard More, Arthur Samuel of the IBM Corporation, Oliver Selfridge and Ray Solomonoff, both of Massachusetts Institute of Technology, and, almost as an afterthought, two vaguely known persons from the RAND Corporation in Santa Monica and Carnegie Tech in Pittsburgh, Allen Newell and Herbert A. Simon. It was to be a significant afterthought.

Presently, no one is quite sure how the Cambridge people got in touch with the Carnegie-RAND group, though there are several possibilities: Oliver Selfridge had given a talk at RAND the previous fall, and had mightily impressed young Allen Newell, indeed, had turned his scientific life around. Marvin Minsky was a consultant at RAND and might have known about the work of Newell and Simon that way. Newell remembers visiting McCarthy at IBM in 1955, where he first met Arthur Samuel too—there were certainly connections.

In addition, others came to Dartmouth for short visits to talk about related work, and among those visitors was Alex Bernstein, then a programmer for IBM in New York City, who was invited to talk about the chess-playing program he was working on. His work was known to Shannon, Rochester, and Arthur Samuel, who himself was working on what was to be one of the earliest and most successful of the game-playing programs with computers, one that played checkers.

Besides serving as a fine moment for us by assembling onstage nearly all the significant characters in the cast of this drama called Artificial Intelligence to meet and be met, and to foreshadow events of the drama to come, and to reveal hitherto unknown events that had already taken place and that would influence the plot mightily, the Dartmouth Conference was also a confluence of several different intellectual streams of the twentieth century. They themselves had flowed from other streams, from the work of individuals in mathematics, statistics, psychology, engineering, biology, linguistics, and the emerging disciplines of management science. If certain scientists were not present at the conference, their spirit was represented by their work, and sometimes by their colleagues and students. I think here of Norbert Wiener and his work in cybernetics, Warren McCulloch and Walter Pitts and their research toward a physiological theory of knowledge, and the computer design of John von Neumann, and to a lesser extent, Alan Turing.[1] These men in turn had been influenced by still others,

---

[1] In a logical genealogy, Turing world be central. He held what were to be some of the central ideas of AI—the symbolic nature of the computer, the necessity to look at comparable functions instead of comparable hardware in humans and machines —very early. But the history of ideas has its own way of doing things and, as it happened, Turing's work had practically no influence on most people at the Dartmouth Conference. For instance, Minsky felt himself much more influenced by McCulloch and Shannon (especially Shannon's early chess paper); Simon considered Turing of no particular influence on his work.

logicians such as Whitehead and Russell, or daring engineers such as Leonardo Torres y Quevedo.

Antecedents were plentiful and varied, as this history has already shown, but they all pointed in the same direction, toward the idea that there was a rigorous and objective way of explaining the human intellect.

Recollections differ. Some of the participants did have the sense that they had gathered for something momentous. Oliver Selfridge, working then at MIT's Lincoln Laboratories, says he had a great sense of something important having been conceived. "Here was a field which was going to be great things. It was to fulfill almost none of its promises in the time scale planned, but the promises are still there."

But the intense and sustained two months of scientific exchange envisaged by John McCarthy never quite took place:

Anybody who was there was pretty stubborn about pursuing the ideas that he had before he came, nor was there, as far as I could see, any real exchange of ideas. People came for different periods of time. The idea was that everyone would agree to come for six weeks, and the people came for periods ranging from two days to the whole six weeks, so not everybody was there at once. It was a great disappointment to me because it really meant that we couldn't have regular meetings.

McCarthy's present recollections seem borne out by the work that followed immediately from the Dartmouth Conference. Though it can be seen in retrospect that the new paradigm, the new model that would dominate for the next ten years, made its debut at the conference, it isn't at all clear that anyone but its discoverers were so persuaded.

A dispute occurred over what the new field should be named. Although the conference was officially called The Dartmouth Summer Research Project on Artificial Intelligence, many attendees balked at that term, invented by McCarthy. "I won't *swear* that I hadn't seen it before," he recalls, "but artificial intelligence wasn't a prominent phrase particularly. Someone may have used it in a paper or a conversation or something like that, but there were many other words that were current at the time. The

Dartmouth Conference made that phrase dominate the others."
McCarthy had not chosen accidentally to call the conference by
that name. During the summer that he'd worked for Shannon,
when they had put together the book of collected papers on some
of the subjects that McCarthy was interested in, to be called *Automata Studies* (Shannon and McCarthy, 1956), McCarthy wanted
to use a term different from automata studies for the papers he
hoped to get for the book, but Shannon objected that any other
phrase was simply too flashy, that the theory of automata would
be sober and scientific. McCarthy went along with that, thinking
that it probably didn't make that much difference.

"The original idea was that Claude Shannon would be the
name to attract good papers, and I would do the work, but it
ended up that he did the work too," McCarthy now recalls. "One
of the reasons why he did all the work was I was unenthusiastic
about the papers." Most of the papers they received for the book
were in fact about automata theory in the narrowest sense, that is,
mathematical principles underlying the operation of electromechanical systems, and not about the relation of language to intelligence, or the ability of machines to play games, or any of the
other topics McCarthy was becoming more and more fascinated
by. The Shannon and McCarthy volume contained many significant papers and is still a reference in automata studies, but
McCarthy felt he'd learned his lesson. Thus in the proposal and
again at the conference, he argued strongly for the term *artificial
intelligence* to distinguish it from automata theory, though to this
day there are persons in the field, including some of the original
participants in the Dartmouth Conference, who object to it.

"The word artificial makes you think there's something kind
of phony about this," says Arthur Samuel, "or else it sounds like
it's all artificial and there's nothing real about this work at all."
Neither Newell nor Simon liked the phrase, and called their own
work complex information processing for years thereafter. But
artificial intelligence is the phrase that stuck. In my view it's a
wonderfully appropriate name, connoting a link between art and
science that as a field AI indeed represents.

It may be that the concrete results of the Dartmouth summer conference were scant. Marvin Minsky circulated some drafts of what would later be his influential paper "Steps Toward Artificial Intelligence" (1963). Ray Solomonoff, who was then working on models of inductive inference, remembers trying to convince people that they did not have to give the machine hard problems to determine whether it was really thinking. Without a firm notion of what thinking really was, some had felt the only way to show that a machine was doing it was to give it problems that they themselves would have trouble solving. Solomonoff felt, to the contrary, that if he himself were working on a hard problem, he'd have some idea about how to solve it, but that a simple problem offered to a machine allowed you to concentrate on the methods by which the machine solved the problem, and not on the solution itself. And he remembers being particularly taken with an idea of McCarthy's that summer, of the possibilities of expressing any intellectual problems in terms of a Turing machine, that hypothetical machine where a string of symbols goes in to be processed and later a string of other symbols comes out. McCarthy suggested that this process be inverted, that to look at the string of symbols that came out might give you an idea of what had gone in. This notion alternately riled and beguiled Solomonoff for the next few years until he could show that with some modifications it might be a good way of understanding inductive inference.

Alex Bernstein, who had come up to Dartmouth from New York to talk about the chess-playing program he already had underway, remembers hearing McCarthy's plans to begin on a chess-playing program, and listening with interest to his ideas. But when they came to play a game of chess with each other, the equivalent of *mano a mano* in the world of science, Bernstein won, despite the fact that he'd accepted the handicap of playing blindfold. Thus fired up, Bernstein was confident he could get back to New York and produce a program to beat McCarthy to the punch, which in fact he did. Because his visit to Dartmouth didn't coincide with that of Newell and Simon, he discovered only later that he and they had arrived independently at some of the same ideas for the problem.

Minsky toyed with a geometry theorem-proving program, an idea that had attracted Newell and Simon earlier, but which they'd abandoned because of the difficulties of representing diagrams in the computer. Nathaniel Rochester would carry this idea back to IBM and put to work one of the brightest young men in his shop, a new Ph.D. in physics who had gone to work for IBM because it promised better money than academic physics to pay off his graduate-school debts. He was Herbert Gelernter, and he was subsequently to design a successful and celebrated plane geometry theorem-proving program, which also provided an intellectual link between Newell and Simon's first, somewhat awkward list-processing computer language, IPL-V (IPL for Information Processing Language), and John McCarthy's sophisticated later one, LISP.

I asked Minsky if he remembered having a sense of being at a historical gathering during the Dartmouth Conference.

Well, yes and no [he answered]. There was a false sense that people were beginning to understand theories of symbolic manipulation and theories of cybernetics which dealt with concepts rather than simple feedback, and that things were going to be understood around the world on a wide scale. I think we had the feeling that these ideas were beginning to become popular, and maybe that's a historic event. It wasn't really true. It took another ten years before people could tolerate the idea of AI without thinking that it was funny and impossible. It wasn't, really. It did look like the time had come when something very impressive ought to be possible within a fairly short time. In that sense it looked like various of us had come to similar conclusions and maybe the field was ready for another stage of recruiting people and attempting large projects. None of the large projects worked very well. The field wasn't really ready for groups of more than two or three people to work on a given project. Very few things are. I think you can only get a large group of people to work on something when the general plan is very clear and it is just a matter of details that can be chopped up. But what happened here was that the ideas really were changing very quickly, and if you got two or three people committed, doing parts of the thing, then two months later you'd probably have to tell them, "Sorry old man, we can't use that because so-and-so has discovered a trick which can do that in five minutes and a real problem is over here instead." And that's still happening today.

McCarthy's disappointment, however, was unequivocal. "I was simply measuring the distance between what I had hoped to accomplish and what we did accomplish, and it was pretty large." McCarthy's disappointment reflects a persistent problem with artificial intelligence—that making machines think, designing computer programs to behave intelligently, was far harder than anyone in 1956 thought it would be. Over two decades have imbued the field with more modesty than it had in its infancy, but the fact remains that the problems continue to be harder than anyone expected. I often think of this when I hear complaints that the effort to make machines behave intelligently can only diminish our human self-esteem. It seems to me the contrary: that the efforts of the last twenty years, some successful and some failures, should inspire in us nothing less than awe at the potent elegance of the human brain.

The Dartmouth Conference proposal is a fine illustration of this point. For example, Minsky described some work he had been doing on the notion of the neural net, and then stated his goals for the summer:

> The important result that would be looked for would be that the machine would tend to build up within itself an abstract model of the environment in which it is placed. If it were given a problem, it could first explore solutions within the internal abstract model of the environment and then attempt external experiments. Because of this preliminary internal study, these external experiments would appear to be rather clever, and the behavior would have to be regarded as rather "imaginative."

It was ten years before that concept was even remotely realized.

Nathaniel Rochester hoped to answer the question, "How can I make a machine that will exhibit originality in the solution of its problems?" And McCarthy thought it desirable

> to attempt to construct an artificial language which a computer can be programmed to use on problems requiring conjecture and self-reference. It should correspond to English in the sense that short English statements about the given subject matter should have short correspondents in the language and so should short arguments or conjectural arguments. I hope to try to formulate a language having these properties and in addition to contain the notions of physical object, event, etc., with the hope that using this language it will be possible to program a machine to learn to play games well and do other tasks.

Shannon planned both to apply information theory concepts to brain models and to study the synthesis of brain models by developing a parallel series of matched and gradually more complicated environments and brain models that would adapt to those environments.

No one was naive enough to expect to finish everything during the summer of 1956, but neither did anyone expect to map out his professional life for the next twenty years, which was almost the case with Minsky and McCarthy, and Newell and Simon. However, Rochester and Shannon found themselves more interested in other things, and remained no more than sympathetic spectators of AI in the years to come.

An appendix to the conference proposal, written in the spring of 1956 (some months after the original piece), outlines the work that Newell and Simon planned to do at the summer conference. It describes their progress to date on a chess machine as well as on another machine,[2] which employed humanlike procedures to discover proofs in the propositional calculus, and a less well-developed idea of learning theory. Perhaps it was their practical experience with the immense difficulties of these problems that made

---

[2] Artificial-intelligence workers continually use *machine* when they mean what an outsider would call a *program*. In the hardware sense, the *machine* is always the general-purpose digital computer, and the same piece of apparatus can be programmed, of course, to do a multitude of different tasks. No one seems to know how this convention started. Possibly it derives from Turing's use of *machine* to describe an abstract procedure.

their appendix sound more cautious than the rest of the proposal. "We are a long way from even knowing what questions to ask or what aspects to abstract for theory. The present need is for a large population of concrete systems that are completely under- stood and thereby provide a base for induction." And that, as it happened, was just how artificial intelligence was to proceed in the future.

Claude Shannon had been thinking about machine intelli- gence for a long time. His brilliant master's thesis of 1937 at MIT, an application of Boolean algebra to the study of switching sys- tems in engineering, is an early example. Boole, whose work is described in Chapter 2, had intended to formalize the "laws of thought" with his algebra. The hope that if Boolean algebra could be used to express the behavior of electrical switches, then con- versely human thought might be expressible in the behavior of electrical switches occurred not only to Shannon but indepen- dently to others.

Shannon remembers a conference at MIT during the early part of World War II when Mauchly was talking about his com- puter, and others talked about work they were doing. Shannon says, "You may say that [the subject of this conference] isn't artifi- cial intelligence, that it's a different thing, but I see that as every- body striving to find the farthest reaches of computers. We real- ized that this was a lot more than an adding machine, a much more general and powerful tool than that."

In 1950 he published an article in *Scientific American* called "A Chess-Playing Machine" (Shannon, 1950). Here he pointed out that the new machines could not only carry out numerical calculations, but were so general and flexible that they could "be adapted to work symbolically with elements representing words, propositions or other conceptual entities." This insight escaped many people, who continued to regard computers only as giant calculators and missed the fact that the computer was in some sense misnamed, for it could not only compute, but also manipulate symbols of many kinds in different ways. Shannon's report is a lucid representation of the problems any chess-machine builder must face, including the inescapable fact that brute-force methods—the popular idea of a computer being able to explore every possible move and countermove — simply will not work. Shannon calculated that a typical chess game, played that way, has about 10^{120} possible moves. "A machine calculating one variation each millionth of a second would require over 10^{95} years to decide on its first move!" Thus, Shannon's hypothetical machine would play good but not perfect chess, based on a method of evaluating certain positions numerically. The chief weakness of the machine would be that it could not learn from its mistakes, a problem for which Shannon saw a theoretical, but not a practical, remedy. He too addressed the problem of whether such a machine could be said to think, and concluded, like Torres, that the definition of thinking was much too fluid to say for certain.

Three years later, in 1953, Shannon took up these ideas again, in a paper called "Computers and Automata," which appeared in a technical journal (Shannon, 1953). It was a survey of current ideas largely meant to provoke research: "We hope that the foregoing sample of nonnumerical computers may have stimulated the reader's appetite for research in this field. The problem of how the brain works and how machines may be designed to simulate its activity is surely one of the most important and difficult facing current science." Indeed, the list of questions he poses in 1953 (Can we organize machines into a hierarchy of levels, as the brain appears to be organized, with the learning of the machine gradually progressing up through the hierarchy? Can we program a digital computer so that [eventually] 99 percent of the orders it follows are written by the computer itself, rather than the few percent in current programs? Can a self-repairing machine be built that will locate and repair faults in its own components?) provided a program for the entire field of computer science for decades to come. What comes across now is the excitement Shannon must have felt as he was writing; there's an enthusiasm, almost a fervor, in this piece that is seldom found in scientific writing. These were high times.

It was also in 1953 that Minsky and McCarthy came to work for Shannon during the summer at Bell Labs. He remembers them all *digging around in Bell Labs trying to find people who were working on this kind of thing. There was one character at Bell Labs who had a sort of robot telephone exchange, where in place of crossfire switches and rotary switches and so on they had an old-style plug and cord thing that a woman or man would sit at. But he had a robot sitting there which would reach for these cords and move them up. It was kind of a fascinating idea. It sounds weird but it had certain advantages that you could get out into three-dimension things, and cords could go over other cords very easily. You had access to an enormous number of places where these cords—well, that's another story, but that was the kind of robot thinking of that period.*

All this activity brought Shannon to edit the automata book with McCarthy and to join with McCarthy and Minsky and Rochester in sponsoring the Dartmouth Conference. The latter was for him no more than another step in a long, continuous series, though after the Dartmouth Conference his interests went in a somewhat different direction. He resumed his work with information theory, and played about with ideas of a minimum universal Turing machine.

One late afternoon in the winter of 1975, Shannon and I sat talking about all these things, looking out over the bare trees and the lake behind his house toward the distant skyline of Boston. "It really is fantastic that a very simple thing could produce the most complicated things in the world. Unbelievable. Sometimes you see little ants running around with obviously very complex behavior patterns," he said thoughtfully, echoing Simon's example of the ant as a simple mechanism producing complex behavior in response to his environment. "They're able to survive and live in this very hostile environment we have, and reproduce and eat and do everything they have to do. And they only have a few hundred nerve cells in them. It seems utterly incredible because if I had to do that with a few hundred relays, I really couldn't—and I'm pretty good at relays. But somehow it can be done. And these results with Turing machines perhaps show something about, some reason for that." He's shy and quietly spoken, with memorable blue eyes. I thought about him pedaling his unicycle through the halls of MIT when he'd been a faculty member there, and commuting on it too, to the surprise of his Cambridge neighbors, who still mention it many years later with wonder.

A little while later we went into another room where he showed me what remained of his famous "mouse," an early electronic creature that found its way through a maze and seemed to "learn" by trial and error. It had been intended for study of telephone-switching systems. The mouse would remember the solution to the maze, and when the barriers in the maze were changed, it would remember only that part that remained the same, and learn the new part by trial and error. Along with the mouse, Shannon had built a chess-playing machine too, a real-life special-purpose machine (unlike the other so-called machines, which are in fact programs, intended to adapt the behavior of general-purpose computers to the task of playing chess). It was called Caissac, after Caissa, the muse of chess, and it played a variety of endgames, computing the advantages of various moves, flashing a light in the square of the move it wished to make. But these machines, this approach, proved to be unextendable, which may be one of the reasons Shannon ended his flirtation with AI and contented himself with being only an interested spectator.

Most people, then, don't remember bringing home very solid intellectual souvenirs from Dartmouth. And that brings us to a fascinating puzzle. For two scientists had arrived on the scene with what no one else had and everyone else yearned for—a working and genuinely intelligent program. That alone should have earned them special attention from their colleagues. Perhaps more important, it was a program embodying the new paradigm, the information-processing level of modeling, which would dominate research in artificial intelligence in the next decade. Why wasn't this information-processing level of modeling, as invented by Newell and Simon, recognized at once for what it was?

The program was the Logic Theorist, which was able to prove theorems in Whitehead and Russell's *Principia Mathematica*, a feat of intelligence by anybody's standards. Its inventors were three scientists connected with the RAND Corporation in Santa Monica and Carnegie Tech in Pittsburgh, and two of them, Allen Newell and Herbert Simon (J. C. Shaw was the third, not present at the conference) brought the very first printouts of the first intelligent computer program to Dartmouth with them. They displayed their work with pride, delight, and not a small sense of one-upsmanship. How they'd succeeded in this coup, by completely ignoring the paths everyone else was on of physiology and formal logic, to arrive at the Logic Theorist, is the subject of the next chapter. The important point here is that these masked men had galloped out of the West with a virtual bandolier of silver bullets. They alone had managed to do what everyone at Dartmouth had faith was possible but had been unable to accomplish: they had made a machine that could think.

Their work was certainly received with interest. But the evidence is that nobody save Newell and Simon themselves sensed the long-range significance of what they were doing. For instance, in a paper published in December of that year, another version of the notes he'd circulated at Dartmouth, Minsky mentions the work of Newell and Simon, but names his own major influences as Shannon, Solomonoff, and Selfridge (Minsky, 1956). A later version of the same paper (Minsky, 1959) now at least cites the papers of Newell, Shaw, and Simon, but seems to regard their Logic Theorist and their General Problem Solver as no more than instances of programs that worked, in the same way that Bernstein's chess program worked, or Samuel's checker program worked, but not at all as models to be extended and generalized to other tasks or even as a general theory of intelligent behavior. Indeed, Minsky seems not to have come round to this view until 1961, when, resigned that his hardy perennial of a paper was not going to transform itself into a book, he added a large section to it that acknowledged the work of Newell and Simon as a major method of accomplishing artificial intelligence, named it all "Steps Toward Artificial Intelligence," and handed it over to two of Simon's former students who were putting together a collection of reports of working artificial-intelligence programs. Nearly all these reports were themselves dominated by the Newell-Simon model, even the ones from MIT written under Minsky's supervision. The collection would be known as *Computers and Thought* (Feigenbaum and Feldman, 1963).

Now one possible way to account for the sense of disappointment in the conference expressed by both Minsky and McCarthy may be the fact that each of them had come to a stopping point well short of accomplishing their goals, and no new good ideas were readily apparent at the conference to take them further. Neither Minsky nor anyone else had been able to extend beyond the trivial the neural model of human cognition promoted by McCulloch and his followers. McCarthy's hope of inventing a formalism to describe human thought, a calculus ratiocinator, was looking more and more impossible. There had to be at least momentary chagrin when they discovered that two other scientists, invited as an afterthought and coming from quite a different background, responding to quite different cues, had arrived at Dartmouth with the prize everyone sought so avidly, a computer program that exhibited intelligent behavior.

This *fait accompli* must have been greeted with a mixture of joy and vexation. Surely there was joy that an existence proof in the form of a running program showed it could be done; and surely there was vexation that the popular routes were dead ends, and would be hopelessly so for at least the next twenty-five years. Newell and Simon and their colleague Shaw had been first with the most.

These days Minsky certainly believes that Newell and Simon had the right idea. He remembers them as being slightly stand-offish during the Dartmouth Conference, because they were so well ahead of everyone else: they'd already implemented their list-processing language, their Logic Theorist really worked, and they were well on their way to getting their first version of the General Problem Solver to work. But Minsky's recollections of why the Newell-Simon model failed to capture imaginations at once has less to do with professional jealousy than with human perceptions of what the model was all about. He remembers Newell and Simon promoting their model all right, but not to workers in artificial intelligence so much as to psychologists. Instead of saying that any intelligent program must have, say, a goal-seeking control of the character exhibited in the Logic Theorist, and later in the General Problem Solver, Newell and Simon seemed to Minsky—and perhaps to others at Dartmouth—to be addressing psychologists, saying that the Logic Theorist and the General Problem Solver were good models of how humans behaved. Minsky wasn't sure if that was true and, furthermore, even if these were good models of human intelligence, he wasn't necessarily convinced that human and artificial intelligence needed to resemble each other. This last was a very strong theme in AI research in the early years. This impression that psychologists were talking mainly to psychologists was surely reinforced by the fact that Newell and Simon wouldn't even call what they were doing artificial intelligence, but insisted on calling it complex information processing. But Minsky speaks for himself:

By the time of Dartmouth, I was thinking about several topics, such as geometry, and when Dartmouth began, my mind exploded into activity, [I started] writing down and elaborating the structure of the proposed geometry machine that eventually became Gelernter's project. I certainly don't want to denigrate his large and productive effort. But probably the important event in my own development—and the explanation of my perhaps surprisingly casual acceptance of the Newell-Shaw-Simon work—was that I had sketched out the heuristic search procedure for the geometry machine and then been able to hand-simulate it on paper in the course of an hour or so. Under my hand the new proof of the isosceles-triangle theorem came to life, a proof that was new and elegant to the participants—later, we found that proof was well-known, and attributed to Frederick the Great, presumably erroneously [see Chapter 9, note 3]. The others checked the hand-simulation and agreed that the machine would indeed soon find this proof. Who could predict that it would be a long time before machines would find as beautiful a demonstration for a harder theorem?

Well, there are a couple of points here. First, you see, it was clear to me that the elaborate and heartbreakingly tedious construction of computer programs was not the only path toward understanding—or even correct anticipation of what programs would do. Second, I already considered the idea of heuristic search obvious and natural, so that the Logic Theorist was not impressive to me. On the other hand, I did find the basic ideas of the General Problem Solver [Newell, Shaw and Simon's next effort] quite impressive, but that was a couple of years later. In GPS [General Problem Solver] they introduced the beautiful difference-method idea, and I was impressed, but largely at my stupidity, since it was a very tiny step beyond the "character-method" matrix idea in the memo I wrote during Dartmouth.

But Minsky agrees that for most people, there's a large difference between an idea on paper and a scheme that is implemented and really works:

There's no way the outsider can tell whether I was right. I'm obviously not saying that everyone should have been able to tell, and should have instantly agreed with me! I'm just explaining why, from my very own viewpoint, it seemed natural for me to regard the beautiful demonstration of Newell, Shaw, and Simon as a fine and pleasant confirmation of ideas that I was already done with. I too was already finished with the Logic Theorist and trying to discover GPS!

In any event, Minsky remembers more floundering after the Dartmouth Conference for a clear direction for research. As a theoretician who hates doing experiments, he himself put a lot of effort into the precursor papers to "Steps Toward Artificial Intelligence," and also into kinds of mathematics that he hoped would be useful to AI, such as recursive function theory, theories of formal languages, and even discrete mathematics. What was needed was something like complexity theory, which didn't then exist, and it was some considerable time after its invention before it began to yield results that were actually of interest to people in AI. Minsky himself would go on to develop the idea of structural description in the pattern-recognition area, a critical idea, though not identified with any particular program. He would work with his student Thomas Evans, pursuing the description theory in terms of analogy, and with another student, Daniel Bobrow, on even higher levels of linguistic description. With the arrival of Seymour Papert at MIT, whose concerns were centered on developmental intelligence in children and human visual perception, Minsky's interests were led in those directions:

My interests never again came much to coincide with Newell and Simon—that is, never much at the same time. Again, I did not realize until much later the great joke; at almost every stage, I was in various ways more concerned with human psychology, they with artificial intelligence—but neither of us would have agreed at all with that description.

McCarthy, of course, doggedly pursued his own mathematical formalism, and worked in computer languages and time-sharing, which we will see more of later.

Whatever the reasons Newell and Simon failed to receive a general and immediate acclaim, the whole business raised hackles. A meeting of the Institute of Radio Engineers was to take place at MIT in early September following the Dartmouth summer, to be attended by many who were active in computing research, and a report on the activities of the Dartmouth Conference was scheduled there. It was to be given by John McCarthy, but Newell and Simon objected strenuously to that. McCarthy recalls, "They felt, perhaps quite correctly, that the situation was anomalous, the conference being reported on by people who hadn't actually done anything, when they had."

Says Minsky,

The unfairness was that they had a well-developed project that they'd been working on a long time, pretty much full time, and we'd been working much more casually and much more as generalists for a shorter time, and wanted to share the stage with more or less equal authority, which wasn't very nice. We were reporting speculatively about what we thought we wanted to do, while they were talking as scientists who had worked quietly. for a long time and prepared their results and done a lot of experiments to validate them. Newell must have thought it was unfair, and in retrospect it was, for like Darwin and Wallace, Darwin had done all this work and Wallace had gotten this bright idea, but they both got equal attention at the time.

Simon says now,

Well, we allowed as how that wasn't going to happen, and so poor Walter Rosenblith, who was supposed to chair the session, walked around with us on the MIT campus, we strolled down Mem Drive and so on, negotiating this. We were not feeling at all good about John, and Rosenblith was trying to be in a neutral corner when we didn't think there was any neutral corner.

In the end, Newell and Simon were satisfied with what happened. McCarthy gave a general talk and then Newell and Simon presented their work in its particulars, which appeared as a paper in the transactions of the conference, the first widely published announcement of the Logic Theorist (Newell and Simon, 1956).

So, if the most important piece of science—perhaps the only one—to emerge at the Dartmouth Conference was in fact done before that summer by Newell, Shaw, and Simon, maybe the value of the conference was elsewhere. There are different ways of measuring the significance of a scientific gathering. Science, being a human endeavor, is a social as well as an epistemological enterprise. From time to time human beings seem to need to take a census of the clan; in any event, we certainly feel the need to band together, even if no more than temporarily, with others of like mind. Thus we organize clubs and professional associations and political parties; we publish magazines for fellow joggers, stamp collectors, and purchasing agents. We take strength from the fact that other people share our beliefs, even our passions, though they might not share each detail. I think something of the sort took place at the Dartmouth Conference. Artificial intelligence, if it wasn't quite a legitimate scientific field, had at least emerged as an entity about which one could ask questions: Was it a science? High jinks? Both?

Arthur Samuel too remembers it all as historic, though for a different reason. He sees the Dartmouth Conference as having defined the establishment in artificial intelligence. "It was very interesting, very stimulating, very exciting." Then he adds, "I'm sort of a loner myself, unfortunately, and I've always objected to this in-group running things while you're on the outside. And that was fostered by that meeting, I think. Not deliberately, but meetings of that sort tend to do that, and that's my one objection to what's been done in the field of artificial intelligence. It's always been run as a sort of closed group."

Samuel perceives himself at the fringes of that group, though most people would place him comfortably toward the center. Nor could he think offhand of significant work that might have been overlooked because it was not done by one of the original group members or their students. Samuel acknowledged the need for a small group of persons involved in their work who exchange information informally, the aptly named invisible college that dominates most intellectual disciplines. To depend on publication is simply too slow and frustrating. But when such a group becomes clannish, it does tend to ignore contributions from outsiders, and the clan often forgets—simply forgets—to tell interested outsiders what's going on.

So perhaps the most influential result of the Dartmouth Conference itself was the social patterns it set. Though Arthur Samuel was a participant in 1956, he's one of many who have raised the question of professional nepotism. Accusations of clannishness have persisted since 1956, and they aren't without foundation.

At the 1977 International Joint Artificial Intelligence Conference, twenty-one years later, the invited papers were by Edward Feigenbaum and Harry Pople, both former students of Simon, by Simon himself, by McCarthy, Minsky, and Douglas B. Lenat, a former student of Feigenbaum and then on the faculty of Carnegie Mellon University, Only three other invited presentations came from the outside. Participants in the panels also showed a preponderance from the big four—Carnegie Mellon, MIT, Stanford, and Stanford Research Institute—with the representation from other laboratories being sparser than might be expected in a field that had grown from the ten Dartmouth pioneers in 1956 to nearly a thousand registrants in 1977. Contributed papers came from a much wider range of geographical centers, however, and the total proceedings, after some diligent refereeing, turned out to be eleven hundred pages in two volumes. (In addition to the spiritual fathers and sons, a genuine father and son appeared: Oliver Selfridge spoke on the history of AI, and his son Mallory Selfridge coauthored a paper on modeling the linguistic processes of a child, which gave him the unusual pleasure of citing his father's earlier work.)

Now the argument can be made that AI is Big Science, in the sense meant by Derek J. de Solla Price (1963), and Big Experimental Science at that, requiring large, expensive computer installations to carry out research. The major source of financial support in this country through the 1960s and early 1970s was ARPA, the Advanced Research Projects Agency of the Department of Defense. Unlike many government funding agencies, ARPA does not use the peer-review system,[3] but disburses funds based on its own judgment of the best people doing the best projects related to its mission. That judgment has been to concentrate resources, and has thereby enriched the four main centers, which are again, Carnegie Mellon University, where Newell and Simon work; MIT, where Minsky works; Stanford, where McCarthy works; and Stanford Research Institute, which is heavily populated with former students of these Dartmouth Conference participants.

A man who considers himself outside the field, but a fascinated and friendly critic of AI, is Professor Lotfi Zadeh of the University of California at Berkeley. He suggests that AI is possibly more cliquish than other fields because most of the funding has come from one source instead of small amounts from many sources. This pattern causes a certain insecurity, he believes, which shows itself in resistance to ideas that emanate from places outside. But there are other problems which might cause this insecu-

---

[3] Where independent specialists in a given field are asked to review and recommend whether proposals should be funded by a given agency, such as the National Science Foundation or the National Institutes of Health. Even this system has recently been severely criticized for being an old-boy network.

rity, and hence this cliquishness, he goes on to observe. Some outsiders make very clear their feelings about the impracticality of AI, and what they see as its excessive salesmanship.

I don't share that view [Zadeh says]. I think AI is extremely important, and I'll always feel that way. But there are many people working in much more mature areas of computer science who look at it all suspiciously. Then, too, people in AI are very young, on the average, young and quite brilliant. [He stops and sighs.] I think there are more brilliant people working in the AI field than in any other part of computer science. Brilliance is an abundant commodity there. So being brilliant, they are, let's say, not respectful of the older people in other parts of computer science. And that creates certain antagonisms.

But nobody really knows what might have altered things. If the claims hadn't seemed extravagant, they wouldn't have been very interesting, and wouldn't have attracted all those brilliant but arrogant youngsters Zadeh points to. (AI insiders have a different view of their recruitment: they can't imagine why thousands more haven't flocked in.) Nobody really knows whether funding from several sources, relying more on the peer review system, would have changed matters much either. Perhaps it's in the nature of new scientific fields to be dominated by their founders, and the institutions those founders call home, no matter where the money comes from, until time alone changes things. In AI, time is indeed changing things: Texas, Yale, Rochester, the Bolt Beranek and Newman Corporation in Cambridge, Massachusetts, and Xerox Palo Alto Research Center all have thriving AI groups; more are on their way.

Professor W. W. Bledsoe of the University of Texas says, "I really think the so-called establishment in AI is a beautifully fair one. Now, I'm sort of outside it, though in some ways I'm not, because I do get funds. But I think it's been pretty darn fair." He goes on to say that sometimes he's been hurt by reactions to his own research—he works in theorem proving, which often gets caustic comment from, say, Minsky—but it all reminds him of the great nineteenth-century mathematician Carl Friedrich Gauss, who was once presented with some results by an unknown mathematician, and who dismissed them as "another monstrosity." Gauss later had to eat his words.

Bledsoe continues, in a Texas drawl that manages to be simultaneously emphatic and mellow:

But you see, Gauss has to behave that way. Is he going to read every paper that comes along? If somebody he trusts can convince him, well, that's different, he should pay attention then. And the AI establishment is the same. People come out of nowhere and they're recognized if they're good. But they've got to pay the price to get into the club. I don't want everybody knocking on my door saying, "I've got good results." Please leave me alone. In fact it gripes me that anybody in the world walks into Minsky's office about when they please and gets his attention. And they do, incidentally. He just loves to talk to people, and just wastes his valuable time. A lot of it. So there ought to be some price to get in.

An insider's view of the in-group comes from John McCarthy:

I think we don't talk to each other as much as we should. We tend to have these separate empires which exchange ideas by means of ambassadors, in the form of graduate students. And when we do get together we tend to discuss politics, though we don't even spend enough time doing that to have the maximum defensive effect against the bad things that might occur. If we were really going to understand what everyone else is doing, we'd have to spend a lot of time together—none of us has an excessive talent in understanding other people's points of view. There's a tendency after starting a discussion to say, ah yes, this suggests something that I want to work on, and the real desire is to get off alone and work on it.

In any case, each of the four main centers was to evolve its own distinct style, set by the personalities of the men who dominated it. A good way to taste the flavor of the resultant style of the 1960s and 1970s is through a whimsical comparison of AI and the garment industry.

Consider MIT haute couture, the *Women's Wear Daily* of the field. No sooner do hemlines go down with enormous fanfare than they go up again, the provinces growing dizzy with trying to keep pace and usually falling behind. MIT thinks itself stylish, but outsiders have been known to call it faddish. Carnegie Mellon, on the contrary, represents old-world craftsmanship, attending to detail and using the finest materials. These qualities presumably speak for themselves in gowns you can wear to a dinner party ten years from now and never fear the seams might part. But classic can be stodgy: if Queen Elizabeth of England bought artificial intelligence, she'd surely buy at Carnegie Mellon. Stanford has two ateliers. The first is the Levis' jeans of AI: sturdy, durable, democratic; worn by socialites and welfare clients alike; and mentioned proudly by everyone in the trade whenever questions of practicality or utility come up. The other is *Nudist World*, incorporating After Six; this shop is visionary about the formal wear of the future, but meanwhile remains naked. Finally, Stanford Research Institute is Seventh Avenue. Maybe those models are knockoffs, but hardly anyone can afford haute couture, and except for the jeans people, who else is going to bring AI into the real world?

Samuel was right. The Dartmouth Conference had set some interesting patterns.

But what were these men doing there at all that summer? What on earth would motivate these ten to devote their professional lives to the organization of large laboratories that were dedicated to building machines either to mimic the human brain or to behave intelligently, by hook or by crook? Several motivations have been offered by armchair psychologists.

There is, for example, the desire to be as gods. Bledsoe, for one, scoffs at this. "Look, did we become gods when we made machines to dig ditches for humans who used to do it with shovels? We'll need more success than we've had to feel godlike. I was working in pattern recognition when my wife's child was born. He's now nineteen. In nothing flat he was recognizing patterns, just went whizzing right by us. Our machines are still creeping— no threat to *him*." Well, maybe.

Then there's the urge to have offspring without the help or interference of a woman. This suggestion publicly surfaced most recently in 1973 in a report written by Sir James Lighthill at the request of the British government's Science Research Council intended to evaluate the state of research in British artificial intelligence. Sir James found the womb-envy idea dubious, but couldn't resist bringing it up. Since his report is largely negative, it seems a bit malicious of him to mention it at. all.

The Freudians have had their turn, seeing in the creation of such beings a yearning to desexualize or cleanse procreation, counter-pointed by the Oedipal drama. Others have detected an urge to divide the self, to make a doppelganger that would carry away the evil in one's soul, leaving of course the residue of good (Plank, 1965).

These all are blunt paraphrases of more sophisticated analyses that don't do justice to their original subtlety. And I can't really dispute any of them, except to say that a single-minded view seems impoverished; all these reasons may be operating, but certainly there are others. As I'll argue later on when I come to talk about the critics of AI, we each undertake projects for a multiplicity of reasons.

But perhaps the main reason is also the most obvious one. To know intelligence well enough to be able to build a working model of it is surely one of the most intellectually exciting and spiritually challenging problems of the human race. To do so is to know ourselves as we've always yearned to, to make us a part of nature instead of apart from it, in Herbert Simon's felicitous phrase. Such knowledge implies a solution of the mind-body problem, which has eluded the most intense human efforts for over two thousand years. And such a model promises to be an extension of those human capacities we value most, our identifying properties, which we sum up as our intelligence or our reason; the thinking machine would amplify these qualities as other machines have amplified the other capacities of our body.

The effort toward artificial intelligence might even bring us face to face with intelligences that surpass those of their creators, and that can move in to solve some of the persistent, even lethal, problems humans have created for themselves but aren't quite smart enough to solve. "In the long run," Bledsoe says softly, "AI is the *only* science."

These considerations constitute the optimistic view, and it pervades AI. A less optimistic view says that while indeed we might be able to create problems we aren't smart enough to solve, an artificial intelligence might just fall into that very category.

Chapter Six

—◆—

The Information-Processing Model

A persistent rumor at Carnegie Mellon University, founded as Carnegie Institute of Technology, is that old Andrew Carnegie was skeptical of any institution of higher learning, and so the buildings were designed to be converted effortlessly from campus to factory if the college failed. In fact the rumor is untrue, but the cheerless yellow brick and the long, dark sloping halls of those early buildings continue to keep the rumor alive; it's passed on as gospel from one student generation to the next.

But not all the buildings are like that, and it happens that most of the early work in artificial intelligence at Carnegie Tech took place in the Graduate School of Industrial Administration, a building in early 1950s International Style, which could easily pass for the city hall in some frugal Great Lakes municipality.

Herbert A. Simon, the Nobel laureate who is now Richard King Mellon Professor of Computer Science and Psychology at Carnegie, and who helped to found GSIA and served as one of its early administrators, describes the school:

GSIA got off the ground in 1949. We felt like we were going to have the first business school that had academic respectability, scientific respectability, and we didn't think it needed to run like dead-headed, old-fashioned business schools, or we wouldn't have been there. None of the people who came in were from a business-school background. We came in with the understanding that we were going to build a

137

different kind of business school, that we were going to experiment and see where these new ideas—operations research, or management science, as we preferred to call it, and organization theory—where they led. So the simulation stuff came into GSIA by accident, so to speak. It came as a result of Al Newell and my collaboration, which had started out at RAND, and not because anybody had planned that GSIA was going to get into individual cognitive psychological research. I just decided that that was bigger and more important than anything else I was involved in, and after 1955 I began to wind up my other commitments and devote more of my time to that. When we got some graduate students who got interested in it, we said, well, there's no reason a business school can't give them degrees.

Edward A. Feigenbaum, now a professor of computer science at Stanford, recalls his work at GSIA.

I was an undergraduate senior, but I was taking a graduate course over in GSIA from Herb Simon called Mathematical Models in the Social Sciences. It was just after Christmas vacation—January 1956—when Herb Simon came into the classroom and said, "Over Christmas Allen Newell and I invented a thinking machine." And we all looked blank. We sort of knew what he meant by thinking, but we didn't know. We kind of had an idea of what machines were like. But the words thinking and machine didn't quite fit together, didn't quite make sense. And so we said, "Well, what do you mean by a thinking machine? And in particular, what do you mean by a machine?" In response to that, he put down on the table a bunch of IBM 701 manuals and said, "Here, take this home and read it and you'll find out what I mean by a machine." Carnegie Tech didn't have a 701, but RAND did, though actually it was the Johnniac[1] that Newell, Shaw, and Simon were working with. But Herb chose the 701 as an introduction for us. So we went home and read the manual—I sort of read it straight through, like a good novel. And that was my introduction to computers.

---

[1] Named, over his mild protests, for "Johnny" von Neumann. An earlier machine at Princeton had also been named for him, but that one somehow didn't stick. RAND'S Johnniac did, as did the Los Alamos Scientific Laboratory's MANIAC.

The Newell-Shaw-Simon team, a familiar one to any student of computer science, had formed in the early 1950s at the RAND Corporation in Santa Monica. In retrospect, their collaboration seems almost inevitable, given their intellectual predispositions, research interests, and the spirit of the times.

Paul Armer, who was head of the computer science department at RAND during much of the time that Newell, Simon, and J. C. Shaw were associated there, describes the atmosphere of RAND in the early 1950s:

I think a good deal of RAND's success in the early days was due to the research philosophy of the Air Force, which said to RAND management, "Here's a bag of money, go off and spend it in the best interests of the Air Force." And then RAND management divvied that one large bag into a number of smaller bags and said to each department head, "Here's a bag of money, go off and spend it in the best interests of the Air Force." In this environment, you could hide failures, and consequently you were much more willing to bet on long shots, which had maybe one chance in ten of paying off. Actually, the number that turned out well was much higher than that. Anyway, if the probability of success is one in ten, but when one succeeds it pays off at odds of a hundred to one, then you're really going to do well in that game because it's so stacked in your favor. I think that's exactly what was happening with RAND. When a fresh Ph.D. was hired, John Williams, then my boss at RAND, would say, "Okay, I Just bought your time from you for a couple of years and here it is. Go off and spend it in the best interests of the Air Force." That strategy only works if you have very good people, and in the early days of RAND, since it was just about the only game of its kind in town, it attracted some very good people. And once it attracted some, it became very easy to attract more.

One project RAND decided upon was to study man-machine interactions by simulating an air-defense direction center, modeled after the center at McCord Field in Tacoma, Washington. In this model, the machines were radar sets and fighter planes, and the men were plotters who had to trace on the surface of a large lucite screen the location and direction of aircraft spotted by the radar. If the craft were unknown, a decision would have to be made as to whether a fighter should be scrambled to go out and look at it. To scramble a fighter for every unidentified object was costly; to miss an enemy plane would be costlier still. It was a tense, hectic job which no one was performing very efficiently. The RAND experiment was to show that greater attention to organizational factors would improve the performance of human workers substantially. The experiment was ultimately so successful that the Air Force asked RAND to train all its plotters and spotters, which required that a separate educational program be set up. By that time Allen Newell had tired of the project and wanted to do something else.

Newell had come to RAND as a full-time staff member in 1950, after a year of doing graduate work in mathematics at Princeton, an experience that convinced him he wasn't temperamentally suited to be a pure mathematician.

I was a problem solver [Newell says], and I wanted problems you could go out and solve. I simply couldn't understand what motivated pure mathematicians to go on working, looking at the structure of some mathematical logic. So Princeton and I just passed in the night. They sort of acted like they'd be glad to have me back when I left at the end of the year, but what I'd found out was that I couldn't have led that life for anything. None of their concerns are my concerns, and I learned that, and it took me a year to learn it, and then I got out of mathematics.

By coincidence, he had entered Princeton in math at the same time John McCarthy did, but though he knew who McCarthy was, they seem never to have exchanged words.

Newell had written to RAND because he heard work was being done there in game theory. He was interviewed around the Christmas holidays and offered a contract stipulating that he would return to school at the end of a year. The arrangement even included transportation from Santa Monica back to Princeton.

I was kind of interested in game theory, but really, it simply seemed to me like a good place to move. I didn't know a lot about RAND except that it was just sort of bubbling around. Meanwhile, at Princeton,

Allen Newell and his favorite cryptarithmetic puzzle (Carnegie-Mellon University)

I'd got a job as a research associate for Oskar Morgenstern just to have something to do. He was starting up a project in logistic models, little stochastic models of whatever was supposed to be interesting about freight cars running around railroad yards—and I'm not sure there's anything interesting about that. So when I went out to RAND I got involved not in game theory but in this logistics stuff.

Newell too was struck with the freedom of RAND, where ideas could come from the bottom as easily as the top. It seemed to him a sort of exploration of a new way of scientific life, of doing research in a setting that had the advantages of a university, namely freedom of choice in research and smart colleagues to do it with, and none of the disadvantages of straitened budgets or burdensome teaching. He felt for a long time, he says, that RAND represented a real shift, a place where the scientific action was going to be. Places like RAND supplied higher risk and higher resources, and the salary differential was very large. Professors in the universities had lost ground over the war, and when RAND offered 50 percent raises, it got nearly whomever it wanted. For the first three or four years that Newell was associated with RAND, the talent was flocking there.

RAND held open court during the summer. They'd have fifty consultants in—it's just absurd against the background of modern-day circumstances. But let's say they wanted to build models of bombers going through stages of attrition to bomb targets. They'd get a bunch of consultants to come in and work on it, the best algebraists in the country, to come in and spend their summer thinking about bombers going through antiaircraft barriers, with attritions, and the various probabilities, and ask them whether they could find a closed form solution to these things. Fifty or sixty of these guys wandering through during the summer. It made RAND look like the wave of the future. You put everything on the line in terms of research that you did, but you got supported with a lot of resources, and you worked full time at it. That was where the scientific action was going to be. But it wasn't. It turned out, in fact, that universities have resources for survival well beyond these little institutions. But I really felt the other way.

Newell's work in logistics sent him to take a look at something in the Pentagon called the Munitions Board, actually then responsible for logistics in the Department of Defense. He spent about a month sitting in the colonels' offices and talking to them:

And I wrote a little document called "The Science of Supply," an attempt to understand the supply system. It led me to the strong view that an abstract mathematical model was really not going to do the job. That is, organizational factors made large amounts of difference. That sort of got me interested in organizations and I went on from that.

Newell is a large man, perhaps six foot two or three, with a pear-shaped face that is continuously radiant with the delight he so obviously takes in his work. His clothes tend to billow or cling, depending partly on the stage of a diet he is at and partly on whether he bought them in the 1950s when baggy pants were fashionable, or in the 1960s when they weren't. As he talks, he seems to regard his life with the same delighted awe with which he regards his science—it's a series of events, of phenomena, that when examined might yield a pattern, but where inconsistencies and anomalies just add to the fun. He'll make a statement and then interrupt himself: "What does that mean? Well, let's see what that means." And then he'll begin a chain of speculation that leaves a listener amused, and slightly awed by the degree of his rationality.

This is not to say he imposes patterns where none exists. Once I was puzzling about what seemed to me to be a contradiction between the high value he places on simplicity of explanation (a value shared by Simon) and the fascination he feels for what he calls complex systems. "Are we talking about a matter of degree?" I asked. "No," he said breezily, "that contradiction exists." Philosophy doesn't like contradictions, and I must have moaned a little, because he began to laugh. "I can live with that. Why can't you?" He believes that if science and philosophy are at variance, then philosophy goes.

One doesn't have to have a consistent philosophic underpinning for what one's doing. That's sort of the last thing one has to have. I guess I've said that before. Philosophy is in the service of science, and not vice versa. Consequently, philosophy is something to be discarded daily and picked up afresh; you use what pieces of it you want and forget about it. And there's no necessity for the sort of statement that says, well, if you do thus and so you must believe in a rational universe. I don't have to believe in a rational universe. [He chuckles.] I mean, if it's not, I'll find out soon enough. It's someone else's compulsion to worry about that. All I have to know is what to do next, and one has to have plans insofar as they seem material to the game at hand. So I don't worry. I mean, I worry a little bit about ill-structured problems, but that's because it's interesting to try to come to grips with particular little members of a class like that right at the boundary.

He is playful as he speaks, grinning, shrugging, throwing his arms into the air, scratching what remains of his sandy-colored hair. But he is perfectly serious about the message.

Science [he'll argue], is an optional game. You pick up what you wish to pick up and you leave go everything else, thereby placing a bet. But except for that, there's no moral imperative necessary to pick anything up. So I don't feel bad not thinking about ESP, let's say. I'm completely unmoved by people who come on saying, "But this is important—you have to think about it." They try and produce a moral imperative. I see science again as a thing which is full of bets but not full of moral imperatives. All that happens is that I'll be consigned to the dustbin if I go pick up the wrong thing to worry about, and that's a risk I'm prepared to take. And do, in fact, in much more important ways than whether I spend time thinking about ESP or understanding the critics of artificial intelligence. There are all kinds of other things which I don't pick up on, which I probably ought to much more.

Perhaps there are reasons for that. He saw me once just after I'd written a letter to the editor of a Pittsburgh newspaper in which I quarreled with a columnist who insisted that computers were nothing but big dumb beasts.

"Aw, what raises you above threshold on things like that?"

I said I believed that computer scientists hadn't done a very good job of clarifying the issues they were best equipped to clarify, and that if they would do so perhaps they would calm public apprehensions about computers.

"What makes you think that? What makes you think that if people really understand what we're doing, they won't be scared stiff?" He grinned his matchless grin.

Newell grew up in San Francisco, the son of a professor of radiology at the Stanford Medical School, which was then in the city and not in Palo Alto with the rest of the university. Newell makes a strong case that his father was enormously influential on him.

He was in many respects a complete man. We used to go up and spend our summers in the High Sierra. He'd built a log cabin up in the mountains in the 1920s. And my father knew all about how to do things out in the woods—he could fish, pan for gold, the whole bit. At the same time, he was the complete intellectual. At Stanford Medical School he had the reputation as one of the great teachers. I've had characters come up to me all over the country and say, "Oh, you're

Bob Newell's son," and go off and tell me what a fabulous teacher he was. He has a string of publications—sort of middling, not great science. But within the local environment where I was raised, he was a great man. There was a standard saying in our family about Newell men, and how they were somehow so much greater than the women. And all the gals used to fawn over my father in this sort of intellectual way. My father knew literature, all the classics, and he also knew a lot of physics. He was extremely idealistic. There are two or three papers he wrote to other members of the medical profession, lecturing them on ethics—for a flavor, there's one called "Mink Coats and Cadillacs" which essentially states the fundamental dysfunction of ostentation. He used to write poetry. Crummy poetry, limericks, but still . . . ."

And Newell says that he has adopted almost all his father's basic attitudes toward life; it is the exceptions that bear examining. Newell is ambitious and wants to go into the archives with the scientific greats, and to that end, he is willing to forego some things his father considered extremely important.

My father didn't want to be great in the scientific sense. For instance, he viewed having friends as infinitely more important and so he cultivated his friends—rationally, I might remark—which is really something. He had millions of friends and he would spend all kinds of time with them. And that precluded what I would think of as extreme scientific achievements. I'm driven by a kind of ambition he wasn't driven by, and I've also retreated strongly on the issue of how universal I am.

Once when Newell and I were talking about scientific partnerships and I asked him to describe Simon's role, he did so, then went on to talk at length about his father again. Finally, I asked him if I should make the corny inference. No, he said. But the similarities are not without interest.

Herbert A. Simon has a distinguished reputation in the fields of political science, business administration, psychology, and computer science. He has made significant contributions to philosophy and economics, and was awarded the Nobel Prize in Economics in 1978. At various times he has been a member of the

Herbert A. Simon
(Carnegie Mellon University)

President's Science Advisory Committee, the Governor's Milk Control Inquiry Board of Pennsylvania, and the Board of Trustees of Carnegie Mellon University. He is fluent in a handful of languages, gets along in several more, and can read still others. Unlike Allen Newell, who hates to read fiction, Simon loves it, and is the only person of my acquaintance to have read all of *À la Recherche du Temps Perdu*—in the original—twice (though perhaps that selection isn't so odd for a man who's spent so much time thinking about thinking). I don't know whether he writes poetry, but he plays the piano, and Marvin Minsky of MIT, himself an accomplished composer and pianist, jests about challenging Simon to a sonata contest, the rules, I suppose, to be improvised as they go along. Simon is said to have a formidable temper.

Simon came to RAND as a consultant in the summer of 1952, two years after Allen Newell's arrival. He was already well established as a political scientist and an economist, and a few years earlier had moved into a new field, examining human behavior in organizations. In 1947 he had published *Administrative Behavior*, a seminal book that went into a second edition in 1957, and a third in 1976, and has been translated into German, Italian, Spanish, Portuguese, Japanese, Dutch, Korean, and Swedish.

Administrative Behavior proposed a theory of human behavior in organizations that fell between what Simon considered the extremes, on the one hand, of the Freudians, who attributed all cognition to affect, and on the other, of economists, who attributed to man "a preposterously omniscient rationality." This latter especially was a theory, Simon wrote tartly, that had "reached a state of Thomistic refinement that possesses considerable normative interest, but little discernible relation to the actual or possible behavior of flesh and blood human beings" (Simon, 1947).

His own view was an expansion of ideas he had been playing about with since his undergraduate days at the University of Chicago, when he had studied the organization of recreational services in his home town of Milwaukee:

It happened that in Milwaukee the recreation department was poised neatly between the school board and the city government. And I went up there and interviewed a lot of people, and I found that there were certain points where they tended to have rather serious disagreements, particularly with respect to what part of the budget should be spent on maintaining playgrounds as against what part should be spent hiring recreation leaders. One could guess then very readily who was going to be on which side of that by what position he occupied in the organizational structure. That phenomenon fascinated me, because here were very reasonable people arguing about these things, and each making out a convincing case, and the case they were making seemed to depend largely on their organizational position. I wanted to understand more about that.

Simon developed the theory that a system—a firm, or a municipal board, or a government agency—which had to make decisions or choices about courses of action would probably do so by means of a process that was, in some broad sense, a reasoning process. The process would be one of drawing conclusions from premises, and it was therefore the premise, rather than the decision, which served as the smallest unit of scientific analysis. People in different places in an organization would start from a different set of premises—the salesman from the premises that it was most desirable to sell and meet market competition, the production engineer from the premise that orderly, efficient production was most desirable, and so forth—and therefore one could predict the kinds of decisions people would make by examining the likely shape of their particular perspective. This view is simpler than that explaining human behavior in organizations directly in terms of goal conflict, or in terms of various kinds of emotional mechanisms, yet it accounts for a large part of organizational behavior. Moreover, it is free of the value judgments that are so often implicit in explanations that focus on emotional aspects of behavior. Most important in a scientific model, it is a good predictor of real life.

Administrative Behavior was specifically cited by the Swedish Academy of Sciences, along with subsequent books Simon would write, as containing ideas that drastically changed the way modern business economics and administrative research were done. But Simon himself considers his work applicable not only to business and economics, but to decision making in general, which, he explained to a reporter, "cuts catawampus across the disciplines."

Simon recalls how he came to RAND:

My work in organizational studies was pretty well known, and one day Bob Chapman and I guess Bill Biel, and probably John Kennedy, all scientists at RAND, came around and we had a long discussion, and I agreed to be consultant to this new lab which had been set up, a social psychology lab really, intended to study air-defense systems. So in the spring of 1952 I went out to Santa Monica. I was familiar with computers—I'd wired some boards in my time, and I'd given lectures to businessmen on the implications of computers for business. But that air-defense lab was really an eye opener. They had this marvelous device there for simulating maps on old tabulating machines. Here you were, using this thing not to print out statistics, but to print out a picture, which the map was. Suddenly it was obvious that you didn't have to be limited to computing numbers—you could compute the position you wanted, a spot to appear on a piece of paper. You could print pictures, with things that weren't even a modern computer, just old card calculators.

It was this device that suggested to Simon the capacities of the computer to manipulate nonnumerical symbols as well as to calculate—a profound insight.

"Al Newell and I in various ways were trying to understand the behavior of the human plotters and tellers in this air-defense set-up, and he had already developed a language—the phrase 'information processing' was already part of his vocabulary by '52 when I arrived out there—and I began to map this onto my decision-making, decision-premises ideas and the like." This was the language that Simon had developed for describing human behavior in organizations, and which he had published a few years earlier in *Administrative Behavior*. "It seemed to me that Newell's views and mine had much in common. So we more and more found this *gemütlich*, and began to work together, trying to use information-processing ideas to understand the ways in which these air-defense personnel were operating."

When Newell moved from the study of logistics to the study of organizations, he had not jumped directly into studying and simulating something so complex as an air-defense direction center. With $200 from John Williams for supplies, he had begun more modestly by hammering together a plywood table with barriers on it so that those seated at it could not see each other, but could only communicate by a primitive set of toggles and lights. This bit of carpentry had been done at the workbench in the computer laboratory, where Newell could look over and see RAND technicians putting together their fancy new computer, the Johnniac.

The idea of the table was to assign some sort of task, and watch how a small group of people interacted in the course of trying to accomplish this task, given the fact that they could communicate only in limited ways. It turned out to be a frustrating experience for the experimenter. As his subjects, he had chosen some of the consultants attached to the laboratory—they were there, and it was part of the RAND ethos that along with doing their own work, they would participate in whatever experiments they were invited to by the RAND staff. But distinguished logicians and mathematicians such as Stephen C. Kleene, Lloyd Shapley, and Melvin Hausner, and a young mathematician on the RAND staff named Ruth Wagner, did not behave like ordinary organization types. Instead, they would think silently through the assigned task, discover the best solution, and do it.

And I kept trying to enrich the situation [Newel says], so that what would happen would be organizational behavior rather than this highly intellectual behavior. The tasks became more and more complicated so that these guys couldn't simply figure out the problem. But it was hopeless. They were too smart. Actually, it had a very fortuitous effect, because the frustration led me to insist that simple tasks were not the right environment for studying organizational behavior— you had to make the task environment much richer, much more realistic, and you'd get genuine psychological behavior only out of environments that were too rich to allow the thinking human to think his way through and understand all the possibilities. And so we went from that little bitty one to the forty-man organization with a total simulated input, the air-defense direction center.

In Newell, eleven years his junior, and a graduate school dropout, Herbert Simon found a young partner who not only shared his ideas about the possibilities of studying human behavior in organizations in a scientific way, but also had a useful set of metaphors drawn from the computer for describing this behavior. This tension between the simple and the complex, which Newell had discovered at his plywood table and Simon by scrutinizing the way people really behaved in organizations, was to inform their scientific work for the next twenty-five years. Were thinking processes simple or complex? If you could get a computer to behave in certain situations the same way that people did, forgetting the same sorts of details and making the same kinds of intuitive leaps, what did that say about people? Or about computers?

These questions led Simon to a much more direct comparison between the mind and the computer than he had felt justified in making up to then. Some years earlier, in a talk at an Econometrics Society meeting, John von Neumann sounded a caution he made often, that the brain-computer analogy must not be pushed too far. Simon had argued then that, nevertheless, computers were interesting systems with a hierarchical structure, which sounded like something that might be important in understanding the structure of human thinking. Now, using the language to which Newell had introduced him, and pushing his own ideas further along, the brain-computer analogy seemed very fruitful to him, and he began looking at it hard.

It is important to remember that for Simon the mind as logic machine had preceded the computer as artifact:

When I first began to sense that one could look at a computer as a device for processing information, not just numbers, then the metaphor I'd been using, of a mind as something that took some premises and ground them up and processed them into conclusions, began to transform itself into a notion that a mind was something which took some program inputs and data and had some processes which operated on the data and produced output. There's quite a direct bridge, in some respects a very simple bridge, between this earlier view of the mind as a logic machine, and the later view of it as a computer.

Finally, provoked by another von Neumann talk given that very summer at RAND on the immense difficulties of building a chess-playing machine, Simon wrote an appendix to a paper he had just prepared on a behavioral model of rational choice, showing how principles embedded in that model could be used to make a chess machine. Though the paper is often cited and has been reprinted in anthologies, the appendix was never published. It is important, though, because it represents Simon's first efforts to apply his information-processing ideas on the similarities between the brain and the computer to a real-life intellectual problem.

We've already seen that the ancient yearning to invent a double of the human brain had revived—lustier than ever—with the invention of the digital computer. Lurid predictions about giant electronic brains filled pulp science fiction and the Sunday supplements, guaranteeing that the reaction would be just as extreme. Norbert Wiener and John von Neumann felt obliged to deliver lectures warning against too facile an analogy between the digital computer and the human brain, and they were joined by many other distinguished scientists and engineers who assured the public and each other that computers were only high-speed morons, incapable of intuition, originality, or any other variety of intelligence. That we possessed no universally agreed upon definitions of intuition, originality, or for that matter, intelligence, seemed not to matter. Like beauty, these concepts were in the eye of the beholder, and one person's intelligence was another's diligence or luck.

To be sure, some of the earliest computer pioneers, among them Alan Turing, W. Ross Ashby, Christopher Strachey in the United Kingdom, and Claude Shannon in the United States, had already proposed ways that computers might play chess, and Turing's engaging little essay "Computing Machinery and Intelligence," first published in 1950, actually addressed the question of whether a machine can be said to think.

But in 1952 sober scientists ran a severe risk to their credibility when they moved beyond the realm of speculation and into actual work intended to simulate any sort of human thinking. A fair number of scientists today—not to mention members of the general public—continue to hold the view that human mental processes cannot be simulated by any machine, or at least not by the means presently proposed by workers in artificial intelligence. Existence proofs—programs performing tasks that, done by humans, would be considered intelligent behavior—merely serve to relocate the boundaries of intelligence. I will explore the reasons for this opposition further on. Now the important point is that such feelings were at least as strong in the early 1950s. Those who thought otherwise were few, isolated from the moral support they might have offered each other and without any concrete results to give them courage to persevere. That said, it seems fair to add that none of these circumstances seems to have made the least bit of difference to most of them.

To move from speculation in science to the more solid tasks of theorizing, modeling, and verification is always a major step. In a simple sense this move requires a new vision of the phenomena being explored, a rearrangement of the data in some fresh way so that new patterns and structures are revealed. That fresh view had come to Simon when he saw Newell's calculating machines at RAND put to use depicting a radar screen instead of merely calculating numbers. It is by no means an obvious view, even now. Many people, including a large number of computer users, still do not see the computer as anything more than a very high-speed number calculator. Newell had been very unusual in detecting the nonnumerical capabilities of the computer so early in his—and the computer's —career. And like the issue of simplicity versus complexity, the notion of the computer as a processor of symbols would inform Newell's scientific work, growing richer and deeper as time went on.

I've never used a computer to do any numerical processing in my life [Newell says now]. The first task I ever did on a computer had to do with simulating an environment—using card-programmed calculators, so this was long before digital computers. We turned their output into the picture of the radar display, with X's printed where the blips were on the scope, mapping planes coming in and out. It was a crude technology, and if we could have dealt with fewer planes we could have used some electronic analog devices. But we had hundreds of planes to deal with [the issue of complexity again]. We used to joke about the fact, back in 1952, that there wasn't a single multiplication in this whole program.

So this device had fascinated Simon, and opened his eyes to the possibilities of a computer as an information processor, although three more years were to pass before Simon and Newell began to perceive what they now call the *symbolic-functioning capabilities* of computers, symbols here signifying objects with access to meanings —designations, denotations, and all the information one might have about a concept such as a pen, or courage, or quality. This view would come to be central to their later work, and in their opinion, as central to understanding mind in the twentieth century as Darwin's principle of natural selection had been to understanding biology in the nineteenth century.

Simon grew up in Milwaukee. His father was an engineer, trained in Germany, who had come to this country as a young man.

He had the attitude of a scientist [Simon recalls]. He really didn't like cookbook engineers. He was involved in designing control systems and built some of the gun turret controls for World War I battleships, and light controls for theaters, and things of this sort. It later became servomechanism engineering. I knew perfectly well what he was doing. I don't mean I ever went into it, but I knew what he was doing. And towards the end of World War II, when the word servomechanism became popular, the idea of feedback was one of many cybernetics ideas in the air which were obviously relevant to what I was after. And I remember asking him for a reference to a good book to get me started in the servomechanism literature. But it really wasn't until that time, or a little later, that I realized that his whole life had been spent in what you might call protocybernetic work, and that it was just a direct ancestor to this whole business. And until the last year or two of his life—he died in 1948—we never had a conversation about this. He used to tell me about his work, but that was about his work, and I used to tell him about what I was doing, but that was about what I was doing, and I don't think the thought crossed either of our minds, certainly not until about 1947 or 1948, that these had any relation to each other. And I don't really understand that now.

Although he is over sixty, Herb Simon is less than half gray, and nearly unwrinkled. He is vigorous and trim, probably a result of his daily walk between the campus of Carnegie Mellon and his big old Pittsburgh house a mile or so away. Simon passes my house on his way to and from work, and I have seen him trudging through deep snow in galoshes and a wonderful knit Peruvian helmet, bright and betasseled, or in a light cardigan and sunglasses during the hottest parts of the summer. But mostly I think of him passing by in his dark overcoat and beloved black beret, walking swiftly and purposefully past my garden hedge. He has been making that walk for a quarter of a century, approximately two round trips to the Pacific coast, he guesses.

I like to walk. I just like to do it. I try to think about things when I'm walking, but I'm a terrible daydreamer. I seldom can keep a coherent line of thought, and going from one block to the next, I go off to thinking about something else. Still, I find the cudchewing just an important part of mental activity for one's research. So I guess an awful lot of cudchewing gets done, but usually in short spurts—you see things on the street, or you think of something else, and there's nothing to get you back in context again. Maybe if I carried a sign in front of me saying this morning I'm thinking about X—. [He stops and grins.] I'm willing to have my crotchets, but I'd feel a little self-conscious about doing that.

There is an interesting contrast between Simon talking in private conversation and lecturing in public. As he speaks to you directly, he is soft-spoken and unemphatic, almost shy. Most striking, he insists on a response from you: at least a nod, or a smile, but best if you speak up and make it a genuine dialogue. Along with Seymour Papert of MIT, he has a splendid gift for making you believe that he regards your intellect seriously. Not until you go away does the intellectual glow diminish, do you remember that the Mendelian shuffle has not dealt you brains that come anywhere near his.

The public speaker is something else. His voice is robust and excited, his eyebrows knit. He frowns, licks his lips, shows his small, slightly parted teeth. He is a restless lecturer, moving from one side of the platform to the other, sometimes a hand in a pocket, sometimes grasping both sides of the podium. When he speaks of the brain, he unconsciously touches his own cranium. The intellectual energy pours out of him, as if it could light the city tonight. Beforehand, I have chatted with him, and seen detailed notes in his hand for the talk. They surprise me because this is to be a very elementary talk to a general undergraduate audience, and surely the issue of whether machines can think is a topic Simon can speak on in his sleep (and maybe does). Later, when I hear him answering the questions he has heard a thousand times before ("But don't computers do only what you tell them to?") carefully, seriously, and without condescension, I guess that this preparation is simply a measure of the respect he has for his discipline and for his audience, whoever they might be. I also suspect that this attention to detail can be intellectually overwhelming, and may account for why some people think of him as unpleasantly, unsettlingly inhuman.

The summer that followed that first fruitful one in 1952, Simon spent only a month at RAND, but he was still excited by the ideas his meeting with Newell had produced. He had already begun to think of problem solving as a more apt substitute for decision making—his notes show that shift in 1950—and he was interested in "a serious rethinking of the whole psychology of the problem-solving process as it applies to administration," he wrote to Jacob Marschak in 1950. By the summer of 1954, just as the radar organization was transforming itself into a training program, Newell and Simon were having some long discussions about computers.

I learned to program the IBM 701 that summer [Simon says]. You know, there's never been a computer since that had as nice an order code as that did. It was very logical and clean. All of the dirty things you put in to use the machine more efficiently weren't in there—it was just a very logical thing. So I learned to program then. Just because I thought that here was an intellectually exciting thing on the horizon, and obviously anybody ought to know about it. You know, you just gotta know about it.'

As it happened, the designer of that elegant machine, Nathaniel Rochester of the IBM Corporation, would be a key member of the Dartmouth Conference on Artificial Intelligence two summers later.

One day during that summer of 1954, Newell and Simon went out to observe a three-day exercise at March Air Force Base:

It was one of those forty-eight hour a day affairs where you go round the clock, which Al loved because he loved to have an excuse for staying up all night, and which I hated for the same reason. Our first conversation, starting out from the parking lot, was about the interpreter in the 701. But further on I can remember us saying, "Well, if we're really going to have a good theory of what goes on in human problem solving, why not simulate it on the computer?" And that was our main topic of conversation going out and maybe part of the way going back.

Nothing happened immediately. This conversation had taken place in the late summer of 1954, and in November of that year, when Simon had returned to Carnegie, Oliver Selfridge, then of Lincoln Laboratories in Lexington, Massachusetts, came out to RAND to report on some work he was doing with G. P. Dinneen on pattern recognition.

Now I'd call it an artificial-intelligence program [Newell recalls]. It was not just a simple pattern-recognition device, but it actually carried out transformations and had several levels of logic to it. I didn't know Oliver at the time at all. We just sat in an office with five or six other people while he talked about this system they were programming, just in order to keep us people at RAND up on it. And that just fell on completely fertile ground. I hate to use the phrase, but it really was a case of the prepared mind. It made such an impact on me that I walked out after a couple of hours and walked into somebody's office—I don't remember whose—and gave them an hour's lecture on this thing. And then I went home that night and designed another system like it, for working on the air-defense center.

The system Selfridge and Dinneen had developed actually aimed at exploring learning (Selfridge, 1955; Dinneen, 1955). Its task domain was visual-pattern recognition—alphabet letters such as A and 0, and some simple figures such as triangles and squares. The program computed certain characteristics of the figure, which yielded values, or numbers. These were then compared with norms, and if the values were sufficiently close to the norms, then the pattern was recognized and labeled as an A or a triangle, or whatever it was. What made it a learning machine was that it could generate new characteristics for itself, and it could eliminate characteristics that had been tried and found wanting. In short, it learned by experience. It didn't learn very well—indeed, it tended to fix itself on one characteristic and improve that one to the exclusion of others. But that wasn't the point. The big point, the thing that had so taken Newell, was that a complex process was underway that was the result of many simpler subprocesses. These simple subprocesses had been organized in a highly conditional and interactive way, and the system showed that, working in concert, a set of simple subprocesses that were easy to understand could lead to genuinely intelligent behavior. This conclusion seems altogether obvious and common sensical in retrospect, but at the time it wasn't at all obvious and it was counter to common-sense notions of intelligent behavior. This assumption—that sets of simple subprocesses could produce a system which behaved in complex ways—was to inform research in artificial intelligence through many, many more tasks of greater and greater difficulty. *Simple* and *complex* were to become relative terms, but the principle never changed.

I can remember sort of thinking to myself, you know, we're there [says Newell]. Those guys, Oliver and Jerry, had developed a mechanism that was so much richer than any other mechanism that I'd been exposed to that we'd entered another world as far as our ability to conceptualize. And that turned my life. I mean that was a point at which I started working on artificial intelligence. Very clear—it all happened one afternoon.

The revelation for Newell was that complex systems of processing information could actually exhibit intelligent behavior. Until then, his focus had been upon building theories of how large human organizations behaved in accomplishing big tasks. Suddenly, for him as a scientist, to build systems that exhibited intelligence became not only a possible, but an appropriate scientific task.

He went home that night—it was a Thursday or Friday, near the weekend—and absorbed himself totally in the new project. He was in a state of high excitement:

I had such a sense of clarity that this was a new path, and one I was going to go down. I haven't had that sensation very many times. I'm pretty skeptical, and so I don't normally go off on a toot, but I did on that one. Completely absorbed in it—without existing with the two or three levels of consciousness so that you're working, and aware that you're working, and aware of the consequences and implications, the normal mode of thought. No. Completely absorbed for ten to twelve hours.

Newell laughs about this turning point:

There are a lot of funny things about that. One of them is that people only change courses when they're ready to, I think, and they don't very often if they're really charging down a trail. They'll often tell you of the blinding experience on the road to Damascus, which showed how Paul did a complete turnaround. But he only did it once, and if he was really capable of doing it, why didn't he do it three or four more times during the rest of his life? From then on, it turns out, he was completely resistant to all the other good ideas that were floating around. So there I was. It was not a turning point in my life, it was an open point—I had all kinds of plans but no deep commitments, and three or four other hot trails I was almost on. But once I started down this artificial intelligence/cognitive-processes path, I haven't changed since. I have no urge to move off this path, though the path itself has changed, in the sense that some features of the mixture of psychology and computer science have changed in it.

The Selfridge and Dinneen program was very influential in one way, but it was a bust in another. It was very exciting to a number of people, and they did almost get it working. But it never evolved in any serious way. It was the basis of another set of ideas that Oliver later had, called Pandemonium.

In the four months following Selfridge's visit, Newell wrote a paper using chess as a vehicle to understand what Selfridge and Dinneen were doing; he delivered it in March 1955 at the Western Joint Computer Conference. He also left RAND for Pittsburgh to work with Simon. Their original impulse had been to collaborate on organizational experiments, an attempt to construct a program—or a machine—that would exhibit intelligent behavior. This earlier work was altogether congruent with Simon's growing convictions that the computer could be made to simulate human thought processes in ways that would yield insights which previous models—mathematical, statistical, behavioral—

had not. The result would be the information-processing model, child of the new scientific paradigm of information theory.

"Al needed to pick up a degree somewhere, and so we figured we could combine business with pleasure," Simon says of Newell's coming to Carnegie. "I was able to work out a way of his getting a degree without doing all the Mickey Mouse, and it seemed a reasonable thing to do, and we wanted to collaborate on this chess machine he was thinking about."

With Paul Armer's help, Newell arranged a means of staying on the RAND staff and going to school in Pittsburgh. Armer says, "I was quite impressed with Al, and because of the RAND environment —whereby if I decided that something was in the best interests of the Air Force, the money got spent and nobody questioned it—we were all very happy with the arrangement." Indeed, Newell continued on the RAND payroll for another six years, until he resigned to become Institute Professor at Carnegie.

In January 1955 Newell brought his wife and small baby back to Pittsburgh, and began one of the richest periods of his and Simon's scientific lives. For Noël Newell, confined by the long Pittsburgh winter and a young baby besides, the adjustment wasn't easy. Allen Newell, on the other hand, was at home right away in the structure of the Graduate School of Industrial Administration.

There was some question about Newell being regarded as Simon's protégé rather than as his partner. He was officially a graduate student and, in any case, eleven years younger than Simon. He had few publications to his credit and was virtually an unknown. But Simon developed strategies to counter that:

We nearly always appeared alphabetically on our joint publications. Al's name came first—people could interpret it as Al being the senior partner or as it being alphabetical, but not the other way around. On our public appearances, our talks, we generally alternated, and we generally didn't appear as a pair. We always argued that we were interchangeable parts and there was no sense in both of us going. Also, I always tried to make sure when I made references myself to the work, it was to both of us. When people sent me manuscripts for criticism and the names got reversed, I put them back in order again. It wasn't really all that much of a problem. The main reason is because everybody who met Al found out he was a big boy—he wasn't anybody's protégé.

Simon recollects the way their partnership worked:

During the very early period it worked mostly by conversation together. It's probably the case that Al talked more at them than I did; it's certainly the case now, and I think it's always been so. But we ran these conversations with a rule which we made explicit. I don't think we knew about brainstorming then, but the rule was that you could talk nonsense, and vaguely, and you weren't supposed to be called on it unless you intended to be talking accurately and sensibly. You could try out ideas when they were half-baked, or quarter-baked, or not baked at all, and try them around and just talk, and listen, and try them again. At various points we would set up goals for one another. I'm sure Al initiated the original chess project. It was pretty much his decision, and my role in it was secondary. We talked about it at various stages, and I'm sure I got him onto Carnap, suggested that Carnap's formalism was probably a good thing. We referred each other to literature. He did most of the active work though, and I would react to drafts and ideas that he had.

Newell returned to RAND during the summer of 1955, and shuttled back to Pittsburgh in the fall. That summer and fall were the critical periods in their decisions to create not only the chess machine, but the Logic Theorist, or LT, the program that proved theorems in Whitehead and Russell's *Principia Mathematica*.

As far as I'm concerned, everything was done by December 15, 1955 [Simon says]. Of course it wasn't really. But during that time Al was looking at the computer side and communicating with Cliff Shaw a great deal, and I was looking at the substantive areas, that is, what problem areas could we do this in and get away with it. I was looking at geometry, and chess and logic—the logic we came on simply because I had the Principia of Whitehead and Russell at home, and I pulled it off the shelf one day to have some problems. I was doing a lot of introspecting about my own problem-solving processes, so I tried to prove some theorems in the Principia, but decided it was pretty hard, so it probably wouldn't be a good thing to try.

In 1957 Simon wrote down his recollections of this period:[2]

When Allen Newell returned to Carnegie in the autumn of 1955, he was committed to the project of programming a computer to play chess. I expressed much interest in the project and stated my intention to continue working on human problem solving .... We agreed to meet each Saturday, and ranged on these occasions over a wide range of topics—particularly discussing problem solving and the chess language he was trying to devise. Al tended to supply ideas starting from the language and computer end, I starting from human problem solving and what we knew of the heuristics there. This is one of the role specializations that, subject to strong qualifications, we have mildly adhered to since. In the course of these discussions, we considered illustrative problems from areas other than chess—including Euclidean geometry, Katona-type match problems, and (I think) symbolic logic. During the third week of October, 1955, we attended the TIMS meetings in New York. I went a day early to see Barney Berelson [of the Ford Foundation] on the morning of October 19. On that afternoon, a beautiful day, I decided to take a walk along the Hudson on Morningside Heights. I don't remember if I had an appointment on the Columbia campus late in the afternoon or not. I pondered as I walked about how one solves geometry problems—the example I had in mind had to do with angles inscribed in circles and semicircles .... Suddenly I had a clear conviction that we could program a machine to solve such problems. I made some jottings on a piece of paper I was carrying and thought very hard about it for a few minutes, the conviction remaining very strong. I think the conviction arose from the fact that I could see the heuristic I was using and how it cut down the search space.

That evening Al and I met in Merrill Flood's hotel room, and after discussion we agreed to try to program a geometry machine before Christmas. We both had strong feelings that evening that we had an excellent chance to succeed. I have a clear picture of the room in my mind, and where each of us was sitting.

---

[2] Simon generously made this account available to me in 1976.

So they began to work on geometry, but it soon turned out that representing the diagrams was very troublesome. Symbolic logic then suggested itself as an alternative, precisely because it involved no diagrams. Simon began to revive his skill in logic by studying the proofs in Chapter 2 of the *Principia*, and meanwhile work was proceeding on the language in which the program would be written. Simon's aide-mémoire continues:

> By the beginning of December I was beginning to have pretty clear ideas about some pieces of the heuristic (e.g., the working backwards in proofs by substitution). I was doing most of the actual work on the proofs, supplemented by our Saturday discussions. Al, after a burst of activity in November or October, was somewhat bogged down by studying for prelims .... Al's notes pick up pretty well from about December 6, by which time we had most of the pieces but little of the organization of the program. During the subsequent week we conferred frequently for short periods—almost daily—and I worked almost every night on the proofs. On Thursday, December 15 (having felt I was getting increasingly close during the week), I succeeded in hand-simulating the first proof using a program reasonably close to that published in the IRE paper the following September [Newell and Simon, 1956]. During the subsequent several days, Al and I worked hard to sharpen up the program, and put it in a form where one could consider coding it for the machine (that is, in the interpretive code). In the above paragraphs I don't want to create an impression of specialization—did Hillary or Tenzing touch the summit first? Most of the actual paper-and-pencil work on developing the LT program was done by me, just as most of the actual work on the language was done by Al. [This of course means Al and Cliff Shaw, Simon says now. A clarification of the working relationship among the three will follow.] We were in closest communication during the whole period, through long association had developed an extraordinary capacity to communicate even our subtleties to each other, and the whole product must be regarded as joint and inseparable. I am firmly convinced that none of us alone had much chance of accomplishing this. As our theory of creativity develops, we may even be able, in a couple of years, to prove this.

Later Simon said,

The reason I mentioned December 15 is because that was the day when very clearly I pushed through the first proof that was a simula-

tion of what became LT. All this time we'd been discussing the succes-
sive approximations, and within a day or two we had identified the
big methods of LT, the detachment method, the contractions, the pro-
cedure, the chaining method, and so on; and I remember one after-
noon over in my office in GSIA—Al and I were there, and I was
outlining that on the board and I was saying to Al, now can you
program that? But our discussions ranged over both the subject matter
and the programming.

Their collaboration also included J. C. Shaw, a senior programmer at RAND, though Simon and Cliff Shaw seldom saw
or spoke to each other. Newell carried out the middleman's role,
mostly by long-distance telephone between Pittsburgh and Santa
Monica. "I thought he was terribly daring, running up those
incredible $200-per-month phone bills," Simon laughs now.
"But, then, Al really taught me how to think big about money."

J. C. Shaw, whose name would appear alongside Newell's
and Simon's on reports of the Logic Theorist, the General Problem Solver, their chess machine, and their series of computer
languages called IPLs, or Information Processing Languages, had
come to RAND the same year as Allen Newell, from a small Los
Angeles insurance company where he had been an actuary. "It
was to be an actuary or teach when I got out of school in mathematics," Shaw says, "so I was an actuary. Then during the war I
chose to be a navigator rather than a pilot, inventing methods
of navigation and teaching. Navigation was also related to mathematics, to computation—modeling, if you will. The whole
business of plotting on charts and so forth is a fine example of
modeling."

Shaw had grown up in California, where his family owned a
paint and decorating store. Shaw's father had started in the trade
as a thirteen-year-old apprentice in Ireland, and even at age
eighty-eight was busily engaged in painting his house. "He was
a first-class marathon runner and made a creditable showing in
the U.S. Olympic trials of 1932 at age forty-five," Shaw says. "I
learned many a lesson in perseverence from him." And Shaw
also remembers his grandfather, a fifth-generation painter, pass-

ing time in the family's paint store in Fullerton, filling the backside of old rolls of wallpaper with digit-by-digit calculations of cube roots.

By 1950, Shaw's own interest in mathematics had vanquished whatever interest he'd had in actuarial tables and had brought him to RAND:

We were in the building at Fourth and Broadway in Santa Monica, which had been used, I believe, by the evening newspaper there, the Evening Outlook. Our department was down in the basement where the presses had been. We had some 604 calculators back in a low-ceilinged room that we called the sweatshop because we literally sweated—took off our shirts. But it was an open place in the sense that every visitor wandered through, and everybody in our department knew who was there. That opportunity to see what was going on was lost a little bit when we later moved to the building at 1700 Main Street.

He remembers meeting Allen Newell:

One day he came down into our offices, and because things were rather open in our basement, no one could conduct business at one desk without everyone else knowing what was going on. He was concerned at the time with simulating—at that point not a complete air-defense direction center crew, that's what it evolved to. He was simply seeking some device that could simulate warnings, and I recollect thinking in terms of punch cards and moving the cards with the holes over red lights underneath the desk!

Shaw is a big ruddy man with sandy-colored hair and blue eyes topped by unruly eyebrows that look as if he had just come in from a windstorm. In an otherwise sober demeanor, those eyebrows seem an irrepressible clue to the extraordinary inventiveness that has characterized Shaw's work. The numerical-analysis department, where Shaw worked, did indeed get involved in implementing Allen Newell's radar simulator. Shaw's function then was partly administrative, allocating manpower in the department, and that meant meeting regularly with Newell to plan schedules. "That was the first I found out how energetic that man is, and driving," Shaw says, laughing. Apart from these administrative duties, Shaw's own interests then were in redesigning RAND'S primitive calculators to be more automatic, more responsive, to applications people were rapidly inventing for them, and helping to make Johnniac work. It wasn't until 1954 that he and Newell began even to talk about designing the chess machine.

It was very painful to try to program anything, to make progress towards a chess-learning machine, because we didn't have an adequate language for communicating. We had done a number of programs in what was essentially machine language, actually using the symbolic assembler that I had written for Johnniac, but it was far too low-level a language to begin to specify the chess-playing program. As programmers, we had a creative task each time with trying to invent a representation in the machine corresponding to what we were communicating fairly loosely in English. The natural direction then was to suggest interpretive languages, higher-level languages, trying to approach something where Al and Herb could specify more completely the complex concepts of chess. But we wanted to do it on the machine. So that involved Al more directly in programming at that point, and we created the information-processing languages, IPL-I and IPL-II. IPL-I was actually a label we put on retroactively to a language that Al and Herb used to lay out the specification of the Logic Theory Machine.

It's hard to believe now [he adds], but we did all the early work on Johnniac with only a forty-column numeric printer—the alphanumeric printer wasn't installed until 1957—so there was a lot of time wasted decoding the printouts.

As language specification was proceeding jointly between Shaw and Newell, it looked to Newell and Simon as if some earlier work they had done on the logic theorem prover might be a more tractable problem than chess or geometry after all. Temporarily, therefore, chess at least was abandoned, and as Simon's aide-mémoire has described, the Logic Theory Machine became the first complete Newell-Shaw-Simon program. On December 28, 1955, Newell wrote to Shaw that the work on the chess machine had come to a momentary standstill because the work on the Logic Theorist was so involving. "In one sense, we're over the hump—we have a machine which can perform one of these 'intellectual' tasks." It had occurred just before Christmas; Newell described himself and Simon as elated. "Kind of crude, but it works, boy, it works!"

Aside from the fact that it worked—that it was proof positive a machine could perform tasks heretofore considered intelligent, creative, and uniquely human[3]—it also exhibited some extremely clever solutions to programming problems that had plagued computer programmers and seriously hampered all kinds of computer applications.

Perhaps the most significant of these solutions was called list processing, which was a technique developed to answer the problem of allocating storage in a limited computer memory. Until that time, the allocation of memory—a very precious resource—had taken place at the beginning of a program's run, and specific blocs of storage had been dedicated irrevocably to specific functions. But the Logic Theorist ate up memory so voraciously that there was no question of allocating storage permanently for any particular function. As information was used and had no further value, it was necessary to recover that storage space for other uses. Shaw and Newell solved the problem by labeling each "word" of storage, and then by making the machine keep a list of all available space that could then be reused, rather like a vacancy list in a hotel. This idea was refined and expanded in subsequent list-processing languages.

List processing introduced the process of recursion to computing, a way of determining the next instruction in a sequence from one or more of the preceding sequences, and it also addressed the problem of data structures. Everyone knows that some ways of arranging information are better than others—a

---

[3] In fact, the Logic Theorist discovered a shorter and more satisfying proof to Theorem 2.85 than Whitehead and Russell had used. Simon wrote this news to Lord Russell, who responded with delight. However, *The Journal of Symbolic Logic* declined to publish an article coauthored by the Logic Theorist describing this proof.

telephone directory arranged by sequential telephone numbers is probably more useful than one with a random arrangement, but it is much less useful to the average telephoner than an alphabetical listing of subscribers. The same principle holds with computers. Some ways of arranging data in a computer are more useful than others, and those ways vary depending on the task to be performed. It happens that data arranged in lists, and lists of lists, lend themselves most readily to the simulation of human thinking processes. This fact is unsurprising, for as Simon points out, one important source of the ideas for list-processing languages was what psychologists knew about associative memory in human beings.

Since Newell and Simon had done some simulating of the Logic Theorist in the winter and spring of 1956, using their students at Carnegie and Simon's wife and children as "the machine," much of the programming for it, and the implementation on Johnniac, was done with an air of verification rather than expectancy.

"But it was still a point of achievement," Shaw says, "when I was able to go into Paul Armer's office and write on the blackboard that P implies P. As simple as that, it was a great event." Then he smiles broadly. "But it was just sweat trying to get the thing to hang together, and on a machine that I'd been involved with at every level, helping to diagnose hardware malfunctions on up through the software levels to fixing the bugs Al and I would discover in the highest level processes of LT. It was a bit of a surprise, by the way, to hear that LT had discovered an elegant proof of Theorem 2.85 of *Principia Mathematica* that Whitehead and Russell had missed. That added a spark to the whole business."

The Logic Theorist confirmed one more interesting fact: though complex systems in computers might do what you tell them to, there is no way of predicting what that behavior will be, apart from running the program. There is no way to be sure that the changes the programmers make actually influence a run of the program in the ways intended. The program behaves reasonably, but not always as expected, and it's impossible to tell whether that behavior is a consequence of what has been specified in the program, or the consequence of a bug. Newell, Shaw, and Simon had not instructed the Logic Theorist to find a better proof for Theorem 2.85, but it had.

Cliff is a very taciturn guy [Newell says]. One of my dominant recollections is going in and talking with him about some of these problems, and going through a whole session and he wouldn't say a single word, and getting up and leaving. This was when I hardly knew him. It's probably the case that the whole scientific enterprise with the three of us would never have worked out if we were all sitting in one place. Cliff found this way of working, with me located miles away, to be just about the right level of controlled interaction for him to flower. And so I operated both by letter and by telephone—by two and three hour-long conversations a week through this whole period— so in fact the three of us never got together, almost.

Shaw recollects their relationship then:

Energy is the thing I remember mainly about working with Al. Energy and brilliance. Long phone calls and long sessions on the teletype were typical. I made a couple of trips back to Pittsburgh, and we would have sessions late into the night at Al's home, into the early morning hours. I felt as if he and Herb were always a few lengths ahead of me in some respects. I felt like I was tagging along behind, trying to get that Johnniac to do what we already knew could be done in the case of the Logic Theorist, or what we had already simulated in the case of chess openings, or what had been simulated in the case of the General Problem Solver. So it was a struggle to keep up. And with Al's energy, it was a good thing he had IPL-V, the programming language, as another outlet, so all that energy didn't descend on me!

Newell disputes Shaw's modesty:

Cliff himself was the genuine computer scientist of the three—I mean in some fundamental way in which I'm not a computer scientist, okay? Cliff was the guy who had developed an assembler, really knew and operated with the machines and so forth. I was very much a middleman—not in the social sense, though that was also true by the way—in the sense that I didn't operate with the machines directly, and I never had. By programmer, you shouldn't think that I was dictating to Cliff what to do. He was the one guy who understood what computers are all about. I've always had sort of a large capacity for a mass of detail in terms of specifying large systems, and Herb has much less tolerance for that. Cliff himself also has a very large tolerance for detail, but he also had all the programming skills and understanding of machines which I didn't have.

So the main work on LT was done quickly, and Herb Simon was able to walk into his mathematical modeling class in January 1956, and declare that over the Christmas holidays he and Allen Newell had invented a thinking machine. By the summer, thanks to Cliff Shaw's programming genius, Newell and Simon could carry up to the Dartmouth Conference on Artificial Intelligence the printout of their first actual working program, the Logic Theorist. As an operating example of the information-processing model applied to a task that requires of humans, at least, imagination and intelligence, the Logic Theorist gave the first justification to the claim that artificial intelligence was a science.

Chapter Seven

Fun and Games

Instead of *homosapiens*, we should probably be called *homo ludens*, suggested the Dutch scholar Huizinga—not because we're the exclusive players of games, which any zoo visitor knows is untrue, and not because we're the only thinking animals. But we've raised each of these activities to such an elaborate complexity that it stupefies us if we stop to think about it.

Games are deep in the heart of us. In the streets of London today, schoolchildren play a game that can be traced back to the time of Nero, and popular books declare that interpersonal behavior can best be expressed as the games people play. From solitaire to the Super Bowl we're nourished on games, those abstract expressions of real life where we know the rules and can test our wits against an opponent or against chance, or watch our agents do it for us. Real life, of course, is never that tidy. Games let us work up to life. For some they even become life, but that's a slightly different issue.

No area of endeavor seems to be exempt. I attend an exhibition where chefs compete with each other to see what can be done with food: legs of lamb become mallards, hams become liberty bells; prizes are awarded, hopes dashed. I've seen the same thing with hairdressers and writers, the conversion of craft into competition. Pure art and pure science, with their *prix de* and

Nobels, enter into the same sort of competition, science all the worse for rewarding only the first scientist to arrive at some scientific landmark, and allowing nothing to the also-rans. Science and art share one more element of gaming: they are abstract expressions of nature; they stand outside it and yet are of it, absorbing their players altogether.

So it's no surprise that the most wonderful of twentieth-century toys, computers, were involved in games from the start. It wasn't just that Alan Turing had always been playful by nature, so with his friends at Manchester set to work in the late 1940s programming their early machine to play chess and tic-tac-toe. A cast of mind must have come from John von Neumann too, who was not only profoundly influential in American computer development, but also had written a book with Oskar Morgenstern published in 1944 called *Theory of Games and Economic Behavior*. In the introduction they described the book as "an exposition and various applications of a mathematical theory of games." In other words, an abstraction of an abstraction, the scientist's *crème de la crème*. When Herbert Simon saw an advertisement for the book—he was standing in line for concert tickets and saw the ad on the back of a journal, upside down under someone's arm—he felt a flush of envy so great he could remember it vividly thirty years later.

Games are models of situations in life, just as physical models imitate, simplify, and express the essence of physical phenomena. In their collection of early papers on artificial intelligence, Edward Feigenbaum and Julian Feldman (1963) introduce the section on the machines that play games this way: "A favorite area of research in artificial intelligence, past and present, is computer programs that play games. Why should one be interested in game-playing, a mere human pastime? Or, as a Soviet acquaintance once put the question to one of the editors of this volume, 'Who allows you to do it?' "

The answer to that question must have come to the Soviet scientist at some point, for by the time of the 1974 meeting of the International Federation of Information Processing Societies in

Stockholm, a Soviet chess-playing program was undisputed champion of the chess-playing machines, successfully beating out American, British, Canadian, and French attempts. The Soviets must also have come to see what Feigenbaum and Feldman declare are the appeals of game playing for artificial-intelligence researchers:

> Affectively, it provides a direct contest between man's wit and machine's wit. On a more serious level, game situations provide problem environments which are relatively highly regular and well defined, but which afford sufficient complexity in solution generation so that intelligence and symbolic reasoning skills play a crucial role. In short, game environments are very useful task environments for studying the nature and structure of complex problem-solving processes.
> (Feigenbaum and Feldman, 1963)

I suspect the first reason predominates. The others are valid enough, but smack of being discovered after the fact, the sober justifications we make to authorities and the suppliers of funds, so we can do what we want. I've seen too many gleaming eyes to believe otherwise—and then there's the evidence of history. The two game-playing programs I'll describe in detail here began as all games begin, with a sense of fun and competition. Curiously enough, they were both created under the roof of an organization whose connection with fun would go unsuspected by most, the IBM Corporation.

I myself first played checkers with my grandfather in England. We called it draughts, and I must have been no more than four or five, for we left England just after I turned six. I tell this because it illustrates two notions associated with checkers—that it's a simple game, and that it's the preserve of children and old men.

These two notions weren't far from Arthur Samuel's mind when, sometime in the summer of 1947, he airily proposed to design a checker-playing machine in order to raise money (the ghost of Babbage stirs) to build a big computing machine for the

University of Illinois, where he was a newly arrived professor of electrical engineering.

Samuel had left Bell Telephone Laboratories in 1946 to teach at Illinois, and his connection with Bell Labs had made him aware that such things as computers were on the horizon. He longed to have one at Illinois. The first dean he convinced to acquire a computer for the university retired a year later without having done anything about it, but his replacement was Louis Ridenour, later a vice president of Lockheed, and Ridenour was more decisive. He persuaded the university's board of trustees to come up with $110, 000—an enormous sum for those days, and in fact $20, 000 more than Samuel and his group had originally asked for. But it was soon clear that the sum was still insufficient to build a computer from scratch. Could they purchase one instead? Samuel consulted John Mauchly and J. Prosper Eckert at the Moore School of the University of Pennsylvania, who'd built the first general-purpose electronic calculator in the United States, the ENIAC, which had only been dedicated in 1946. The Moore School wasn't ready to go into computer production—though Mauchly and Eckert were later to start the firm called UNIVAC—and neither was anyone else. Samuel canvassed industry and government and even spoke with John von Neumann at the Institute for Advanced Study at Princeton. Disappointed, he returned to Illinois with the certainty that to get a computer, they would have to build it themselves. It would take more money than they had, and it occurred to them that perhaps they could fatten the money from the trustees with some government funds.

We would build a very small computer [Samuel recalls], and try to do something spectacular with it that would attract attention so that we would get more money. It happened the next spring there was to be a world checker champion meeting in the little neighboring town of Kankakee, so somebody got the idea—I'm not sure it was mine, but I got blamed with it at least—that it would be nice to build a small computer that could play checkers. We thought checkers was probably a trivial game. Claude Shannon had talked about programming a computer to play chess, and other people had been thinking about it,

so we decided we'd pick a simpler game, and write a program to play
checkers. Then, at the end of the tournament we'd challenge the world
champion and beat him, you see, and that would get us a lot of atten-
tion. [He laughs merrily at this.] We were very naive. I was given the
job of writing the program for it while we were still designing the
computer. So I started. I didn't know anything about programming a
computer or anything, but it was a good place to learn.

The Illinois group invited experts in to instruct them, and
wrestled with the design of the machine at the same time that
Arthur Samuel was trying to write a program to make it play
checkers. By the time of the world checkers championship in
Kankakee, it was clear to everyone on the project that neither the
machine nor the program had a ghost of a chance of being completed.

Nevertheless, the effort had a profound effect on Samuel himself. He was an expert in vacuum tubes and head of a large laboratory at Illinois, but the computer bug had bitten hard. He knew
now that he wanted to build computers. As long as he stayed at
Illinois, he'd be burdened with the administration of the laboratory, and worse, identified forever as an expert in vacuum tubes.
Could he find an industrial firm ready to launch into the computer field, where he could work and teach himself as he went
along? From his days of searching to buy a computer, IBM seemed
the very place. It had the resources to do valuable research and
development in computers, and had been involved in the Harvard
effort to build a computer, the Mark I. After some bad judgments,
the firm was just developing a small computer, the 604, a plug-
board affair. It would be two more years before IBM came out
with its first mass-produced, stored-program, general-purpose
computers, the 701 and 702, in 1951.

In the circuitous way of IBM, Samuel was hired for his own
purposes, which weren't precisely the purposes IBM had in mind
for him. The company was in the midst of terrible troubles with
vacuum tube reliability. They wouldn't admit this to Samuel because they didn't want anyone, not even a potential employee
who was an expert, to know that they were trying with no success

to make their own vacuum tubes instead of using the ones readily available—unsuitable for computers because they were intended for radios and such. Samuel told IBM frankly that he was interested in building computers, but under the impression that he was telling them this to be polite, they nodded, and hired him for his vacuum-tube expertise.

Samuel set up shop in a small laboratory in Poughkeepsie, the forerunner of what would later come to be one of the biggest and best industrial laboratories in the world. Here Samuel learned that his real job was to pull IBM out of its trouble with vacuum tubes. Again because of his Bell Labs connection, Samuel knew that the transistor was on the way, and he tried to persuade IBM not to put funds into developing what was already obsolete.

I didn't realize until afterwards that I came so close to getting myself fired for opposing management. Young Tom was not yet president of the company; old Mr. Watson was still active. But they gave Tom the job of deciding whether this young upstart knew what he was talking about. So I went down and had many talks with him. I hung around the office for about two weeks. He would talk to me in any free time he had, between other things, me trying to convince him that they shouldn't set up to make vacuum tubes.

But once Samuel succeeded in convincing Thomas Watson, Jr., that transistors were the way of the future, the problem remained of producing machines with vacuum tubes that would be reliable enough to market. Watson's impulse was to withhold the computers from the market until transistors were available, but Samuel persuaded him to put out the first 701s and 702s with vacuum tubes made specially for IBM by other firms.

And while all this was taking place, Samuel was still fascinated by his checkers project. To make it more respectable, he began adding features that would make it learn from its mistakes.

When we were designing the 701, I decided I'd rewrite the checker program, which I'd written in this pseudo language for the pseudo machine which we hadn't built yet at Illinois. So one of the first programs that we had to run on the 701, one of the first computers of its type, was this crude checker program. IBM never looked with favor upon my working with it really, because it smacked too much of machine thinking, et cetera, and they wanted to dispell any worry people had with machines taking over the world and all that sort of thing. But as a test vehicle for computers it was a very good program, because it was very complicated and could be set up to run for a long time, and there was some way of checking whether it was really working right or not. So I continued. But it was never my main job; it was always by sufferance.

Bored by and temperamentally unsuited for managing the research laboratory at Poughkeepsie, which was beginning to grow as big and cumbersome as the laboratory Samuel had left at Illinois, he switched jobs. He became IBM's eyes and ears in Europe, reporting back to Americans on what the Europeans, who in some ways were more advanced in the early 1950s in computing, were up to. Curiously enough, his checkers-playing program was often his entree to laboratories where he would otherwise have been unwelcome. "I had a ready-made topic. I could tell them everything I was doing on the checkers problem, and then I would stay over and let them show me what they were doing."

Thus it was that Samuel's checkers-playing program became even more widely known. He worked steadily on it, incorporating one feature after another, and debugging it on the new machines that were coming off the IBM assembly line. There they stood in a row, playing ghostly games of checkers with their programmers in the hours between midnight and eight, being tested to go into the world and do accounting, inventory control, and other sober tasks by playing the game old men play with their grandchildren.

That Samuel's name was becoming widely associated with a checkers-playing program was not only an embarrassment to IBM, it was also something of an embarrassment to Samuel himself. It had been an accidental choice, after all, and he was only too well aware of the popular reputation checkers had for being a trivial pastime. In fact the game is not trivial, as Samuel was to find out over twenty years of work on his checkers-playing project. He never enjoyed playing checkers himself, and despaired of learning anything about the game's general principles by reading literature on the subject. Even worse than most chess literature, checkers literature falls into two extremes: either it is a very complicated expression of the theory of the game, which has little to do with its reality; or else it is a compilation of games the masters have played, with no principles extracted. Giving up on the literature, he invited checkers masters to help him work on the program, but always without success. One came to the Stanford Artificial Intelligence Project, where Samuel had gone when he retired from IBM, planning to work with Samuel for six months.

"We got simply nowhere," Samuel reports. "The more he attempted to analyze his own thought processes, the more confused he got, and he finally quit, saying that he was really afraid he'd become a poorer checker player because of trying to think about it."

But then Samuel had seen the same problem with chess. Earlier on, he'd watched an international group of chess experts gathered at an atomic-energy facility in Italy where a small artificial-intelligence effort was underway. The chess experts had come and talked to the programmers solely in chess terms, and the programmers, who knew about computers but not about chess, misunderstood, or forgot, or were unable to translate the chess insights into computer expressions. Samuel had been called in to advise the two groups, and he was reluctant to tell them that they were all just crazy. Finally, one of the chess players wormed it out of him. "If you keep doing it the way you're doing," Samuel said, "it'll be a thousand years before you'll get any results." He was thanked, and soon afterwards the group was disbanded. Samuel was astonished to read a newspaper account of the break-up, his phrase transformed into the claim that the problem of chess playing by computer would take a thousand years to solve.

Samuel's approach to the checkers-playing program was somewhat different from other learning programs. He did not believe that human learning processes ought necessarily to be imitated, because he believed that the differences between the human brain and the computer were simply too great. "I think you study the way people solve problems to get an insight into what the real problem is, not to get an insight into the method the brain uses to solve the problem. And then you sit down and you say, 'Okay, given the technology available, the speeds, and what the computer will do, how best can we solve the problem?'"

Samuel's program was a celebrated success. It became a better checker player than he was himself (by far), playing at the master's level by 1961. One reason for the program's success is that checkers is a less complicated game than chess, though by no means trivial. Thus it presented more tractable problems of look-ahead (the program's search to examine future possible moves) and evaluation. Another reason for its success has been, quite simply, Samuel's intelligence and tenacity:

I believe the reason why more progress has not been made on chess is that no one person has worked on it as hard and continuously as I have done on checkers. There have been many people who have dabbled at chess and made good starts, and then they get discouraged and quit. So somebody else comes along and says, well, he didn't make any progress because he didn't know what he was doing, so let's start all over again. So they begin again, and get discouraged, and quit.

Key to the success of Samuel's program is that it learned— that is to say, it adapted its behavior to past events. If it encountered a position it had already come across in another game, it made a more accurate evaluation of the position based on the results of the completed game than the chancier evaluation it could make by looking only two or three moves ahead in the present game. It thereby improved its performance with each game. It was able to improve in this way because there are fewer different positions in checkers than in chess, and the situations to be evaluated are simpler.

In the summer of 1962, Robert W. Nealey, a former Connecticut checkers champion and a nationally known player, engaged the checkers program in a game. Nealey commented,

Up to the 31st move, all of our play had been previously published, except where I evaded "the book" several times in a vain effort to throw the computer's timing off. At the 32-27 loser and onwards, all the play is original with us, so far as I have been able to find. It is very interesting to me to note that the computer had to make several star moves in order to get the win, and that I had several opportunities to draw otherwise. That is why I kept the game going. The machine, therefore, played a perfect ending without one misstep. In the matter of the end game, I have not had such competition from any human being since 1954, when I lost my last game.

(Feigenbaum and Feldman, 1963)

Samuel has long maintained that his ignorance of checkers was helpful in getting the program going, that he was never tempted to put in what he himself knew, because he knew nothing. But ignorance didn't help the hopeful designers of chess machines, and in fact, one of the first really successful programs was done by Alex Bernstein, who knew a great deal about chess to begin with, and soon learned more.

Bernstein had a flair for it. He had begun playing when he was nine or ten, and as I was waiting for him one afternoon in his pleasant Brooklyn Heights brownstone, I watched his own nine-year-old patiently teaching a friend some chess tactics.

Bernstein was captain of the Bronx Science High School chess team, a school whose pasteboard playing fields have done as much for American science as Eton's grassy ones did for the British Empire. He recounted the early history of his involvement with the game:

I started playing chess seriously, I guess, when I was in high school. I played chess so much that it affected my grades in college. One year I played chess to the exclusion of everything else and woke up at the end of the term and discovered I had failed two courses. I was going to City College at the time. I failed a physics course and a math course— theory of functions of real variables. It was quite a shock and I gave up chess after that term. I suppose I continued reading about it, but I stopped playing chess. Then in graduate school, although I'd given up math for medieval literature and poetry, I worked as an assistant in the civil engineering department. After that I went into the army, and because of my work at Columbia and what I was doing in the army—working in a special research and development outfit of the Signal Corps—I became acquainted with computers and what they could do.

From the summer of 1953 to the summer of 1955, then, Bernstein worked at the Bureau of Standards, his flagging interest in mathematics revived by a kind woman who had known his father, a well-known European mathematician, and who taught him about the operations of the bureau's computers. Bernstein was given permission to try his hand at simulating the air-defense system protecting Washington—the first missile air-defense system—to see how it would work. This was at just about the same time that Allen Newell was involved in simulating an air-defense message center out at the RAND Corporation. Computer simulation was in the air. But one main difference between Newell's and Bernstein's models was that Bernstein's was a mathematical computer model and Newell's was nonmathematical.

After Bernstein got out of the army he returned to Columbia, but the academic atmosphere got him down, and when the opportunity came to work full-time for IBM he took it. He had been working part time for them, and had become friendly with a young man named Hal Judd. Judd knew Bernstein had played serious chess, and one day suggested they try to produce a chess-playing program.

He was a very poor chess player but an avid one, as most Americans who play chess are [Bernstein recalls]. I guess he knew computers far better than I did at the time, but he really didn't know chess, and he felt that with his expertise in computers and my understanding of chess, perhaps we could produce something. Well, almost immediately after I'd said that I thought it would be a good idea, he was transferred someplace out of New York City. Eventually he wound up working for IBM in Australia. Originally, he'd said he'd work with me at long distance, but it became impossible. Nevertheless, the idea of a chess-playing program really intrigued me, so I decided to see if I could come up with a scheme for producing this program and for getting some backing. It wasn't impossible to work on this problem oneself, but one did need computer time and computer time is very expensive, although at the time there was a good deal of unused computer time during the third shift. Nevertheless, we had to have some permission at least to do it. So we went to Charlie DeCarlo, who was head of the so-called Applied Science Division at IBM. DeCarlo is a mathematician and came out of Carnegie, and he was very sympathetic to the idea, and said he'd support it on a limited basis. Thus I went to work full-time for IBM at what was then the Scientific Center, and later became part of the Service Bureau, in Manhattan. I was given other work to do, but essentially it was understood that half of my time I would be allowed to spend working on the chess program.

Bernstein drew upon not only his own experience with chess, but began to study *Modern Chess Openings*, which came out then every two years, and spent six months going through some five hundred chess openings. He assigned scores to various positions, scores that depended not only on the pieces retained, but also on area control of the board and mobility. He also developed a fourth measure, what he called a "greens area" around the king, meaning that the more squares outward from the king controlled by his own side the better. But after six months of this he gave it up. He couldn't make any sense out of it.

"It was essentially a chastening experience. I was sort of abashed because I had said to people, I can do it." Bernstein was well aware that chess literature contained mentions of chess machines, and certainly science fiction—say, Ambrose Bierce's "Moxon's Master"—provided examples. Baron von Kempelen's nineteenth-century chess machine might have been a fraud, but the idea was intriguing enough to even so serious a scientist as Charles Babbage, who hoped to raise money to complete his Analytical Engine by working up a little chess-playing machine for exhibition. And Torres y Quevedo had actually built two endgame players in the 1920s.

At this time, Bernstein was unaware of Shannon's seminal papers, and did not know that chess had caught the interests of a group at Los Alamos, including J. Kister, P. Stein, S. Ulam, W. Walden, and M. Wells, who were working on a limited 6 x 6 board, rather than the regulation 8 x 8. Nor did he know that Allen Newell, J. C. Shaw, and Herbert Simon together, and John McCarthy independently, were also pondering chess-playing machines. Alex Bernstein only knew that the problem was hot, and though his confidence was slightly shaken by the experience of his first six months on the problem, he was all the more anxious to try again. That it was a classic problem made it even more tempting—the creation of a smart chess-playing machine had come to seem as elusive as solving the four-color problem,[1] and the solution of either would have been a scientific coup.

Bernstein went back to the chess-theory books. This time he found insight in a book called *My System* by Aron Nimzowitsch, an early twentieth-century Russian chess master who had revolutionized modern chess. Nimzowitsch had abandoned the traditional strategy of building up the center game and gradually grinding down one's opponent. Instead, he stressed the notion of imbalance on the chess board—not necessarily immediate control of the center, but a delayed strategy, emphasizing what are called strong points and lead points.

It was now that Bernstein became aware of Turing's work and read at least one of Shannon's papers. When he finally began to see how he might codify some of the principles he felt were essential, he telephoned Claude Shannon at MIT. "I went up to MIT

---

[1] The four-color conjecture has teased mathematicians since the nineteenth century. It seemed plausible that every map drawn on a sheet of paper could be colored using no more than four colors yet in such a way that countries sharing a common border would have different colors, but nobody could prove this mathematically. The conjecture was finally proven in 1977 by Kenneth Appel and Wolfgang Haken at the University of Illinois, using twelve hundred hours of computer time. They have written a lively article on the history of the four-color conjecture and how they proved it as a theorem, which appeared in the October 1977 issue of *Scientific American*.

and spent a day or two with him, telling him what I was planning to do, and he said he thought it was intelligent, and a good way of proceeding. Essentially I felt I'd received his blessings, which was pleasant."

Bernstein also mentioned that he was working on the problem to Dr. Edward Lasker, a well-known chess writer, who introduced him to Stanislaw Ulam of the Los Alamos group. Bernstein had the advantage that the Los Alamos group didn't have, of a machine with a large amount of memory, although the four thousand words of memory the IBM 704 had to begin with were insufficient for Bernstein's program in the end. The 704's memory was to have doubled by the time Bernstein finished his program, and he still came within two hundred words of overflowing memory.

Bernstein's program turned out to be a perceptive combination of his own chess intuitions, what he had learned from Nimzowitsch's book, and happily, some of the things he had learned from his first six months with *Modern Chess Openings*. One of the program's major features was that it eliminated a large portion of the legal possible moves from consideration, and concentrated upon those legal moves that were likely to prove fruitful.

As noted earlier, a popular misconception about a computer chess-playing program is that it can somehow consider all possible moves, and from this omniscient survey, pick the best move. As we've seen, in 1948 Claude Shannon had computed the possible moves in a chess game to be 10^{120}, which means that the heat death of the universe would terminate such a game before it could be fully played out. It had been clear to researchers from the start that some means of economizing was essential. Thus, for example, the Los Alamos group's program considered all alternatives to a depth of two moves (that is, two for white and two for black), computing a score by evaluating mobility and material. To carry out computing on their machine, a MANIAC I, within reasonable times (if an average of twelve minutes per move can be considered reasonable), bishops were eliminated as pieces, as were all special moves such as castling and two-square pawn moves at the opening. Bernstein introduced some rules of thumb to his game that would direct it toward the most immediately fruitful moves, and Samuel had done the same for his checkers program. Such heuristics were that, and no more. They weren't invariably the best strategy; sometimes they missed the unusual and the brilliant, and sometimes they didn't work at all. But they were a reasonable means of pruning away the less likely possibilities of action. The notion of economizing searches for solutions among a host of possibilities would be central to artificial intelligence, with a good case to be made for its centrality in natural intelligence as well.

So Bernstein's chess program selected what seemed to be the likeliest fruitful moves, and these it examined in considerable depth, comparing one to another among a number of dimensions. The program contained a large data base, which allowed it to examine any particular piece or square at any time. In descending order of importance, the program asked such questions as, Is the king in check? If the king is in check, there is nothing else to do. Is the king in double check? If he is, merely to capture one piece that threatens the king will be insufficient; the king must be moved. The next question had to do with material: is there any to be gained, or any in danger of capture? And clearly it is more important to rescue or capture a rook than to rescue or capture a pawn, and this was factored into the program. And so it went.

By now Bernstein had transferred to IBM proper, and he and his group were sharing their machine with the research team that was designing FORTRAN, a computer language that would allow a human to write programs in a language more natural to humans, and that would then be translated by the machine itself into machine code. Each group needed the entire machine, and when one was on, the other had to be off. Since the groups needed huge amounts of computing time, both of them worked the third shift, so an extraordinary amount of computing was going on in the small hours at the corner of East 56th and Madison Avenue in Manhattan. Bernstein had begun working full-time on his project, and he had a number of assistants, some of whom were even officially his. The two groups watched and were cheered by each other's progress.

There came a time to try out the chess machine. Bernstein had the sense to try it out first without an audience. "There was a bug. The very first move the machine ever made was to resign!" It had taken two years of work to get the chess-playing machine going, and for several years thereafter, bugs were still being discovered, not only by Bernstein and his group, but by outsiders who had requested copies of the program, and who uncovered more surprises. "It played, I think, a sort of respectable beginner's game," Bernstein says, "and every once in a while it made a move which was remarkably good."

The success of the chess machine had some unexpected results. To be sure, Bernstein received all the publicity he could have hoped for—besides the usual scientific meetings he was invited to address, he found himself written up in the *New York Times*, and an article in *Scientific American* reached a wide international audience.

Life magazine came asking for photographs of Bernstein sitting at the computer, and wondered if they could get Bobby Fischer to pose too. He was in his early teens at the time, but shrewd enough, and said he would for a fee of $2500, Bernstein recalls.

They said forget it, and asked me did I know anyone else who might be willing to pose for a picture. And I said, I'm sure that Ed Lasker was a very respected name in chess and a charming man and a gentleman, and would not ask for $2500. He said he would be delighted, and they paid us each $1. The funny thing about it is they proceeded to go to some antique dealer on Madison Avenue and rented a chess set—an Indian chess set of the sixteenth century which cost all of $2500 and a chess board which cost $1200, and everybody was absolutely dying in case any of the pieces should fall over. They were extremely delicate filigree ivory, and they were insured.

But the pictures didn't turn out and had to be retaken, meaning that the chess set and board had to be re-rented, the cost well exceeding, Bernstein calculates, what it would have cost to rent Bobby Fischer instead. As it turned out, *Life* never did use the pictures in its magazine, although six or seven years later Bernstein was surprised to discover a picture of himself and Ed Lasker standing in front of their 704 in the Time-Life series on mathematics.

But T. J. Watson, the president of IBM, was not amused. IBM's original, or at least official, justification for allowing Bernstein to use the first 704 for nothing more serious than game playing had been the hope that if he were successful, it would show the world—in particular, businesspeople—that computers could be used to solve problems even as difficult as ones that came up in business. But IBM's stockholders had challenged Watson at the last meeting, wanting an explanation for the money being wasted on playing games.

So here was Bernstein, getting what for Watson was unpleasant notoriety about his chess-playing machine, and Arthur Samuel, reaping a harvest of publicity for his checkers-playing program. Faced with these two very successful homegrown examples of artificial intelligence—soon to be joined by a third, Herbert Gelernter's geometry theorem-proving program—sales executives at IBM began to grow nervous lest the very machines they were trying to sell prove so psychologically threatening that customers would refuse to buy them. Thus they made a deliberate decision to defuse the potency of such programs by conducting a hard-sell campaign picturing the computer as nothing more than a quick moron. Countess Lovelace's dictum, that the machine can do nothing more than we tell it to do (without the qualifications she added about the necessity for experience with a real machine), was raised to a universal truth, and parroted by every sales and service person connected with the company. It came to be a popular idea, a sort of slogan of the backlash, and is offered to this day by people who feel threatened by the idea of machine intelligence. I heard it just recently in Princeton.

What was it about chess that so enchanted the early workers in artificial intelligence? Newell, Shaw, and Simon were to write in 1958: "If one could devise a successful chess machine, one would seem to have penetrated the core of human intellectual endeavor." This statement seems to summarize part of the spirit that moved the creators of chess machines. Chess is the intellectual's game par excellence. But the ironic fact is that plenty of intelligent people play chess without distinction—in this history alone, Turing, Wiener, Minsky, McCarthy and Simon—and champion players cannot necessarily be considered intellectuals. Another irony is that the two earliest and most successful game-playing programs—Alex Bernstein's chess player and Arthur Samuel's checker player—told us practically nothing about how humans perform these tasks.

Checkers and chess are only two of the games computers have been programmed to play, either with each other or with human opponents. Tic-tac-toe and *Go*, to name two polar extremes of simplicity and complexity, have also been programmed, the former easily, the latter with such difficulty that no effort has yet got beyond the novice stage. A host of other games exist in computer form, from the mundane to the exotic, from poker to *Kalah*. *Kalah* seems to have come from the Pacific or Africa, and relies on short-range tactics and precision without much long-range strategy. One artificial-intelligence researcher writes, "Computer programs at MIT and Stanford have probably been world's champions at this game for close to ten years. We cannot be sure, because skilled human players have been hard to locate" (Raphael, 1976). Indeed.

Research in game playing languished in the late 1960s and 1970s, at least in the West. The Russian chess machine may be one of a series of game-playing programs the Soviets are working on, keeping their efforts modestly under wraps until success is assured, or it may be a solitary exercise in national one-upmanship. But here, funding agencies are for the most part disenchanted with games, and sound these days like Mr. Watson's irate stockholders.

With or without official sanction, work goes on sporadically. The Greenblatt program at MIT, one of the strongest chess-playing programs of its time, was done against the advice of Richard Greenblatt's thesis advisor, Marvin Minsky, who told young Greenblatt there was no progress to be made, and cheerfully ate his words when confronted with a *fait accompli*. By 1975, computer chess was Class C, which was certainly superior to novices and beginners, but was not yet approaching championship or Class A chess.

Then, in 1976, a program designed by two young programmers at Northwestern University, David Slate and Larry Atkin, showed itself capable of beating Class B players. Its design was based not on any radical new ideas about how humans play chess, but rather on better, faster hardware and more extensive memory. Called Chess 4.5, it earned a United States Chess Federation Rating of 2070, which is in the expert category, though it earned this rating through its brilliant, Grandmaster-quality tactics, and not through its strategy and long-term planning, which were mediocre.

But here at last was a chess program to take up the challenge made by a young British international chess master, David Levy. In 1968, Levy had declared that no chess program would beat him within the next ten years, and waged 500 pounds on the proposition with Professor Donald Michie, then working on AI at Edinburgh University. Soon Edward Kozdrowickie of Western Electric, John McCarthy of Stanford, and Seymour Papert of MIT joined in the bet, raising the sum to 1250 pounds.

A match was arranged for April Fool's Day, 1977. Northwestern's Chess 4.5 had beaten players with ratings as high as 2100, and though Levy's rating was 2375, he had sometimes lost to players lower than 2100, so if Levy was favored, the outcome wasn't a foregone conclusion. Should Levy lose or tie, he would be entitled to a rematch, but if he won, the computer would not be entitled to a rematch for this session. Also, if Levy won, he could still stand to lose the bet if some computer program beat him before August 1978.

Since Donald Michie was visiting at Carnegie Mellon during the spring of 1977, it was arranged that the game would be played there. Levy would be in a quiet room with a computer terminal and a chess board. Across from him would be Slate, one of Chess 4.5's designers, who would move the chess pieces as the computer dictated its moves. The computer, a Control Data Corporation CYBER 176, was physically situated in Minneapolis, so moves would be communicated to Pittsburgh by a telephone link. Closed-circuit TV brought the match to spectators elsewhere in the building, and commentary was supplied by Hans Berliner, himself a former correspondence chess world champion.

Before a highly partisan audience, then—partisan on behalf of the machine, I suppose it should be added—the machine lost in forty-two moves made in three and a half hours. At the twenty-fifth move, the machine made a serious positional mistake, which the calm David Levy took advantage of immediately and expertly. (In blitz games, where only ten seconds is allowed between moves. Chess 4.5 has beaten both Levy and Berliner, but with respect to the wager these games didn't count.)

As interesting as the game itself was, reactions were equally interesting. The *New York Times* headlined "Chess Master Shows a Computer Who's Boss," and the reporter for the Pittsburgh *Post-Gazette* wistfully misquoted Carnegie Mellon researchers as saying the machine had never defeated a player of tournament quality. A colleague of mine in the humanities misread the misquote, and thought it interesting that a machine had never beat a human player, period.

I suppose we are still "boss" over the computer, as we nervously need to remind ourselves. In any case, Chess 4.5 went on to beat the champion Russian machine in a tournament in Toronto during the 1977 Conference of the International Federation of Information Processing Societies.

It was an improved version of the program, now called Chess 4.7, that played the final games of the wager with Levy at summer's end in 1978. The commentator for the match was George Koltanowski, who wrote a description of the triumph of man over machine in his syndicated chess column. Levy won the match 31/2 to 11/2. After beating Levy in game 4 "in masterly fashion," Koltanowski remarked, the machine then went on to experience mechanical difficulties in game 5. "Computer 4.7 was doing well when, inexplicably, it went berserk. Possibly it had no juice, or somebody pulled a plug. It was dead for over an hour and, as they were playing forty moves in two hours of play, the computer lost on time."

Koltanowski went on, "Too bad, but I am now definitely convinced that it will not take another ten years before the computer will be ready for participation in international tournaments with Grandmasters."

But again, it must be emphasized that despite their skill, these programs make no claims to mimic human behavior. Thus it was dapper Arthur Samuel, after hearing a report on Chess 4.5 and on a new special-purpose machine built by Greenblatt (who had developed the first really strong chess program)—with sixty-four processors corresponding to the sixty-four squares on a chessboard — raised a pertinent question: if all you have is a machine that bests humans by means of speed alone, what really do you have? What have you understood about human intelligence, what core of the human intellect have you penetrated? ("Well," one Berkeley professor mused, "maybe it turns out humans are fast and not deep after all!" These questions are presently unanswered.

But that line of research is now being pursued at Carnegie Mellon. Hans Berliner, who received his Ph.D. in computer science under Newell and Simon, is also working on a chess machine. Deep in the Newell-Simon tradition of cognitive psychology, Berliner claims to be more interested in simulating the cognitive behavior of a chess champion than in producing a flashy winner. Of course, if he succeeds, he'll have both. His approach is considerably more sophisticated than many prior efforts, and takes advantage of what AI workers have discovered about the role of goals in human cognition, and the quite different mental representations of chessboards experts have in their heads from those of ordinary duffers.

What seems to be holding up the world's champion chess machine—whose advent is awaited like the Messiah's in some circles, an instrument to smite down the skeptics and enemies of AI (and conversely, whose no-show would be sufficient proof to those same skeptics and enemies that the true believers are off their collective rocker)—seems to be our inability at present to know what kinds of knowledge championship chess players have that non-champions do not. There then remains the problem of how to capture such ways of knowing in a machine. Many in the field believe that as we come to understand how to model and represent knowledge in a machine, a major and highly controversial topic in artificial-intelligence research at the moment, then we'll have made another enormous leap toward understanding how machines think. The way chess knowledge is embedded in the intelligence of a championship player, whether human or machine, will surely be suggestive about how other kinds of knowledge are structured in the intelligence of agents accomplished at other tasks.

But it might not be. Some very smart people are middling and even poor chess players, as I've said, and expert chess players aren't necessarily very good at other intellectual endeavors. This fact seems to suggest that expertise rests not on powerful general principles of intelligence, but on large bodies of expert knowledge, acquired and arranged to be immediately useful in a given task. So the world's champion chess machine may not be the thing either hoped or feared.

It will be an enormous leap from the primitive days of 1956, but not just yet to the millennium, I suspect. The initial pulse of artificial-intelligence research was inspired by the methods people use to solve problems or to behave intelligently. Many of them could be seen at their clearest in games. But an impressive amount of research to identify and capture those methods in a computer program has shown that these are but one essential component of intelligent behavior. *Acquired knowledge* and its structure now seem to be equally essential, and very difficult to represent, since we presently understand so little about it. Once these problems are resolved, it wouldn't surprise me if a new component of intelligence surfaces, obvious in retrospect, which in its turn must be dealt with. The programs incorporating each of these aspects will grow cleverer and cleverer. When they might be clever enough to be, let's say, the world's chess champion is anybody's guess, and nearly everybody has had a try at guessing. The date predicted for such an auspicious event has varied from ten to a hundred years in the future, depending on the predispositions of the prophet. But the date most often predicted by people outside the field, is simply and unequivocally never. These are the persons who quote the Countess Lovelace (or variations, such as Gödel's incompleteness theorem) and certainly each other as proof that intelligent machines are an impossibility. Science isn't always as polite as a Pall Mall Club, regardless of its reputation for coolness and disinterest. Indeed, the ferocity of attack and riposte in artificial intelligence is remarkable and merits a long look.

Part Three

—

Resistance

Joyous distrust is a sign of health.
Everything absolute belongs to pathology.
— *Friedrich Nietzsche*

It seemed so simple when one was young and new ideas
were mentioned not to grow red in the face and gobble.
— *Logan Pearsall Smith*

Chapter Eight

~

Us and Them

Looked at one way, human history is a continuing series of attempts to define and exclude The Other, the alien. Our culture might have begun, let's say, with small bands of women and children foraging off the land and accepting occasional gifts from those strange Others, the men, who stayed at a cautious distance for the most part. Something, sure enough, has caused men to regard women as The Other—mysterious, threatening, not quite a part of the human species. Then again, we might have been in the charge of one old clan chief, a wise matriarch, let's say, who led us, small brave family that we were, in the endless search for the food we shared within the clan, but not outside it. When some of us settled down to farm, it was The Other who wanted to take away from us the literal fruit of our labors. Against such strangers we must have defended ourselves in savage ways.

We're pleased that these days the distinctions are less rigid and often breaking down or disappearing altogether, and so we should be, since it's now in our general interests that they do. The music I hear on my taxi driver's transistor radio as he drives me from the airport into Warsaw is the same music I might hear in a San Francisco taxi. Clothes, by which you could once tell not only a person's age and national origin but gender too, are no longer a good clue. And we know that the music and the clothing are signs of a profounder convergence. Slowly, painfully, with all sorts of holdouts and reservations, we are getting together, and even have to be reminded sometimes of our differences.

Who, or what, then, will be left to be The Other?

By its very nature, The Other has a different face from ours. The more distant it is, the harder it is to see, and thus the easier to be suspicious of, and feel enmity toward instead of affinity. This may account for why the United States at midcentury found itself making war on enemies living in countries that most U.S. citizens had never heard of. We knew everyone else too well. This seems to say that The Other can only be an enemy, which has most often been the case. But it need not be. We approach the alien with deeply mixed feelings, part terror and part exhilaration: how much of ourselves will we find there after all?

Consider Joseph Conrad's Marlow, traveling down the river into the heart of darkness, in the novel by that name. There he comes face to face with the native Africans in their natural surroundings, and he is shocked by what he finds. "We are accustomed to look upon the shackled form of a conquered monster, but there—you could look at a thing monstrous and free. It was unearthly, and the men were—no, they were not inhuman. Well, you know, that was the worst of it—this suspicion of their not being inhuman . . . what thrilled you was just the thought of their humanity—like yours—the thought of your remote kinship with this wild and passionate uproar."

When we come face to face with the idea of thinking machines we have much the same reaction. What thrills us—in the deepest sense—is the thought of our remote kinship with these contrivances. We are speaking implicitly of power: ours over them (the natural course of events, we believe) and theirs over us (unnatural and monstrous). This potent fear of The Other lives in strange places. Here's a scientist from NASA Ames Research Center talking about building cities in outer space. The effect, he said, is that the first species to reach intelligence and master technology would flash across the galaxy in a few hundreds of millions of years, snuffing out most other intelligent life forms before they could evolve. We could be the first, he added hopefully. We can call such an imperialistic view shabby and outdated, but it speaks of very real fears. And I believe that this potent fear of The Other is one of the things that informs the violent reaction we have to the idea of thinking machines.

Still another fear might be operating here. Earlier I pointed out the contrast between Hebraic and Hellenic attitudes toward imitation human beings, the Hellenic being open and curious, the Hebraic being specifically opposed. That proscription against graven images, which appears by my rough count four times in the first three books of the Old Testament—what does it really mean?

The scholarly commentaries suggest that God knows we humans have a terrible time distinguishing between the real thing and a representation of it, a symbol that quickly takes on the *mana* of what it is only supposed to represent. Nobody will argue with that— whole wars erupt over what somebody considers an offensive symbolic act. This confusion between a deity and its mere representation, the commentaries argue, narrows the mystery in a sense; it makes a *Reader's Digest* Condensed Version of the awesome.

But the confusion does something else as well. The history of doll making or image manufacture, which is, so far as I can tell, nearly universal among humans, is suggestive (even the Israelites somehow squared the Second Commandment with little human-like images they made for magical purposes). Whether dolls are used to assure a good harvest, a plague on an enemy, or the fidelity of a spouse, they all have one thing in common: they are all magic. Put into other words, they partake of the power of the gods.

How much of that proscription against graven images, then, is a fight over territory, an effort to keep separate the sacred and the profane? Mixing both makes us nervous at a deep level; keep that in mind as we talk about thinking machines.

The Second Commandment could also be interpreted as a remedy against our insufferable human chauvinism. Faced with the ultimate in power and knowledge and grandeur, we're incapable of imagining it in any but anthropomorphic terms. If something isn't precisely like us, then how can it be intelligent? (Maybe we ought to snuff it out to ensure our hegemony in the universe, as the scientist from NASA Ames suggests.) Defining intelligence and determining who might have it seem to be problems psychologists and semanticists ought to work out, but the notion that human beings might be just one instance of a larger class of intelligent beings is so provoking that everyone wants to get into the act with an opinion.

And why not? These issues lie at the heart of our identity as *homo sapiens*. To agree that a machine can be intelligent is to open the door to one more Other and share our identity a bit further. We make this opening in terror and exhilaration—how much must we give up of ourselves; what will we gain? With this gesture we cede some of the power or position we suppose we have; we look at a thing that used to be shackled and see it now as free. Faced with an uppity machine, we've always known we could pull the plug as a last resort, but if we accept the idea of an intelligent machine, we're going to be stuck with a moral dilemma in pulling that plug, one we've hardly worked out intraspecies.

An intelligent machine is a psychological and moral puzzle for us. If, as some claim, we've been creating intelligent machines by biological means since the race began, then a computing machine is something different, an artifact of our own hands, whose building blocks are different from ours. More important, such a machine is *an artifact whose value is diminished because we have made it.* Contrary creatures that we are, we house alongside our human chauvinism a fundamental suspicion that we may be inferior, which shows itself in our contempt of the artificial—the manmade—as compared with what we're pleased to call natural. At the moment, the human race is in transit from the time when we were utterly dependent upon the natural world and showed that utter dependence by worshipping it, and through a time when we dreamed imperiously that we could master nature, and got our rightful comeuppance for believing so. We are still deeply unsure of ourselves vis-à-vis nature; a part of it and yet apart, we have still to find out just where we can best stand. Our experience in the twentieth century with partaking in the power of the gods has been mixed, to say the least. Dare we complicate things with this even more presumptuous artifact, the thinking machine? Won't our comeuppance be the worse in proportion to our presumptuousness? To put it bluntly, won't the machine take over?

Related to our mistrust of the human-fashioned artifact is a superstition about the human-fashioned explanation. This superstition holds that to explain something, to remove the mystery from it, is to explain it away. Thus explanations of human thought will somehow explain away—and thereby demean—the wonders of the human intellect. Time and again, workers in artificial intelligence have felt compelled to show this superstition for what it is. "Reverence and mystery shouldn't have anything to do with one another," says Bruce Buchanan of the Stanford artificial-intelligence group and a key member of the DENDRAL project, "but they always have. I admire things I understand." Understanding how and why humans think the way they do, he goes on to say, doesn't mean that those thought processes aren't terribly important and shouldn't be treated with sensitivity. Herbert Simon, in his *Sciences of the Artificial* (1969) quotes the early Dutch physicist Simon Stevin, who took as his motto *Wonder, en es gheen wonder*, or, "Wonderful, but not incomprehensible." And this is Simon's enterprise as he sees it, to explain the wonderful and let it raise new wonder at how complexity is woven out of simplicity.

The arguments against machine intelligence can be stated simply. Arguments as to why machines, in particular digital computers, cannot be said to think sort themselves into four categories:

1. Arguments of emotion.
2. Arguments of insuperable differences.
3. Arguments of no existing examples.
4. Arguments of ethical considerations.

The first category, arguments of emotion, is based on the premise that intelligence is an exclusive human property; for reasons of divine origin or biological accident, human beings are the only creatures on the planet who have or ever will have genuine intelligence. A variation of this argument says that some organisms have a rudimentary intelligence, but still limits intelligence to organisms alone, and rather complex ones at that.

Under the second category, arguments of insuperable differences, the reasoning says that machines can't be said to think because intelligence requires creativity and originality, and no machine has been or can be creative and original. A variation of this is that intelligence requires a special kind of experience, only acquired through interaction in the real world with others of like mind. Intelligence, this argument runs, requires autonomy, and no machine can ever be autonomous. Another argument along this line says that the only so-called intelligent tasks machines can accomplish are one of a kind; that is, even if a machine can play a decent chess game, it won't be able to transfer that savvy to, say, translating an ordinary news story or composing a melody. Intelligence means coping with a variety of tasks successfully and with originality. Even if a machine could do these things, it would not be conscious of having done so, and consciousness is a significant part of intelligent behavior. Finally, mathematicians in the twentieth century (in particular Gödel and Turing) have shown that some problems are for all time uncomputable, or undecidable. Gödel has shown that in any sufficiently powerful logical system, statements can be formulated which can neither be proved nor disproved within the system, unless the system is logically inconsistent. Thus, discrete machines that must behave within a logical system have an insurmountable disability. Human intelligence seems to be so different in method and flavor from the ways in which we are able to program discrete machines that we will never come close to capturing anything like it, with its swoops and curves and intuitions and simultaneity of conscious and unconscious.

In the third category, arguments of no existing examples, the statement is simple. Even if computers are in principle capable of intelligent behavior, nobody has made them behave that way yet. Whether anyone ever will remains to be seen.

Arguments of ethical considerations run something like this: Given that computers might be capable of intelligent behavior, ought we to pursue the possibility? Can we foresee the outcome of such an awesome step? And if we cannot foresee it, shouldn't we take a lesson from some of the other examples in science— good ones and bad—and approach the whole topic with much more caution than we have so far? Shouldn't there be public debate over whether we want other intelligences around, since we now have a choice? Shouldn't we give some thought to the kind of limits we might put on such machines?

Personally, I find the first category, arguments of emotion, quite difficult to deal with. I can't see how those arguments can be answered beyond appeals to personal inclination. The other three categories, however, raise interesting and arguable questions.

It would be nice to report that such questions have been gravely and profoundly debated with all the human intelligence and rationality that could be brought to bear upon them, but they haven't. Of course, the criticism of science is seldom the disinterested affair that a storybook view of science suggests. Ian Mitroff's fascinating study of moon geologists, called *The Subjective Side of Science* (1974), documents the intense partisanship permeating all science criticism, a partisanship that is in fact absolutely necessary for science to be done. If a scientist doesn't believe in his or her own work, even in the face of some counterevidence, science would never move beyond the status quo. What moves it, what opens the universe for us, is a dogged pursuit by committed individuals of a hunch, a theory, a feeling. If this hunch runs against the prevailing beliefs, the theoretician can expect skepticism, even such penalties as being removed from committees or denied funds or a job. If lucky, our theoretician can at least expect attack. If unlucky, he or she will be ignored altogether. It takes an astonishing strength of personality or an unusual disregard for social approbation to be original in science, or anywhere else. More often than not, an astonishing strength of personality is a pain to put up with. As a consequence, scientific arguments transform themselves into feuds that are painful, comical, unedifying, but scientifically energizing.

Given that the idea of artificial intelligence touches a deeply rooted human nerve, and given that it is a field in its infancy that has a large proportion of speculative theoreticians, it shouldn't be surprising that the history of its criticism borders on melodrama.

As I've said, computers and thinking have been associated from the start, for much of the impetus to develop them has come from the desire to devise some means of removing the drudgery from thinking. Charles Babbage was inspired to his Analytical Engine by the tedium of calculating logarithmic tables. A century later, Konrad Zuse worked in his parents' Berlin apartment on the first general-purpose computer with a clear feeling that he was making a machine that would somehow think. At the same time, ignorant of Zuse, Alan Turing was playing with the idea of a machine that would think, alarming his colleagues with the notion that robots would scamper all over the British countryside, making, in that all-purpose Anglicism, a nuisance of themselves. As we have seen, when Turing published his charming essay on machine intelligence ten years later, merchants of computers were deliberately deciding to deny that their products were, as the Sunday supplements were calling them, "giant brains," calling them instead giant morons, capable only of the most repetitious and ordinary tasks, though splendidly suited for those.

J. C. Shaw, of the RAND-Carnegie team, amused himself for a time by collecting declarations made against thinking machines. They range from the thoughtful to the dogmatic. The thoughtful comments want to examine what the real difficulties and differences are; the dogma tends to anthropomorphize—"moron" and "idiot" appear regularly, and one Stafford Beer, a British cyberneticist, even declares that moron is an overstatement to describe a computer. A common theme in these critiques seems to be, "Cheer up, human reader, you are just as unique and special as you ever were; no threat, no sweat." After I read a number of these, our species began to seem a bit like Snow White's wicked stepmother, pathetically anxious to be assured by our magic mirror that we are still the smartest in the land.

Interestingly, most of these critical statements—the thoughtful and the dogmatic ones—assume without question that we know and agree upon the nature of intelligence. The only question to answer is whether machines have it, and the dogmatists declare they do not.

Some twenty years since Shaw made his collection, we are less sure we know just what intelligence is, but few raised that question then. One exception was a statement by L. A. Hiller and L. M. Isaacson in their introduction to *Experimental Music*, published in 1959. Since, they write, the real nature of adaptability, learning, and purposeful behavior will only be decided in the future, we may conclude that for the present computers are still definitely limited to specific tasks in which one type or another of data processing are of a fairly routine nature.

> The question of whether computers will ever be "creative" in the sense that we speak of creative composing is rather similar to the problem of whether they "think." Also we might ask: "What is meant by the term creative?" Being "creative" would seem to depend at the very minimum, like "thinking," on having a computer operate on a self-sustaining basis, and to "learn from experience." Moreover, it seems that what we first consider strokes of insight and manifestations of "creative thought" are, once they are analyzed and codified and particularly, codified to the extent that they can be processed by a computer, no longer "creative processes" in the usual sense.
>
> (Hiller and Isaacson, 1959)

But even Hiller and Isaacson do not go on to ask why a formalization of the creative process robs it of its essence, or, to put it another way, why the creative process—or thinking, or adaptability—is forever beyond our understanding.

Marvin Minsky once talked about this resistance to believing we could ever penetrate the essence of human thinking and creativity:

The interesting thing is, according to my youthful stereotype, it would be musicians and artists who would be hostile to the idea of intelligent machines. But I never got over it, that musicians and artists weren't hostile at all. They'd say, well, oh gee, that's great, what's the idea? How would it do that? And it would be mathematicians and physicists who would get very angry [and still do, he might have added]. It took me a long time to understand that professional artists are not superstitious about creativity; they're very concerned with it, and most of the artists I've talked to don't think much of the theory of talent, and they admit they learn things by looking at other people's work and thinking about it and asking them how you do things, and so forth. Mathematicians don't. Mathematicians never talk about how they think about mathematics, and they worship their creativity as a God-given gift. They're hypocritical about teaching students because on the whole they believe that you can't teach students to be mathematicians—some of them have it or they don't; these classes are a sham—but eventually the real mathematicians come out. And besides they're mathematicians when they're thirteen already, and there's no hope in—

He waved his hand in mock despair.

There is superstition about creativity, and for that matter, about thinking in every sense, and it's part of the history of the field of artificial intelligence that every time somebody figured out how to make a computer do something—play good checkers, solve simple but relatively informal problems—there was a chorus of critics to say, but that's not thinking.

In Cliff Shaw's collection is a rather long quotation from Mortimer Taube, which appeared early in 1959. Even at this distance, the reader must resist the impulse to duck spray from the snorts of derision. After a thorough scolding, Taube declares, "One may wonder why reputable scientific journals publish material of this sort and why it should have an audience beyond the readers of the Sunday supplements."

Taube did not intend to be rational so much as persuasive. Two years later he was to publish a book (which included the statement above) called *Computers and Common Sense* (1961), whose subtitle was *The Myth of Thinking Machines*. Mortimer Taube was disturbed and angry about artificial intelligence.

Generally, Taube's criticisms seem to fall into arguments of insuperable differences between humans and the digital computer, and arguments of no existing examples of machine intelligence. Underlying these is a distinct assumption that intelligence is indeed a special human property, which of course serves as an explanation for why there are insuperable differences and no existing examples.

Taube examined a series of applications computers were being put to in the late 1950s with the hope of raising not only the scientific but the philosophical issues involved. He was curiously selective—ignoring some efforts and citing others that weren't necessarily the best illustration of his points. I take those points to run something like this: Certain things in science are possible and others are not, given certain natural laws. For example, building a space platform is difficult but possible, whereas building a perpetual-motion machine is impossible, given the second law of thermodynamics. Cognition by machine is of the last sort of impossibility, because machine thinking means mechanizing thought processes, which is the same thing as formalizing them, and Gödel's result showed that this is impossible. (Taube does not address the argument that whatever obstacles to solving problems and finding consistency in an axiomatic system Gödel's theorem raises, they are the same for human mathematicians as for machines.)

Taube attacked the McCulloch-Pitts neural model, again citing Gödel, and then went full-tilt after machine translation. Machine translation from one human language to another, which had got underway not long after World War II, had certainly been launched with enormous optimism, more than was appropriate, given the primitive state of linguistics. Thesaruses were compiled, word-for-word translation was attempted; on the whole the effort ended in failure. The *coup de grace* was a report prepared by a committee headed by John Pierce, then at Bell Laboratories, which concluded that a genuine machine translator was simply too ambitious at the moment, given the state of what was known about linguistics and about the use of natural language. Some of these reasons were correctly anticipated by Taube, but he also seemed to believe that there was something special, ineffable, and exclusively human about translation. It's an intuitive thing, he declared, and no machine—no computer—will ever be capable of intuitive behavior. Intuition may indeed be a present mystery, but Taube gives no reason why it will always be so.

Mechanical translation, Taube concluded in triumph, was too expensive even if it could be done. He quotes Donald Booth as estimating that machine translation might cost twice as much as human translations. (But then human translators might just be underpaid. I know one who quit the field because she couldn't live on what she was making.) And as it turned out, by the mid-1970s, machine translation had risen Lazaruslike and was flourishing, especially in Europe, as we will see in Chapter 11.

Another of Taube's targets was learning machines, computer programs capable of learning. Human beings, he conceded, exhibit many kinds of learning, from the simplest sort of rote memorization to the most complex sorts of self-guided invention, and we're not very good even now at doing much more than recognizing that something is taking place, though what we can't quite say. "Machine learning, if it occurs," Taube wrote, "would of necessity be a formal process in the sense of formal mathematics. This, as has been noted, is also true of mechanical translation, and in this case, as in the discussion of mechanical translation, it is necessary to ask whether there is any reason to assume that human learning is merely a formal process" (Taube, 1961).

It's equally necessary for the curious reader to ask just what Taube means by "formal," which by no means need be mathematical, and why he sees hopeless problems in encoding human learning in a machine. To be sure, machine languages were relatively primitive —Newell, Shaw, and Simon weren't even getting alphabetic symbols from Johnniac, and had to decode numbers instead—which might have limited the vision and enthusiasm of even the wooliest prophets. But Taube's insistence on limiting what computers could do to a highly restricted sense of mathematical formalism meant he had to ignore those programs which did seem to exhibit learning, in particular, Samuel's checkers program. In fact he did not literally ignore them. Noting their existence, he complained that they were busy at such trivial tasks as games: why weren't they doing something socially useful? He clearly wanted to have it both ways. And while Mortimer Taube was wondering why money was being spent to design learning machines that couldn't really learn (without defining what would satisfy him as real learning) those scientists who read the book wondered what was really bothering Taube.

The concluding paragraph of Taube's book is a wistful piece of fiction about how science is done: "All noble things, says Spinoza, are as difficult as they are rare. So it is with respect to science and scientific advance in the computer and data processing field. Genuine advances are difficult and rare. And even more rarely are they the product of prophecy and the premature announcement of what someone expects to but has not done."

But on the contrary. The visionaries of science are what drive it, whether they incite the experimentalists to prove them wrong or inspire other experimentalists to prove them right. This process is clearer when Taube's view of the scientific model is examined. He wants models that rest on knowledge already acquired, models based solidly on firm, verifiable facts. But this is a naive view of the scientific model. Scientific visionaries build their models on speculation sometimes very far afield from what meager facts exist, using those very models as instruments of discovery. Can *this* be the case? the theoretician asks, and the experimentalist says, Well, yes, or, Well, no. I oversimplify, for most working scientists have a proportion of both theoretician and experimentalist inside their heads, with one perhaps exceeding the other. A good illustration is James Watson's vivid description of how he played obsessively with the configuration of the DNA model (in this case in wire and wood) until suddenly—and at last—all the pieces fit. By fit we mean that the model accounted for all the data then available on DNA, both from common knowledge of chemistry and more specialized knowledge of structure acquired by X-ray crystallography. The model furthermore was a key to explaining what wasn't known, and a pointer toward where to find more.

The other implication of Taube's words is that science gets done without criticism, which is simply untrue. He acknowledges that some sort of internal controls have existed in the form of peer review, but this sort of control of publication and academic preferment can be evaded by "impatient and ambitious young scientists" who can somehow get government grants and contracts on the strength of their promises, and then publish their slapdash results in "reports and symposia which have not had the benefit of critical review." God save us all from the misplaced envy of the young; Taube was right to suspect it might make him sound sour and uncreative.

Policies differ from agency to agency, but generally speaking, government research grants are reviewed by outside experts. Agencies such as the National Science Foundation screen proposals more carefully perhaps than most journals referee articles, though the process is pretty much the same, with anonymous authorities rendering a series of judgments on each proposal. This anonymous peer-review system has some big drawbacks. Critics have charged that the Establishment in any field, where the experts are drawn from by the granting agencies, tends to favor proposals of known scientists over unknown, the top universities and research laboratories over the lesser ones, with a consequent concentration of resources where they're least needed, thus leading to intellectual stagnation.

The classic case cited is cancer research, where for many years experts who were brought up in a tradition of *seeking a cure* approved awards only to others who were also looking for a cure, while alternative approaches, most notably *prevention*, went almost unfunded. In other words, the major criticism of the peer-review system is that it encourages undue conservatism.

Proponents of the peer-review system concede its faults, but argue that a conservative bias is probably no bad thing considering the amounts of money involved, much of it public money. There certainly are dismal cases of entrenched scientists supporting their own students and disciples to the exclusion of everyone else—this seems to have been true in linguistics in the 1960s—and I've mentioned these kinds of criticisms directed at present-day artificial intelligence, but at the time Taube was writing, the picture he painted of wild-eyed kids on the make while their elders stood back in helpless horror was at variance with the facts.

Taube died not long after his book was published, so we will probably never know the reasons for his choice of targets, or even why he felt impelled to attack. *Computers and Common Sense* contained some ideas it might have been fruitful for a young field to heed. For example, I think Taube was arguing in a somewhat roundabout way for semantic as well as syntactic analysis in any attempts at machine translation, and in this he was quite right. But this idea, like others he had, is buried in such polemical, offensive language—"scientific aberration" is a favorite phrase, along with "dogmatically ignorant," and "jejune"—that any audience he might have reached could only be the already converted.

Taube and his book might be dismissed as a curiosity except for this: he seems to have been the first of a series of critics of the field *whose emotions were as deeply engaged as their intellect.* Fired by his anger, he pretended that science criticism had never existed before he undertook it, and that scientists go on their merry way without accountability to anyone. He seems to have believed that science had fallen from its pure state of "individual activity of dedicated men" to no more than big business, "with its organization men, its dislike of controversy and controversial characters, its share of puffery, and its concern with plans and budgets rather than with scientific contribution"—a harsh statement from someone who was himself an employee of a private firm, not an academic. And we must ask what moment Taube had in mind when he pictured science in such a state of grace as to be done not by smart, ambitious scientists, but by self-effacing, "dedicated men."

In this history alone, we have already seen Charles Babbage in 1830 at pains to convince the British government to support what no private fortune could underwrite. The scientist as leisured gentleman, or even self-sufficient academic, is a pretty but not altogether accurate picture of science as it was done in the eighteenth century. By the nineteenth, science was already getting too expensive and complicated for individuals to manage alone; the growth of the scientific society is partly explained by the economic, not to mention the intellectual, realities of doing science, which made it more and more difficult to accomplish anything without collective action. World War II, which began some twenty years before Taube published his book, should have put to rest once and for all the myth of the scientist as naive, apolitical loner.

It's ironic that one scientist whom Taube attacked with great gusto in his book, a Harvard professor named Anthony Oettinger, would himself write a book criticizing computer-aided instruction (an effort to accomplish some of the more pedestrian tasks of teaching with a computer terminal substituting for a human teacher). In a further disavowal of his youthful folly, and disillusioned by his participation in the failure of machine translation in the 1950s, Oettinger contributed the introduction to a book that was to be an even more detailed and sustained attack on artificial intelligence, a book called *What Computers Can't Do*, by Hubert Dreyfus.

And there's a story.

Chapter Nine

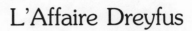

L'Affaire Dreyfus

Hubert Dreyfus, a vigorous and excitable redhead who has the energy to hike from the rim to the floor of the Grand Canyon and back again in one day, is a professor of philosophy presently teaching at the University of California at Berkeley, though he first became interested in artificial intelligence when he was an instructor at MIT. His initial public statement on the topic was a discussion note that he and his brother Stuart, who then was at MIT and now also teaches at Berkeley, inserted into a proceedings volume of a series of lectures held at MIT in the spring of 1961.[1] Thereafter, Hubert Dreyfus did a study of artificial intelligence as a consultant at the RAND Corporation which resulted in a long paper. As its title "Alchemy and Artificial Intelligence" implies, this paper was a free-swinging attack on the field, comparing it with that deluded pseudoscience of the Middle Ages.

---

[1] These remarks, which were not made at the time of the session itself (a talk by J. R. Pierce on "What Computers Should be Doing," and a panel commentary by Claude Shannon and Walter Rosenblith), contained some images that were to endure in Hubert Dreyfus's writings about artificial intelligence: the exaggerated claims and the feeble accomplishments of the field; the comparison between AI and alchemy; and the absurdity of believing you were getting nearer the moon by climbing a mountain (later to be downgraded to a tree). The language is already intemperate—an interesting contrast to the others on the panel, who don't necessarily agree with each other but who don't deride one another's intelligence or motives (Greenberger, 1962). Stuart Dreyfus's initial public statement on AI was a generally favorable review of Mortimer Taube's *Computers and Common Sense*.

Lotfi Zadeh

Published in 1965 in mimeograph form and in a printed version in 1967, the paper was the kernel of a book Dreyfus published in 1972, called *What Computers Can't Do; A Critique of Artificial Reason.*

It is the point of Dreyfus's book that human and artificial intelligence are in fact quite different—in particular, that human intelligence is unique. Not only that, but a great misunderstanding accounts for public confusion about thinking machines, a misunderstanding perpetrated by the unrealistic claims researchers in AI have been making, claims that thinking machines are already here, or at any rate, just around the corner.

Dreyfus is persuaded that in the end artificial intelligence will never work. He sees it as the technological realization of two thousand years of philosophic tradition which has had as its goal the discovery of rules—rules for moral behavior, rules of intellectual behavior, and rules for practical behavior. By assuming that intelligent behavior can be described by rules, Dreyfus believes, workers in the artificial-intelligence field share in a long but by now obsolete tradition. He proposes an alternative view, which is twentieth-century phenomenology as expounded by Heidegger, Merleau-Ponty, and others. He understands the difficulties his readers might have in accepting phenomenology as a way of describing human behavior. What counts as a theory, after all, is presently determined by the very tradition that Dreyfus is rejecting. Nevertheless, the reductionistic models that have worked for the physical world have proved intractable when applied to human behavior, whereas phenomenology seeks a general understanding of perception, language use, and other faculties.

Dreyfus begins by noting what he sees as a stagnation in AI research. That is, after an initial success with some limited problems, AI has not really moved beyond those first successes. This is unsurprising—indeed, inevitable—to Dreyfus, because AI is based upon obsolete biological, psychological, epistemological, and ontological assumptions.

The biological assumption is that the on-off switches in a computer are comparable to the on-off action of individual neurons, and by extension, that human behavior is the result of a computer program.

The psychological assumption, that information processing can somehow describe human cognitive behavior at a certain (though by no means the deepest) level of detail, seems to Dreyfus wrong because it is based on a misunderstanding of Shannon's original ideas of information theory. It is a hybrid notion that takes Shannon's idea of content-free information and forces content into it. Here, too, he questions the generality of the results achieved, particularly by Newell, Shaw, and Simon's General Problem Solver, and also whether those results account for a range of phenomena.

The epistemological assumption of artificial intelligence represents a confusion between the rule one is following to do something and the rule that can be used to describe someone doing something. As his example here Dreyfus uses the movement of the planets, which can be described in their orbits by the formalism of differential equations, but this is merely a model of planetary movement.

Such modeling is mistakenly borrowed from physics. The human world cannot profitably be viewed that way, and the way computers perceive the information they are given—in a piece-by-piece fashion—is simply not the holistic way in which humans perceive the world.

The last point relates to other arguments Dreyfus makes about the inability of a computer to match certain essential human functions: the "fringe consciousness" humans have which allows them to zero in on the important aspects of, say, a chess game, without losing total awareness of other possible moves. In this a human being is different from a computer, which must use a sort of counting-out procedure, and thereby sometimes misses the unusual or the unexpected. A computer, moreover, has no tolerance for ambiguity, Dreyfus writes, which is one of the reasons why machine translation failed. Humans can usually figure out the meaning of ambiguous statements merely by their context. Finally, the computer must do a sort of trial-and-error search instead of immediately sorting out the essential from the inessential, as humans do.

In his alternative model, a phenomenological view, Dreyfus stresses the role of the human body in organizing and unifying our experience of objects, the role of the situation in providing a background against which behavior can be orderly without being rulelike, and finally the role of human purposes and needs in organizing the situation so that objects are recognized as relevant and accessible. For example, phenomenologists posit an outer horizon in perception which remains indeterminate yet allows the zeroing in on the inner horizon, *not* as indeterminate as the outer horizon. Thus, in Dreyfus's example, when we humans perceive a house, the act of perception slides back and forth between the outer and inner horizon, perceiving the whole and then the parts, and then a renewed, richer perception of the whole house. A machine has no inner horizon and must process from the details to the whole, the opposite of humans.

And what we expect to find is a large component of what we will find. In a sense, Dreyfus is his own best argument for that. As we shall see, a great deal of what he found in artificial intelligence is what he expected to find.

Dreyfus argues that artificial intelligence cannot deal with the interdependence we see between the parts and the whole in perception. Some workers hardly even recognize the problem, and those who do are misled by the constraints of the computer. The lack of success in robotics is inevitable, for the human experience in the human body cannot be duplicated by a digital computer—the way humans use motor skills takes place in the domain of the phenomenal.

As for orderly human behavior that is not rulelike, Dreyfus quotes Wittgenstein and Heidegger, who see human behavior to be in some sense like regulated traffic. Though stop lights and pedestrian walks regulate traffic, they do not guide the totality of traffic movement by prescription. The traffic movement has a contextual regularity; it is never completely rule-governed, but is as orderly as necessary. If I understand Dreyfus correctly, when he comes to talk about *the situation* as a function of human needs, he is saying that life is what we humans make it, and nothing else. Since a computer is not in a situation—that is, it has no purpose and no needs except those programmed into it—it must treat all facts as possibly relevant at all times. Computers, unlike human beings, have no means to discriminate between the important and the unimportant, while simultaneously maintaining generality.

Humans manage this, Dreyfus states, because they are somehow at home in the world, a world prestructured in terms of human purpose and concerns. He cites Wittgenstein's suggestion that the analysis of a situation into facts and rules (which is where the traditional philosopher and the computer expert think they must begin) is itself only meaningful in some context and for some purpose.

Dreyfus divides intelligent behavior into four domains, which are progressively more complex, and which are different—that is, discontinuous—in their requirements of the agent that will perform them. Finally Dreyfus asserts that because machines cannot penetrate into the most complex domains of human intelligent behavior, the limit to what AI can achieve is near.

While it is my intention here to report rather than refute, readers interested in what AI has to say about some of these topics should refer to other sources. For example, the robotics literature abounds in material on perception (see Winston, 1975, for an introduction), and even Simon has addressed the problem in several papers (1967, 1969, 1971, 1972). Dreyfus's attack on the biological assumption came long after the on-off correspondence between neurons and computer switches was abandoned as a model. The plausibility of his interpretation of the information-processing model is, like his interpretation of the epistemological assumptions of AI, a matter of judgment. The counting-out procedure he ascribes to computers seems to me mistaken, and the problem of dealing with ambiguities is presently yielding to attack in both language and vision efforts, as Chapter 11 of this book suggests. (This problem is still a troublesome thing for humans, for that matter.) Newell's work on production systems (1973) is an introduction to some AI answers to the problem of systems that have contextual regularity but are not completely rule-governed.

Unsurprisingly, Dreyfus's work was not greeted with cheers by AI workers. They objected that he was basically ignorant of what computers could already do, that his ignorance led him to misunderstand anything but the simplest empirical evidence (for example, can a computer program play chess or can't it?), that he could not grasp current work or its implications and assumed that if a program wasn't already doing something, then it never would. Some who have read him argue that he is simply a man trying to promote his own particular brand of philosophy and using artificial intelligence as a scapegoat.

Seymour Papert undertook a point-by-point refutation of Dreyfus's RAND paper. In an unfinished memo called "The Artificial Intelligence of Hubert L. Dreyfus; A Budget of Fallacies," published by the MIT group where Papert works, he stated some of the reasons why he felt obliged to respond.

> I have been told [writes Papert] that it is irrelevant to refute defamatory charges against Simon since other people really have made false claims about achievements in Artificial Intelligence. I have been told it is a waste of time to show that there is nothing but muddle in Dreyfus's explanation of why machines can "play checkers" but cannot "play chess" since attempts to make a computer translate Russian really have encountered difficulties. I have been told that only a pedant would object to the technical nonsense that pervades every paragraph of Dreyfus's papers about Artificial Intelligence since his real purpose is to provide insight into the rich subtlety of human intelligence. I have been told that his arguments must be read as literary conceits with deep "humanistic" content.
>
> I think it does matter. I sympathise with "humanists" who fear that technical developments threaten our social structure, our traditional image of ourselves and our cultural values. But there is a vastly greater danger in abandoning the tradition of intellectually responsible and informed inquiry in the futile hope of an easy resolution of these conflicts. The steady encroachment of the computer must be *faced*. It is cowardice to respond by filling "humanities" departments with "phenomenologists" who assure us that the computer is barred by its finite number of states from encroaching further into the areas of activity they regard as "uniquely human."
>
> (Papert, 1968)

We are about to go on an excursion into the sociology of science, and examine not only whether the substance of each side's position is arguable, but also perhaps what has made them take such a stand. Papert's language is hot; so was the language of Dreyfus in the document Papert was responding to. Dreyfus riposted by including Papert among the gratefully acknowledged in the introduction to his book, leaving an innocent reader to believe Papert had helped in the usual sense.[2] But none of this can be understood without looking at the circumstances of why Hubert Dreyfus has so persistently and passionately attacked artificial intelligence, and why the artificial intelligence community has responded with less than sweet reason at all times.

It is certainly the case that the claims have been optimistic in artificial intelligence. Surely the most famous, or infamous, depending on your point of view, were made by Simon and Newell at a meeting in 1957. (Simon gave the talk but the predictions were the work of both men. Relying on the precedent of Genesis 27:22, a footnote to the printed paper tells us, "The voice is Jacob's voice, but the hands are the hands of Esau.")

In ten years, they predicted, a digital computer would be the world's chess champion, would compose music that had considerable aesthetic value, and would discover and prove an important mathematical theorem; finally, most theories in psychology would take the form of computer programs or of qualitative statements about computer programs.

What had happened? It was to be twenty years before chess programs climbed into "A" class chess, the category below masters level, which is impressive but not world's champion. Though theorem proving has had some bright success, and further work goes on, none so far can be considered important in the sense Simon and Newell seem to have intended.[3] On the other hand, information-processing models have permeated cognitive psychology, bringing with them an award to Simon from the American Psychological Association for Distinguished Scientific Contribution in 1969.

---

[2] More abundant thanks went to the employee at MIT Press who was smuggling to Dreyfus galleys of a new book by Marvin Minsky so that Dreyfus could frame his objections based on the latest material.

[3] Dreyfus has some fun with W. Ross Ashby's claim that the geometry theorem-proving machine designed by Herbert Gelernter had discovered a fascinating new proof of the *pons asinorum* (that the base angles of an isoceles triangle are equal). But Gelernter told me he was as surprised by Ashby's slip as anyone. In the course of preparing the program, Gelernter had

People have used our 1957 predictions as ammunition against the whole idea of artificial intelligence [Simon said to me on one occasion]. They happen to be pretty good predictions—it takes a fair amount of explaining why, but they are. If they'd come out just that way, we wouldn't need to explain anything. Well, let me give the setting of it first. I was going to talk at an Operations Research Society of America conference, which was in Pittsburgh here, and I wanted to talk about our work, but I wanted to talk about it in a way that was relevant to operations researchers. And I thought a relevant way to talk about it was to try to do some assessment—I say I on this, but almost all of these things are we, Newell and I. I actually gave the talk, but Al and I, of course, worked this thing out together. My intent was to give an assessment of the implications. And so the way of doing this was to try to be concrete—try to give some for instances of the kinds of things you could expect to happen, because you can talk about this in the abstract until the cows come home and it's very hard for people. I was trying to make the implications as concrete as possible. So I took four things that seemed to me to be plausible extrapolations of what was going on then. I'd done some social prediction before that and since. I guess I should add that to the premises; this was not a new belief in making prophecies. I'm quite interested in the problem of how you make social predictions and of the importance of trying to make them under certain circumstances. I'd been engaged in a major effort of this sort earlier when I worked with the Cowles Commission in doing a report on peacetime uses of atomic energy back in 1950. So I thought it was important to do this for this field, and we predicted then, because chess was already underway, a chess-playing program that would be world's champion in ten years, and a musical composition with serious aesthetic content in ten years. The reason for that one was that Hiller and Isaacson had already produced the Illiac Suite, which was not trivial and uninteresting. So that

---

immersed himself for months in the plane-geometry literature. He had certainly come across this unusual proof, but had expected his theorem-proving machine to use the proof normally taught in high school geometry courses. The surprise was that the program unexpectedly discovered the proof for itself anyway. This is the same proof that emerged in Minsky's hand simulation of his geometry machine at the Dartmouth Conference. See Papert (1968) for a slightly different version of the Dreyfus-Ashby dispute.

was almost there. The third was that most psychological theories would be stated as computer programs, and since we were going to do that, that seemed a reasonable one to say. Chess, music, psychological theories, theorem proving—each arose out of work that was already beginning. And ten years seemed like a reasonable time in terms of what we thought would be the effort applied to those.

We made no prediction about natural language, in which we were far too conservative, because at that time that was very far away to us. And that moved much faster than we expected. So, at the end of ten years, we didn't have our chess champion, but we had chess-playing machines. There we just vastly underestimated two things: first, how little, how few man-years would go into this; and second, how much very specific knowledge had to get poured into it. Maybe we left out some other things, but those are the only things I'm willing to admit we left out!

Marvin Minsky of MIT, who does a fair share of prophesying himself, is perfectly at ease with the Newell-Simon predictions:

The claims that the AI people made, I think, were correct, because AI is possible and we will eventually learn how to make machines that are smarter than even the wildest speculations said. But the estimates were short, though not characteristically more exaggerated than in other fields. Simon might have thought that chess programs were well enough along so that in ten years there would be a world champion, and he was off by perhaps twenty years. But that's a pretty small factor. The trouble, perhaps, is that the usual prediction in old science is a hundred years. People are saying we'll be able to make an artificial cell in perhaps a hundred years. Nobody expects to make a biochemical cell in the next ten.

Then Minsky leans back and grins.

Perhaps Simon underestimated how hard chess was, but his real mistake was an amusing one, I think, since he's been the head of a school of management. Simon's big mistake was in his estimate of how many people would work on chess. When he made that prediction, I think he thought that three or four years from then, there'd be hundreds of people all over the world working very hard on chess programs. As it happened, they themselves were the only ones to work hard on chess in that decade. McCarthy worked casually, assigning good problems to bad students. My laboratory discouraged it.

I remember having a conversation with Herb about it once while walking on the beach at Santa Monica. And he said, "Well, if they don't do it in another ten years, I'll have to do it myself." I said, "Well, aren't you annoyed that they're not doing the right thing, all going and doing the wrong thing?" And he said, "No, it's sort of nice. It leaves more of it for us to do." And that was the first ungracious thing I ever heard him say. I was very shocked at that. I might have thought such things to myself but the idea of admitting that you didn't mind that people were on the wrong track because you would have more fun yourself seemed sort of shocking. But that's the way it was, and still is, on the whole.

I think probably if the idea of a chess player had swept the world in 1957, and a hundred projects had started to get at each other's throats around the country to make the best chess program, then by 1967 there ought to have been much more powerful chess programs than there are today.

Concerning the second prediction, Simon says,

Again, on the music thing, essentially the prediction was correct. But even then, much less labor went into this than we expected. The biggest mistake we made was an overestimate of how much this field was going to fascinate people and trap them into working on it. We just couldn't understand how anybody could stay out of it, and they managed to stay out of it in droves.

I asked why. We had been talking about Thomas Kuhn's model of scientific revolutions, and Simon now took that as his metaphor.

There are probably more timid people in the world, even in science, than one likes to believe, people who like to do things in well-structured environments where there already is a paradigm to work in. There probably are more normal and fewer revolutionary scientists than one likes to believe. Even among my students, there are people who wouldn't march up to things like this because they knew another kind of problem that was well-structured and they knew at the end of the year they'd have a Ph.D. thesis, and what would they have with this wild stuff? So not only did we very much overestimate the number of people who were willing to work in relatively unstructured spaces, but we also underestimated the extent to which the computer-science culture was going to be colored by the mathematics culture during the early years, and heuristics never appealed to mathematicians—there weren't any theorems in it!

Simon and I got back to the public reaction to his predictions. Among many other things he does, Simon is a scholar of the history and philosophy of science, and has published extensively in the field. With his interest in how people behave in organizations, he has paid a great deal of attention to how people behave in organized science:

If you go and look at other sciences, which maybe aren't so personally threatening to people, claims are made all the time. Look at the canons of behavior in astronomy today. You know, someone can go around with the smallest scintilla of evidence and make a new kind of universe that expands or contracts or is permanently in one state or another. Cosmologists go around doing this all the time, and they're regarded as good scientists in astronomy because that's part of the mores of that field. Ditto for geologists—plate theories of the world, for instance. These all go way ahead of the evidence, and in some fields this gets institutionalized as acceptable. Biologists on the whole are much more careful, in that sense of careful.

I thought of how the astronomer Carl Sagan's speculations about life on Mars have brought pain to geologists and biologists alike. Some of them do acknowledge the necessity for a Sagan to drive the field, though, in the hopes of either proving or disproving his assertions. Theoreticians often play this role in science, a point that is nicely documented in Ian Mitroff's *The Subjective Side of Science* (1974).

Simon continued: "So if people from a field which does less speculating look at a field where this is done, so to speak, they say, oh here's a bunch of publicity seekers, and so on and so forth."

"Well, I'm not sure," I answered. "It seems to me you put your finger on it earlier when you said the animosity people feel towards machines imitating humans has a lot to do with it, because nobody really cares whether cosmologists make these outlandish claims— one can be amused or not amused, whatever."

"Yes. Well, this is the reason that I don't believe the difference here lies in the behavior of the people in the field—people in AI aren't somehow wilder and more speculative than other scientists. I think the difference lies in the field itself and the feelings people outside have about it."

"But you're aware of this enormous animosity?"

"Oh yes, I certainly am. It even includes people who aren't very far from the field, and occasionally it includes people who are converted out of the field. I think Tony Oettinger is an example, and so is Joe Weizenbaum, though they're quite different kinds of guys. Weizenbaum feels it's wicked. Oettinger, or [Yehoshua] Bar-Hillel, for example, they think it's hopeless. There are two guys who tried real hard, and the particular things they tried didn't go—which is always a good proof that it can't! Joe, I think, is a different case."

But controversy is very much a part of the gestalt of science— Mortimer Taube to the contrary—and scientists expect to have to defend their hypotheses. That said, the acrimony seems higher in artificial intelligence than elsewhere.

In the introduction to *What Computers Can't Do*, Hubert Dreyfus dwells on chess, which he sees as emblematic of all the inflated claims of AI. In 1958, Newell, Shaw, and Simon had indeed announced a program that played chess, though they hedged their claims as to how well it played. This was because their main interest was not in producing a killer machine, but in simulating how human beings played chess. Their work on chess had begun in 1955, but had languished while they turned their attention to programs that simulated other kinds of human problem solving. Drawing from this experience, they returned to the chess problem, and used it in combination with some of the features other researchers had developed. "Perhaps the only common characteristic of the other programs that is strikingly absent from ours—and from human thinking also, we believe—is the use of numerical additive evaluation functions to compare alternatives" (Newell, Shaw, and Simon, 1958).

Newell, Shaw, and Simon also reminded their readers of something essential. Earlier chess programs had been written in machine code, including even Bernstein's, which required sophisticated techniques and nearly all the memory available in his computer. The NSS program was, as they put it, "already beyond the reach of direct machine coding; it requires a more powerful language." They then reported the development of their series of such powerful languages, the IPLs, which were to have a significant impact on computer programming in general. Then and now, the three realized that here was one of the major achievements of their research, that whether the chess program played well or not was comparatively less important.[4]

This is what Seymour Papert was driving at when he declared that Hubert Dreyfus knows how to evaluate only the grossest empirical evidence—does the machine play chess or doesn't it?—without understanding the aims of the research. Machine coding had permitted an impressive enough level of achievement; we sometimes need reminding that for all their comparative simplicity, nothing like the computer programs of the mid-1950s had ever existed in the world. They might not be up to human standards of performance, but there was nothing else like them. Higher level languages were going to permit much more sophisticated achievements, and hand in hand with hardware improvement,

---

[4] The growth of high-level computer languages is a complete story in itself. Perhaps the easiest way to understand it is to compare computer languages with natural languages. Machine code might be compared to the language of traffic signs: stop, yield, do not enter, and no parking are fairly direct messages, admitting little complexity or ambiguity. The lyrics of popular songs are generally richer, and allow a greater expression of emotional range. But to be able to grasp the extraordinary complexity of ideas and effects in the language of James Joyce or Ludwig Wittgenstein requires effort, tenacity, and even special training. Nevertheless, the same human brain can cope with all three kinds of language. In a sense, so it is with the machine, the machine code standing as the simplest one-to-one sort of language, while more powerful languages (which are automatically translated by the machine itself from highest to lowest level) are capable of dealing with concepts that the machine code cannot deal with directly.

which included faster operations and much larger memories, or storage, more ambitious projects were planned and executed.[5]

It's important to remember this progress in reading Dreyfus's book, for he implies that Newell, Shaw, and Simon simply abandoned their chess machine because they couldn't make a go of it. It would be more accurate to say that developments in software—some invented by Newell, Shaw, and Simon themselves—and hardware both had overtaken the early NSS machine. To try and improve the machine as it was would have been like feeding a horse enriched oats in the hope it could thereby compete with an automobile.

But how had Hubert Dreyfus got himself into the position of archcritic and exposer of artificial intelligence in the first place? Several stories are told.

Herb Simon remembers that talk he gave at MIT in the spring of 1961, when both Hubert Dreyfus and his brother Stuart were in the audience. The Dreyfus brothers took strong exception to some of the things Simon was saying about the elementary perceiver and memorizer he and his former student Edward Feigenbaum had designed:

The Dreyfus brothers were so exercised by that they asked permission, and what's worse, received it, to insert a half page of discussion into the proceedings volume. It was not a discussion that took place during the meeting, it was an afterthought they had. And it was a nasty little diatribe about this preposterous stuff that was being peddled. Then Stuart managed to get Hubert a consultantship out at RAND for a summer. And he went out to RAND and wrote that "Alchemy and Artificial Intelligence" thing, which got peddled as a RAND report.

---

[5] Marvin Minsky gives several examples of work his students have done which, in the late 1950s and early 1960s, would have been enormously complicated. Now these programs are understood and reprogrammed in simpler ways in more powerful languages, and stand as a special case of more general efforts. For example, Thomas Evans's analogy program was to become a special case—easily programmable—of a subsequent thesis by Patrick Winston. Gelernter's high school geometry theorem-proving program, which took several work years to design in the late 1950s, could be done in a few days by a graduate student in the late 1960s, thanks to a high-level programming language called PLANNER, designed by Carl Hewitt at MIT.

He'd had no connections with RAND before or since; he had no technical background for this at all. But the fact that he was a consultant at RAND immediately gave him credibility. I was about to say I don't mind being criticized; of course I mind being criticized. But you know that's fair game, and I can play it the way the politicians play it. But what I resent about this was the RAND name attached to that garbage. That was really false pretenses.

In fact Stuart Dreyfus had been instrumental in getting RAND to hire his brother for the summer of 1964. Paul Armer, who was then head of computer science at RAND, recalls that in addition to supporting Newell, Shaw, and Simon's work, he was also responsible for hiring Hubert Dreyfus:

It struck me and the idea grew on me that I ought to mix a philosopher in with these guys, that there were indeed philosophical questions here. I thought maybe the project would gain from having a philosopher around for a while. And Stu Dreyfus, Bert's brother, was then at RAND, and so he suggested his brother. His brother had a good-looking resumé, and when I chatted with his references, they were all high on him. And I was only investing three months in him, brought him out in the summertime. It seemed like a good gamble. You know—win a few, lose a few.

Had Dreyfus been presented as an impartial critic? Armer remembers that he was, and is sure that if he had known Dreyfus had already appeared in print in 1962 explicitly attacking artificial intelligence he would not have chosen him to deal with the philosophical problems raised by the idea of thinking machines. Even more interesting, Dreyfus remembered himself as arriving at RAND in 1964 with a completely open mind until I reminded him of the 1962 statement he had published with his brother. There is not venality here so much as a passionate commitment to an idea: as tenaciously as Simon and Minsky and others hold on to the idea that whatever the evidence *against* it, the evidence *for* machines—computers—someday being highly intelligent is overwhelming; so Hubert Dreyfus holds that whatever the evidence that machines can perform intelligent tasks, the evidence *against* their ever being able to be really, humanly intelligent, is overwhelming.

When Dreyfus finished his paper, some time after the summer was over, Armer read it and didn't like it:

I thought it was lousy philosophy. And, further, he had decided he was really going to attack. I think you can disagree with somebody without in some sense questioning their motives. But I also thought it was poor philosophy. Not that I was a philosopher, but I thought it was bad. I had a big squabble with some other people in the department who liked the paper and who kept pinning me to the wall with my own pronouncements about censorship and how bad it is to have yes-men around. A fellow who was my administrative assistant at the time, Robert Ryanstat, still at RAND, a psychologist, he just thought it was a great paper and he was the one who kept using my own arguments on me about how at RAND one didn't just sit on the output of another scholar because one didn't like it; if you could really show that there was something wrong with it, well, that was one thing, but just because it came to a conclusion you didn't like was no reason not to publish it. So eventually it came out. I suppose I delayed publication on it for nine months or so.

Dreyfus was aware of the attempted suppression. He recalls that "Alchemy and Artificial Intelligence" finally did come out as a RAND memo, the lowest level RAND publication. He said, with amusement,

And it became the best seller of any such paper that RAND put out, which didn't make them a bit happy. I don't know whether there have been best sellers since, but in those days it was a popular one. And such is the underground in the sociology of knowledge that everybody all over the world that might have been interested in it heard about it and read it. I've got a folder about that thick of letters from people all over the world, Russia, Japan, the United States. And it's funny, because I thought that if that nonpublished, nonadvertised paper had that much of a response, just think of what would happen when this book is published. But the funny thing is, there was practically no additional response when the book was published; this paper reached the real audience without any advertising, without RAND'S *even wanting to sell it.*

"Alchemy and Artificial Intelligence" indeed caused an uproar, partly because it came out under the RAND imprimatur, and partly because it attacked so viciously, and, some thought, inaccurately. One who held the latter view was Seymour Papert, who himself had had a fair amount of training in philosophy.

Speaking of the paper, which was later expanded to become a book, Papert felt that in fact Dreyfus never really came to grips with the important issues:

One third of the book is gossip, and has nothing to do with AI. As for the rest—well, if you take any Dreyfus quotation and you go back and see where it came from, look at it in context, it's always wrong. And it's not that occasionally he misses the point because he didn't see the context. There just aren't any exceptions, and so that must either prove something about Dreyfus, or that it's very hard to find real examples of bad things that were said by leading AI people. I think it says something about Dreyfus. You have to take sentences as he found them in a very literal-minded way.

Which is another way of saying what Dreyfus himself had said in his book: that a large component of perception is what you expect to find. And Dreyfus clearly had gone looking for statements to fit his passionately held preconceptions.

I think maybe in a deeper sense that's the problem of people who have never programmed a computer [Papert went on], and who want to think about whether AI is possible. It really seems outrageous that it should be possible. You tell them that a program is being written to do so-and-so and this is how it does it. Then they think of a slightly different problem. Any programmer would see instantly how to extend the program to cover that problem. People like Dreyfus can't. So any modification of the problem seems to put it outside the realm of what could be programmed, and that seems to be an impossible weakness in AI. Unless you yourself can play with the directions of extension and see at least vaguely how it can be modified to go in this way or that way, there's just no way in which you can begin to judge what it can and can't do.

How had Papert come to be pulled into the Dreyfus dispute? For he ended up writing a refutation of Dreyfus's paper that in fact was requested by Paul Armer, but RAND'S attorneys felt nervous about publishing it because it contained what they thought might be libelous material. It was eventually brought out as a Project MAC report, with no lawsuits ensuing. About this paper Papert says now,

It was really a serious distraction which I shouldn't have done. I was thinking of writing a paper or a little monograph—I'm still thinking of doing it—on the difficulties of accepting the idea of AI. This means you consider what it is about the idea of machines being intelligent that makes it hard for people to understand or accept that idea. Incidentally, I think an important part of the training of students in AI is for them to work through this kind of difficulty, accept that they do have conflicts and it does clash with many aspects of our culture, so that even if you're convinced intellectually that it's possible, you really have to work through all these resistances. Otherwise you're always going to be hung up about some aspect. Anyway, that was a plan and I was making a draft of it when Dreyfus came along. So I got this probably bad idea that a good way to do that would be to take a real subject—let's take this guy who seems to be having these real difficulties, and deal with him as an interesting subject for study. It's a nice, seductive idea, but it wasn't in the end really so, largely because Dreyfus doesn't really come to grips with the issues: he stays too near the surface.

Edward Feigenbaum, of Stanford, says,

What artificial intelligence needs is a good Dreyfus. The conceptual problems in AI are really rough, and a guy like that could be an enormous help. But I can think of— [and here he stopped to count] one, maybe two philosophers who have the grasp of what AI and computing are all about, and also know philosophy. And both of those guys are interested in their own projects. We do have problems, and they could be illuminated by a first-class philosopher. But Dreyfus

*bludgeons us over the head with stuff he's misunderstood and is obso-
lete anyway—and every time you confront him with one more intelligent program, he says, "I never said a computer couldn't do that."
And what does he offer us instead? Phenomenology! That ball of fluff!
That cotton candy!*

A scientist might well conclude that phenomenology is cotton candy. I myself was troubled by Dreyfus's alternative model
of human behavior because it seemed to me to offer nothing
beyond certain assertions based only on intuition and some observation—rather like declaring that the earth must be flat because that's the way it looks and feels. How could his model be
helpful in teaching people to do better arithmetic, or spelling,
for example?

He agreed that his alternative cannot on principle offer any
kind of scientific understanding of these things, that in fact he
believes that "human intelligent behavior is a sort of something-
or-other which we cannot have a scientific theory of." Thus all
social sciences are, for Dreyfus, as wrong-headed as AI. This is
not an attitude widely held in universities.

Dreyfus's stand touches on all four of the basic arguments
against machine intelligence—arguments of emotion, insuperable differences, no existing examples, and ethical considerations.
The latter is a small element of Dreyfus's argument, and I will
address it later when I come to speak of a critic whose main argument is from ethics.

It is surely clear by now that something besides intellectual
dispute is at issue here. Perhaps it can be seen most clearly in the
great chess match.

That he never said a computer couldn't do that were nearly
Dreyfus's very words at the end of a chess match between himself
and the MacHack program at MIT. His original statement in
"Alchemy and Artificial Intelligence" had been ambiguous: no
chess program can play even amateur chess. Did he mean *ever*?
That is certainly the sense understood by *The New Yorker* in its
June 11, 1966, "Talk of the Town," which reported on Dreyfus's
paper with even more than its usual smugness. Subsequently mated

by MacHack, Dreyfus wrote in his book, "Embarrassed by my exposé of the disparity between their enthusiasm and their results, AI workers finally produced a reasonably competent program. R. Greenblatt's program called MacHack did in fact beat the author, a rank amateur." His footnote to this statement goes on to explain that he did not mean a computer could *never* play even amateur chess, that he was giving a correct statement of the state of the art at the time he wrote, 1965. On the other hand, Richard Greenblatt might dispute the assertion that he conceived MacHack in a flush of mortification over any exposé.[6]

In any event, both parties seem to have found the opportunity irresistible. Smiling, Papert recalls,

I organized the famous chess match. That was beautiful. He was— well, it wasn't all pathetic and sad because he was quite convincing. He was going to beat it very easily. And that also said something about him, something almost naive. We didn't know. About halfway through we all thought Dreyfus was going to win.

Herb Simon says of that chess game,

In that paper of his, Dreyfus said some really nasty stuff about chess because he was looking at our NSS chess program at RAND and he knew that a ten-year-old had beaten it. Then along came MacHack, which was much stronger, and somehow Dreyfus was induced to play it, and it walloped him. One of the things he was arguing in the document which preceded his book was not only that a chess program would play very bad chess, but also that it was going to play mechanical, nonhuman chess. And if you look at this game, it's a wonderful chess game because it's a cliffhanger. It's two wood-pushers, you know, fighting each other, and they have these momentary bursts of insight in which they get a fiendish plan to trap the other guy, usually two moves deep, and alternately the guy almost falls into the trap or he doesn't. Dreyfus was being beaten fairly badly and then he found a move which could've captured the opponent's queen. And the only way the opponent could get out of this was to keep Dreyfus in check

---

[6] In fact, MacHack was an easy extension of a kriegspiel chess program Greenblatt was then working on.

with his own queen until he could fork the queen and king and exchange them. And the program proceeded to do exactly that. And as soon as it had done that, Dreyfus's game fell to pieces, and then it checkmated him right in the middle of the board. So it wasn't mechanical at all; it was a typical game between humans with these great moments of drama and disaster that go on in such games. It was wonderful.

And gratifying. The results of the game were printed in the bulletin of the Special Interest Group in Artificial Intelligence of the leading computer society in the United States, the Association for Computing Machinery, with a headline drawn from "Alchemy and Artificial Intelligence"—*A Ten-Year-Old Can Beat the Machine— Dreyfus*—and a subheadline that read, *But The Machine Can Beat Dreyfus.* Dreyfus protested hotly, saying he'd never said computers could *never* play good chess, and this defeat didn't change any of his major assumptions. The language of his protest is extremely aggrieved; someone less emotionally involved would have let the AI people have their laugh, and probably have joined in. Herb Simon made one of his few public responses to Dreyfus. It is an open letter entitled "Cool it, Friend!" Simon writes,

> Dear Professor Dreyfus,
>
> I was a little touched by your recent letter to SIGART, protesting the comment on your defeat by MacHack, Greenblatt's chess program. A writer who employs the juxtaposition "Alchemy and Artificial Intelligence" can hardly plead ignorance of the uses of rhetoric, or cry "foul" when an editor implies something by juxtaposition of a chess score with a quotation from one of the players. Such writer could not even, in good conscience, protest a return in kind of rhetoric that he began five years ago, and has continued, with escalations, ever since.
>
> What are the facts? A man who exhibited great zest in writing that a "ten-year-old novice" had beaten a particular chess program was himself beaten, and beaten roundly, by MacHack. Neither fact by itself proves much about the present or future of chess programs, but the two facts may interest and arouse emotions in persons already passionately committed to conclusions (pro or con) on these matters. To protest amused comment on the MacHack victory shows either a

desire to apply the rules of rhetoric asymmetrically, or such deep emotional involvement as to cause blindness to the asymmetry. You should recognize that some of those who are bitten by your sharp-toothed prose are likely, in their human weakness, to bite back; for though you have considerable skill in polemic, you have no patent on it.

The discussion of the philosophy and status of artificial intelligence would benefit from de-escalation. Since you have contributed some of the most vivid prose on the subject, may I be so bold as to suggest that you could well begin the cooling—a recovery of your sense of humor being a good first step. You see, the real humor in the Dreyfus-MacHack game, as any chessplayer who plays it over will tell you, is not that you were beaten. The humor is that the Greenblatt program exhibited in this game many of the same human failings that you did (failing to see obvious impending mates, for example) and still clobbered you—by the skin of its teeth. It was a real cliffhanger, in which one fringe unconsciousness was outdone by another. MacHack behaved not like an "omniscient computer" (to quote you out of context) but like a frail and sometimes desperate humanoid—even, shall we say, as you and I.

Below the letter, the editor of the *SIGART Bulletin* had put a spacefiller:

(The following is extracted from a set of notes prepared by Prof. Seymour Papert of the MIT Artificial Intelligence Group.)
1.5 *Computers Can't Play Chess.*
1.5.1 *Nor Can Dreyfus.*

This defeat by no means silenced Dreyfus or cooled his rhetoric. His book, the expansion of his "Alchemy and Artificial Intelligence" paper, was published, and while it seems more restrained than the paper, it doesn't yield an inch in general conviction that computers simply cannot think.

Indeed, if Dreyfus is so wrong-headed, why haven't the artificial intelligence people made more effort to contradict him? Except for Papert's unfinished memo and the reviews of his book—in particular one by Bruce Buchanan of the Stanford artificial intelligence effort, who himself received his doctorate in philosophy—AI people generally avoid confrontations with Dreyfus. "He's just too silly to take seriously," one researcher told me. "His opinions were fixed from day one and all the evidence in the world that he's wrong won't convince him. And he sure as hell doesn't convince me." That Dreyfus was invited to make a keynote address at a general computer conference a few years ago outraged the AI community. "That kind of platform gives him an authority and credibility he's simply not entitled to," Allen Newell told the organizers of the conference. But Dreyfus's invitation was symbolic of the animosity a large segment of the computer science field feels toward artificial intelligence, expressed in conversation and often in print.

AI researchers dismiss Dreyfus's charge that the field has stagnated. "What that means is that the papers are now too hard for him to read," laughed one MIT researcher. "Not that he could ever evaluate what he *could* read." It is true that some present programs achieve an impressive display of behavior that in a human we would call highly intelligent. Two examples are special-purpose programs for diagnosis in internal medicine (at this writing being tried out on the test cases presented by the *New England Journal of Medicine* to its subscribers, and so far having done perfectly in fourteen out of fourteen tries) and evaluating mass-spectrogram data (at the Stanford Medical Center). These are special-purpose programs. If they achieve or exceed human performance in one field, they still cannot travel to a conference and deliver a paper on their successes. But what is a reasonable expectation for a scientific field approaching its thirtieth birthday?

How [wrote Seymour Papert] is progress measured? How much progress is to be counted as refuting Dreyfus's statement? How does one assess the importance of work on fundamental or specialized technical problems? Dreyfus does not even try to face these questions. He merely asserts pontifically that there is stagnation.

There is no stagnation. The crudely empirical criterion of observing performance of machines suffices to demonstrate steady progress. But even if Dreyfus had bothered to find out how well modern programs actually perform, he would have missed a far deeper point, which I shall introduce through an analogy with another branch of engineering. The innovators in aviation at the beginning of the century worked by building whole airplanes and flying them. The problems of super-

sonic airliners and atomic aircraft are being solved now by people who could no more construct an airplane than fly themselves.
Artificial intelligence follows this pattern like any other area of science or technology. The sign of its maturity is the emergence of specific technical problems. But the amateur observer sees this maturation as "stagnation."

(Papert, 1968)

Newell and Simon, who are prime targets of Dreyfus, concluded that any formal rebuttal would only propagate Dreyfus further. Science, Newell believes, lives because one scientist takes up the work of another, either to disprove or corroborate it. Much of the science done simply slips through the net, catching nobody's attention. Therefore, good or bad, the only way to propagate scientific work is to make formal reference to it. Lack of such reference means oblivion. This strategy also suited Newell and Simon's views of gentlemanly behavior.

During the period immediately after "Alchemy and Artificial Intelligence" appeared, people often raised Dreyfus's charges during the question-and-answer periods following lectures Newell and Simon gave, and so they in fact prepared answers. But aside from that, and from Herb Simon's open letter to the *SIGART Bulletin*, the two have refused to answer Dreyfus directly. At times they have reconsidered this decision, especially as Dreyfus has refused to go away and has become something of a fixture in the AI community, but they are so far sticking by their decision.

Papert's response is an unfinished document, largely because Papert lost interest in it, feeling that with a finite life it was more interesting to do science than to defend against what he considered an intellectually irresponsible attack.

Thus, aside from some reviews of *What Computers Can't Do*, which were necessarily too brief to address any but the grossest exceptions they took to Dreyfus, workers in artificial intelligence mainly take no public notice of him. Robert K. Lindsay, in his review in the prestigious journal *Science*, was amused to point out that Dreyfus felt compelled to use empirical evidence to shore up an argument against the empiricist tradition, but like Bruce Buchanan, Lindsay hoped that some of the issues raised might be examined seriously by workers in the field—not Dreyfus's charges against the work, but his proposals for alternative ways of looking at problems. I have already noted indirect responses to issues Dreyfus raised.

With or without the advice of Dreyfus, AI programs are growing more complex, flexible, contingent, tolerant of ambiguities, and situation- and goal-oriented than they once were. My impression is that this progress has taken place piecemeal and in response to tough given problems, and owes nothing to Dreyfus. He may have been correct in some of his criticisms, but like that earlier critic Mortimer Taube, his derisiveness has been so provoking that he has estranged anyone he might have enlightened. And that's a pity.

Meanwhile, Hubert Dreyfus goes on, gleefully sitting on panels, giving talks, and writing reviews he hopes will discredit the artificial-intelligence community.

To have made something of a cottage industry out of attacking a particular scientific field might strike some as infra dig. Most of us have our hobby horses; we beat them and go on. There seemed to me something almost unsavory about a scholar devoting so much time and energy to such an attack. By now it has endured for over fifteen years—through several papers, a book, and heaven only knows how many verbal exchanges. When I finally met and spoke with Professor Dreyfus, an idea occurred to me that will please neither him nor the field he so relishes attacking, and it is that they have more in common than they think. As surely as Herbert A. Simon was firmly convinced from the beginning that a scientific explanation exists for what we call mind, Hubert Dreyfus was from the beginning equally convinced that one does not.

Things aren't quite this tidy, of course. Simon does indeed remember having a lifelong antipathy for thinking of mind as something mysterious and unanalyzable; he decided very early to do social science as science, and so has sought models that would confirm this point of view. Dreyfus began as a student of physics at Harvard; he switched to philosophy as a graduate student, and then grew disenchanted with what he took to be the sterility of conventional philosophy. His real apostasy came when he abandoned conventional philosophy for phenomenology. Only gradually did he see that AI, if it were successful, would be a confirmation of the conventional philosophy he detests.

"AI is a symptom," he said to me "and I've generalized it to all the human behavioral sciences. The idea that science and technology can be generalized to everything is something to really worry about and be concerned with—that's my rational reason for what I do." Which certainly puts Dreyfus at the antipodes of Simon's position.

"But," Dreyfus continued, "I never asked myself, 'Why do I get so upset with people like Papert, Minsky, Newell and Simon?'—and I *really do* get upset. It's really puzzling. I'll have to think about that. I'm always asking myself, 'Why do they get upset with me?' and in a way that's even more obvious than why I get upset with them; I started it. But I've never thought about it. I have a sort of Manichean way of reacting to things, the good guys and the bad guys, but why I picked them as the bad guys and then spend so much energy worrying about them—well, I'll ask myself that and try to find an answer." As we were leaving his office, he mused some more on the question. "I've given you the rational reasons. But you're asking a personal question. Maybe I attack in them what I dislike in myself, an excessive rationality."

It is Us and Them. And that's the worst of it: our remote kinship with the machines, when we long for them to be The Other, and they are instead another face of Us.

Dreyfus thoroughly enjoys ridiculing artificial intelligence, and seldom passes up an opportunity. The final footnote in his book refers to a statement by Feigenbaum and Feldman that intelligence is a continuum, and that there is no proof that an insurmountable hurdle lies somewhere along that continuum as an obstacle to high-level artificial intelligence. Dreyfus transposes this idea to the success alchemists have had with the baser metals, and their hopes then to achieve the philosopher's stone. Well, yes. Perhaps. But.

Dreyfus's assertion that somehow the human body is key to intelligence, and that without it intelligence cannot exist, sounds strangely to me like the claims of nineteenth-century physicians, based roughly on the same kind of evidence and certainly with the same happy complacency, that women couldn't think because they had female bodies, and that the male body was essential to real cognition. (Some allowed as how the female body was perhaps capable of thinking but was drained by all the necessities of menstruation, childbirth, and so on.) On such authority, Augustus DeMorgan, Lady Lovelace's tutor, denied the evidence of his own eyes and discouraged his pupil from further mathematical adventures in order to protect her child-bearing capacity.

A probe into this nineteenth-century dogma reveals that the main reason physicians made such statements was that women were clamoring to get into the medical schools—for the not unreasonable purpose of providing better health care to one another than they'd received at the hands of a masculine profession—and physicians saw a threat in this. With this in mind, it's not farfetched, it seems to me, to suggest that philosophers, whose domain after all is *thinking*, have every reason to protest the loudest that machines are incapable of thinking. In Dreyfus's case, this conclusion is altogether fair: he is the first to admit that a nonhuman intelligence, in particular a thinking machine, would destroy all that he holds intellectually near and dear.

In other words, for Hubert Dreyfus no amount of progress in the field of artificial intelligence will ever persuade him that he is wrong, just as for Simon and his colleagues temporary failures of prophecy are no reason to quit trying. Which camp has the growing evidence and which the diminishing is only partly a matter of counting up. Predisposition, or world view—call it what you will—have more to do with opinions on this scientific question than evidence. I've been amused to discover the same thing within myself.

As we'll see later on, some of the elements Dreyfus insisted were necessary for intelligent behavior, and that didn't exist in the very early intelligent programs, have subsequently been incorporated into intelligent programs as higher-level programming languages have made this possible. But if Dreyfus was right with some of his ideas, he has already been shown to be wrong in saying that no machine could have such properties.

The funny thing is that both sides might turn out to be more or less right. It may indeed be that human intelligence in *complete detail* cannot be realized on a computer, serial or parallel. But that doesn't preclude the possibility that machines may eventually exhibit intelligent behavior that would make a sage swoon with envy.

Ah well, Dreyfus has said publicly that he would accept Turing's test—that is, if, in a blind test, a machine could fool him, or even leave him in doubt 60 percent of the time about whether he was encountering natural or artificial intelligence, he would shut up. Such a test might be an interesting encore to the great chess match.

Part Four

—

Realizations

As it asketh some knowledge to demand a question
not impertinent, so it requireth some sense
to make a wish not absurd.
— *Francis Bacon*

Chance does nothing that has not been
prepared beforehand.
— *Alexis de Tocqueville*

Chapter Ten

Robotics and General Intelligence

When most of us try to picture an artificial intelligence, of course we think first of a robot. This history has been full of them, make-believe humans who clanked their way through our dreams, our stories, our films and plays. Some have been nobler versions of ourselves, some ignoble. The robots who win the war against the humans in *R.U.R.* look forward to a life of bliss in a socialist workers' paradise, while other robots in other stories stand for the indestructibility of the machine compared with the all too frail destructibility of human flesh, one more gloomy reminder of our personal mortality.

Why build a robot?

The reasons are numerous. It was a robot which explored the surface of Mars for us, a handy example of the fact that robots can go where—and do what—humans cannot. (However, the extent of this robot's debt to artificial intelligence is in dispute. Its routines were preprogrammed and, once in action constantly monitored by humans. This represented impressive control engineering, but the robot was not an intelligent machine. On the other hand, Charles Rosen of SRI called the Mars robot "Son of Shakey," for all that the Jet Propulsion Laboratory had borrowed from the SRI robot named Shakey.) In the context of the Mars robot, then, robots are one more tool, one more extension, of the human body and mind.

As I've pointed out earlier, the building of robots has been said to be motivated by all sorts of psychological needs, particularly those of males, who cannot themselves give birth biologically. Again, this point surfaced most recently in the 1972 report of Sir James Lighthill mentioned earlier, in which it was recommended that support for artificial intelligence in Britain be terminated.[1] Oh, not because it represents a desire to give birth, Sir James hastened to say; in fact, he didn't believe it for a moment, but just thought he'd mention it. I don't believe it for a moment either, unless it is meant in the larger sense that we all seek immortality—or at least propagation of our own presence beyond our immediate circle and lifetime—by the art we create, the science we discover, the good works we do.

No, I'm putting my money on a bet that says we build robots for the same reasons we do other kinds of science and art, for the immense satisfaction of knowing something significant about ourselves that we didn't know before, of having our suspicions and guesses about ourselves confirmed or laid waste—simply put, of seeing ourselves in a new way. Without doubt, robot building gets back to the human race. I come again to a point I have made continually, that the building of artificial intelligences is very much a part of our long romance with ourselves as a species. Unlike art, however, artificial intelligence contains in it the possibility of our transcending the species and knowing something about intelligence elsewhere. That is one of the things that makes it a science.

---

[1] The Lighthill report generated much controversy at the time, especially because temporarily, at least, it threatened to end funding of AI in Great Britain. Sir James Lighthill, a distinguished physicist, had been commissioned by the Science Research Council of Great Britain to evaluate the state of AI and recommend whether further funding should be given. Charitably speaking, the report seems to have been done in a hurry—in such a hurry, in fact, that rumors immediately flew that its main purpose was personal vendetta, with Sir James as hatchet man for others who were nettled by AI and some of its practitioners. Since the report coincided with, and surely exacerbated, the dissolution and reorganization of the Edinburgh University AI laboratory, the rumors seemed true. But Bernard Meltzer, still at Edinburgh, has doubts as to any conspiracy, though he's one of several who complained to me about how the report was done. In any case, AI funding seems to be scrutinized now somewhat more carefully than it was pre-Lighthill, but it continues satisfactorily,

We're rather casual about how we use the word robot. It comes from the Czech word for servitude, or slavery, and was introduced into English by Karel Capek in his play *R.U.R.*, which took the London season by storm in 1921. We use it to designate all sorts of machines that do stand-in work for us, from the automatic car-wash device to that rather more complicated instrument that prowled Mars. In artificial-intelligence research the robot that matters is an intelligent robot, one that will cope with novel situations essentially by figuring them out—by comparing them with situations it has encountered before, by generating a set of reasonable alternative courses and choosing the most appropriate, or even by falling serendipitously into an unexpected solution and recognizing it as such. If it all sounds familiar, it should. It's what human beings do all the time.

So intelligent robots must have a general capacity for dealing more or less successfully with a variety of situations, which makes them different not only from ordinary preprogrammed robots, but also from other artificial intelligences designed to deal well with only one task environment, however complicated. It puts them right in the middle of that stream of effort we have seen at least since Leibnitz—the urge toward a universal calculus, a universal set of rules for reasoning. It's the urge George Boole followed when he set up his algebra or laws of human thought in the nineteenth century, the urge culminating in Whitehead and Russell's *Principia Mathematica*, which expressed all mathematics in terms of a single logical calculus.

This same urge informed the spirit—though in a decidedly nonmathematical way—of a program designed in 1957·by the Carnegie-RAND group, Newell, Shaw, and Simon. This program, the General Problem Solver, came quickly—almost simultaneously, as science goes—after their Logic Theorist, the program that had proved theorems from Whitehead and Russell's *Principia*.

The philosophy behind the General Problem Solver was clear. It was not to be task-specific. Once Newell, Shaw, and Simon had demonstrated to themselves with the Logic Theorist that computers could indeed do tasks that required intelligence, that the information-processing level of abstraction was more than a metaphor—was in fact an explicit language for theory building and for simulation, in the same way mathematics might be used by a physicist to describe physical events—they began to worry about generality. Their working assumption was that human beings brought some general processes to bear on a whole variety of tasks, whether getting to the grocery store or solving a mathematical puzzle, and the General Problem Solver was designed to identify and make explicit those general processes by demonstrating them in a variety of environments.

Both Newell and Simon had a long-time interest in human problem-solving methods. Newell had been a student of George Polya at Stanford (Polya had also taught von Neumann in Europe), and Polya was well known for his attempts to demystify problem-solving techniques mathematicians use, which he had gathered together in a little book called *How to Solve It*. It was from Polya that Newell and Simon borrowed the term heuristic. Simon, for his part, had spent many an evening as a graduate student at the University of Chicago discussing with his friend Harold Guetzkow how to spell out the specifics of problem solving.

Somehow Newell and Simon heard of O. K. Moore's experiments at Yale, where Moore had subjects "think aloud" as they were solving various kinds of puzzles. These studies tied into similar experiments underway at RAND in 1955 and 1956, and Newell and Simon sat down with transcripts of these tapes, called protocols, with the hope of analyzing them in such a way as to cull the problem-solving techniques from them and simulate those techniques in some kind of computer program. That people might be able to express in words some of the things that were going on in their minds as they solved problems was what Simon calls "a sketch of a sketch" of what consciousness might be all about—the ability of an intelligent system to be aware of things external to it, and report, at the same time, on things internal to it.

Newell says, "As soon as we got the protocols they were fabulously interesting. They caught and just laid out a whole bunch of processes that were going on. My recollection differs from Herb's, who remembers the history of GPS as being more diffuse. My recollection is that I just sort of drew GPS right out of Subject 4 on Problem D1—all the mechanisms that show up in the book, the means-ends analysis and so on."[2] From the Logic Theorist, Newell exhumed a technique that he and his colleagues hadn't even realized was there, a matching process; now he made it central to GPS just as he saw it in the human protocols.

GPS did indeed codify a number of problem-solving techniques that humans have used without necessarily putting name to them. Among these techniques are what is called means-ends analysis, planning, and selective trial-and-error.

Means-ends analysis, to take one central technique, is a cycle of operations that works something like this. We look at where we are. We compare this with what we want. If they are the same, we have solved the problem. If not, we ask what will reduce the difference between where we are (or what we have) and what we want. We then apply successively methods suggested by heuristic rules as being likely to reduce that difference, each time beginning again at step one. It happens that different kinds of heuristics work for different situations, so although the original versions of GPS were able to handle a variety of tasks (so long as they were specified in a fairly rigid format the experimenters had to back down on how much specialized knowledge was needed for

---

[2] Simon notes,

"The protocols were gathered in May 1957, and then transcribed. In July we held a Summer Institute at Carnegie for some social psychologists. During one week, we broke up into subgroups, taking the protocol of S4 on P1, and each subgroup was to induce a program simulating the subject. Al and I were working either with separate subgroups or individually, apart from all of them. We were probably not communicating much during the week—can't be sure. On 7/6/57, I produced on a single sheet of paper (preserved) a primitive, but clearly recognizable GPS, which I presented as my solution. I don't recall Al's. Al recalls, equally clearly, that he presented his solution at the end of the week, which was also GPS.

"We had been working so closely, of course, that it is quite possible that we both independently arrived at the same analysis of the protocol. I simply don't remember either any communication during the week, nor do I remember any 'aha!' when we discovered our identical solutions. I am afraid that we may never have the true explanation of what exactly went on during the week. I do know that during the ensuing years we never thought or talked of GPS as having been discovered by only one of us. Why???"

solving problems. It was more, much more, than at first they had realized, and this was to be a continuing problem in all general-purpose intelligences.

GPS sounds almost ridiculously simple, but in fact we see just such reasoning in everyday life. Let me give a homely example. I am hungry. I want to be full—or at least not hungry. What's the difference between being hungry and not hungry? One answer is food. But I am trying to lose weight, and food adds weight. I could play tennis, which not only kills my appetite but also burns calories in the bargain. But I have work to do that keeps me indoors. If I want to satisfy my empty stomach and still do my work and not gain weight, can I think of something else to fill my stomach? I brew another pot of tea. And so on. From an initial state, the problem, one moves by means of operators to various intermediate states, and at last, one hopes, to a final state that is a satisfactory solution to the problem.

A word about problem solving. It is not intended as a synonym for all thinking. In their book *Human Problem Solving* (1972), Newell and Simon call problem solving a subspecies of thinking, concerned explicitly with the performance of tasks. Under these circumstances, learning is viewed as a second-order effect, behavior that *improves* the performance of a system *already performing in a given situation*.

I have before me a self-help book that promises to help me manage my time better: its technique is pure GPS. Another book, called *The Universal Traveler* (Koberg and Bagnall, 1972) and subtitled *A Soft-Systems Guide to: Creativity, Problem-Solving and the Process of Reaching Goals*, is both a charming and useful handbook, drawing from many different sources, as its rich bibliography attests. But its methods, its underlying model, and much of its language is GPS. Its authors are primarily interested in solving architectural and environmental-design problems, but they understand that the techniques are more general, that they can be adapted to all sorts of fields once one views the situation as a journey to be made, a problem to be solved. When Newell and Simon predicted in 1957 that psychology would grasp the information-processing model as a useful way of explaining and understanding human cognitive behavior, they did not expect to find it filtering down to popular self-help books in less than twenty years. In fact, the very view that techniques exist for improving one's creativity is still repugnant to some people. They'd be outraged if their physician still practiced medicine as it was done in Galen's time, but they hold that creative behavior, on the other hand, is mystical, unknowable, and therefore inaccessible to improvement. You've got it or you don't. The GPS point of view—the entire assumption of artificial intelligence as a field—is contrary to this belief.

GPS was successfully tried on a number of tasks, among them logic problems and chestnuts of puzzles such as the Tower of Hanoi and the missionaries and cannibals problem.[3] Newell and Simon wrote about GPS in *Science* in 1961 (Newell and Simon, 1961). They were very careful to note its limitations, as they did whenever they talked about it (for example, the first paper describing it, a mimeograph from Carnegie Tech, says that GPS has "pretenses to generality," a phrase Newell invented out of the conviction that they were probably on the right track, but it was by no means certain). Nevertheless, they could assert that GPS was a computer program capable of simulating, in first approximation, human behavior in a narrow but significant problem domain. It provided unequivocal demonstration, they went on to say, that a mechanism can solve problems by functional reasoning.

The original GPS went through several versions. Hubert Dreyfus says in *What Computers Can't Do* that GPS was aban-

---

[3] The Tower of Hanoi is a platform with three upright pegs on it. Stacked on one of these pegs are disks of graduated size, the largest on the bottom, the smallest on top. The object of the puzzle is to move the disks one at a time until the stack is transferred from one peg to another, without ever allowing a larger disk to cover a smaller one. With only three or four disks, the problem is easy; seven or more makes it hairy. Simon has such a puzzle sitting on the bookshelf in his office at Carnegie. In the missionaries and cannibals problem, three missionaries and three cannibals must cross a river in a boat that will accommodate only two people at a time. If ever the cannibals outnumber the missionaries, the cannibals will dispatch the missionaries with gusto. How do they cross the river intact?

doned because it failed to be a genuine general problem solver, but this oversimplifies matters. The techniques of GPS are embedded in one after another of the more sophisticated computer programs that consciously trace their lineage back to the original, a common example of the evolution of any sound scientific idea. For example, ten years after the original was developed, George W. Ernst was reporting a new, improved version of GPS that would handle twelve different kinds of problems, thanks to improvements in internal representation, while at the same meeting, Saul Amarel of Rutgers suggested a complementary approach, which would reformulate problems in a nested sequence of transformations. And fifteen years later, Simon's former student Laurent Siklóssy of Texas reported on his own work, which combined ideas from GPS and from an alternative approach to a general-purpose artificial intelligence, John McCarthy's so-called Advice Taker. If we count such descendants as *The Universal Traveler* and its sibling efforts, which are beginning to appear in classrooms and workshops all over the country, then GPS is alive and well indeed. Though Newell's recollections should make it clear, I want to reemphasize that GPS was not a collection of new techniques. It is the first program ever developed as a detailed simulation of human symbolic behavior; as such it clarified—and through that clarification made more useful—a handful of procedures human beings had been using all along for solving problems. To say that is not to diminish the profound insights of the creators of GPS. We honor Newton because he gave us a language, a means for understanding some major aspects of the physical universe, not because he invented that universe.

But in the minds of some, GPS also stood as a good example of a bad idea. It outraged the poet Adrienne Rich, and she wrote a harsh poem dedicated to GPS (see page 392). For Joel Moses, now a professor at MIT and a member of the-next wave of AI researchers, it seemed to him a thoughtless direction for AI to take, a quagmire that prevented many people from seeing that specialty and not generality was needed for intelligent behavior. In her own way, Rich was saying the same thing.

At about the time that the General Problem Solver appeared in print in 1957, John McCarthy, who had moved by then from Dartmouth to MIT, was wrestling with a similar problem. How, he wondered, could you have a program that would solve a variety of problems, and furthermore take advice in order to improve its performance? So he proposed some ideas for a program called the Advice Taker, a program that would have common sense—that is, it would deduce from what it was told, and what it already knew, the immediate consequences of any actions it might take.

McCarthy's fascination with intelligent machines has already been described. It was his disappointment with the automata studies he and Claude Shannon had edited that caused him to search for some better way of expressing intelligent behavior in machines. His own contribution to the automata-studies volume had been what he now calls an unsuccessful approach to artificial intelligence, an attempt to make a Turing machine behave intelligently. The Turing machine was unsatisfactory for representing human behavior, McCarthy concluded, because although in principle such behavior might be represented, changes in behavior that are small from an intuitive human point of view don't necessarily correspond to small changes in the Turing machine. It was a defect he recognized even before he published the paper, but it was the best he could do at the time.

McCarthy had been until then a pure mathematician, but a summer at IBM in 1955 gave him a better acquaintance with computers, and he marks that time as the point at which he took leave from mathematics and entered computer science and artificial intelligence, the term he coined. As graduate students at Princeton, he and Marvin Minsky had basically agreed that artificial intelligence was a worthwhile project to work on, though they were rather vague about how. And McCarthy too had been provoked by one of John von Neumann's talks, and wanted to explore the idea of a finite automaton as an intelligent agent. But it was an idea he kept working on, trying to improve it before he published it, and he was chagrined to see others rush into print with less careful analyses. It was about this time too that McCarthy began working on the programming language LISP, making the ideas inherent in the list-processing languages of Newell, Shaw, and Simon, and of Gelernter at IBM, more elegant, cleaner, and more powerful—that is, able to do many more things.

It happens that LISP didn't catch on for some time. McCarthy attributes its late blooming to the fact that it could do things powerfully, all right, but at the time of its invention, nobody really wanted to do them. The simple programs most people were aspiring to were actually easier to program in machine code, and not until aspirations rose did people realize that LISP existed and would provide a representation by which they could accomplish more complicated tasks. LISP, with its offspring, is still the language of choice in most AI research.

Meanwhile, McCarthy was also busy promoting time-sharing, through which the capabilities of a single central computer are shared by a number of users in a way that looks to the users as if they are getting custom service, but which is really a trick based on the mismatch between the slowness of human reaction and the speed of the computer. Thus, when I telephone for an airline reservation, the reservations clerk queries the computer (by means of a remote terminal where the clerk is sitting) about whether seats are available on the flight I want. There may even be an intermediate step, where the clerk can answer my request for a late afternoon flight to San Francisco by finding out from the computer which late afternoon flights to San Francisco exist for me to choose from. So far as the clerk and I are concerned, we are the only people making inquiries of the computer, but in fact, there may be hundreds of clerks all over the country making inquiries at approximately the same time, and the central computer (or multiplexed sets of them) cycles through each of us, serving us in turn.

This idea was originally John McCarthy's:[4]

It was one of those ideas that seemed inevitable in the sense that when I was first learning about computers, I was a little surprised that even

---

[4] It happens that J. C. Shaw had the same idea during the mid-1950s at RAND, but never got around to implementing it or even writing it up.

if that wasn't the way it was already done, it surely must be what everybody had in mind to do eventually. It turned out it wasn't, and I promoted it as something for artificial intelligence, for I'd designed LISP in such a way that working with it interactively—giving it a command, then seeing what happened, then giving it another command—was the best way to work with it. The word time-sharing is used in communication as one of the ways of sending several signals over the same line, so that's where the word came from. My ideas on the subject were rather modest with regard to hardware, but I agitated for time-sharing at MIT, and we got a grant from the National Science Foundation to do it.

McCarthy turned over the details of implementation to others, and went back to thinking about what had inspired LISP and the notion of time-sharing in the first place, his program with common sense, his Advice Taker.

There are people who consider McCarthy's efforts on behalf of LISP and time-sharing to be diversions from his serious work in artificial intelligence, but in fact they seem to me to be all of a piece. The common way of using computers to squeeze maximum efficiency from them is to run continuous batches of problems through them. For many uses, this is perfectly fine. A department store needs to bill its customers only once a month; the grinding out of census statistics will not be improved by any human interference during the grinding-out process.[5] This batch processing is what John McCarthy had found at IBM in 1955, and which no one seemed to have had any plans to change. But how could you give advice to an intelligent program if you couldn't get at it during the time it was going through its problem solving? It was rather like being coached in tennis over the telephone the night after you'd lost a crucial match.

---

[5] It also happens to be a splendid excuse for human rigidity. We lose a paycheck, say, and we're informed with great solemnity that "because of the computer" we'll have to wait until the next pay period, and make whatever explanations we can to the landlord and grocer. We protest and fume to no avail in the face of such insensitive system design, especially those of us who know very well the blame is in no way "the computer's."

So the Advice Taker was to improve its performance by having statements—advice—given to it in real time, telling it about its environment and what was wanted of it. It was automatically to deduce for itself a sufficiently wide class of immediate consequences of anything it was told and already knew. Such a program must have certain specifications. For example, it must allow for interesting changes in behavior that are expressible in a simple way, relatively speaking, in the way that genetic change is basically simple but provides all the variety of flora and fauna that we see. It has to have a concept of only partial success, for on difficult problems decisive successes or failures come very infrequently. And so it goes.

The Advice Taker shared with GPS a penchant for relative simplicity, and it certainly was planned to be as general as possible. Unlike GPS, however, no Advice Taker exists.

I'd heard the term Advice Taker so often before I spoke to McCarthy that I told him I was surprised to learn that it was still a proposal. "No," he said, "it doesn't exist. Because in order to do it, you have to be able to express formally that information that is normally expressed in ordinary language. As far as I'm concerned, this is the key unsolved problem in AI. I uncovered the problem in 1958 and it's still unsolved." McCarthy has himself made several attempts to invent a formal language that would be able to express the events of everyday life: it's the one scientific problem he's stuck to, among the variety of others he's taken up. But in his view, the general problem has simply not been attacked by enough good people to solve it, and he believes that until they do artificial intelligence will remain somewhat stuck.

Not everyone in AI shares McCarthy's pessimism. Indeed, in the mid 1960s, logician Alan Robinson published a paper on what he called the Resolution Method, a highly efficient way of proving theorems in the first-order predicate calculus. It seemed as if McCarthy's dream of a uniform problem solver had been realized, and a rush to Resolution was on. Several of McCarthy's graduate students went to work applying Robinson's method to the world of facts, among them Cordell Green, whose QA3 program behaved as a sort of General Problem Solver. A group at the University of Edinburgh also took up Resolution, and it became a topic of great interest for the AI group at Stanford Research Institute. James Slagle, whose SAINT program, a simulation of a freshman calculus student, had been one of the pioneering AI efforts at MIT, directed his group at the National Institutes of Health in a concentrated effort that produced MULTIPLE, one of the best of these mathematical single general-search methods.

But all this effort eventually collapsed. It seemed that the Robinson method generated search spaces as large as ordinary heuristic methods. There was no way by which the Resolution Method theorem provers could use real-world facts to constrain those spaces. Edward Feigenbaum, McCarthy's colleague at Stanford, and one who had not been enchanted by the Resolution Method, put it this way: "It's very awkward to translate your knowledge of a task domain into predicate calculus, and the difficulty of doing it is exceeded only by the awkwardness of how it looks after you've done it." And it was Feigenbaum who provided a small note to the social history of science. Asked how Robinson reacted to his sudden rise and fall in AI, Feigenbaum laughed:

He underwent an unwanted spectacular rise to stardom in AI research, unwanted because Alan is a logician and views his activity from the point of view of that peer group. He wasn't necessarily concerned with his reputation in this very strange peer group called artificial intelligence. But here he was, propelled to the front ranks, and suddenly felt heavy obligation to extract the AI researchers from the pit into which they were falling, the pit of the combinatorial explosion.[6] He understood this, but was really helpless to do anything about it since he was a logician who invented a method, not an AI re-

---

[6] The combinatorial explosion takes place when growth is exponential, for example, as Malthus feared population growth would be, therefore outstripping the world's food supply, which only increases linearly. Thus, two parents have more than two offspring, who in turn each have more than two offspring, who in turn.... Problem solving encounters the same difficulties, when one alternative path to a solution leads to several more branches, each leading to further branches. Thus, Claude Shannon was able to calculate his 10^{120} possible moves in a chess game where every move and its consequences were explored.

searcher interested in formalizing the world's knowledge. Finally he gave up, decided he was really sorry he'd got people into this trap, but he couldn't do anything about it. As AI moved away from the Resolution Method, he moved back to logic and resigned his position on the editorial board of the AI Journal, and retreated from the whole scene.

Not everyone was enchanted. Newell believed that the Resolution Method was no improvement over GPS, the eclecticism of the AI group at Stanford enabled it to evade the trap that devotion to one method would have led to, and the MIT group was openly hostile. But the Edinburgh researchers' passionate fling with the Resolution Method would be costly to them later, in the opinion of several observers.

Yet McCarthy still longs for a formal language that will express the facts of common knowledge, a rather lonely position in AI just now. But then he's a man who has always been driven by extraordinarily high intellectual standards for himself. But those high standards have also perhaps accounted for his relatively short list of scientific publications, and his dissatisfaction with other people's work. They even account, he once said to me, for his long silences in ordinary discourse, because if he can't think of anything to say that's worth saying, he keeps quiet, which can be hard on the partner in the colloquy who is left to wonder whether he or she has said something stupid, offensive, or both.

Like most highly gifted people, McCarthy is really interested in doing only what challenges him. He has been the despair of funding agencies because he couldn't be bothered to write progress reports. Thus, no one in Washington knew whether his scientific work was moving ahead on schedule, or whether the large collection of programmers, hackers, and other assorted AI groupies who find the Stanford AI Laboratory congenial were spending twenty-four hours a day playing Space War, sitting in the laboratory's unisex sauna, or frolicking on its waterbed. Some of this waywardness was remedied by hiring a project administrator to oversee nonscientific activities. Also, McCarthy's introductory artificial intelligence course at Stanford in the 1960s was so vague it was known among the graduate students as Uncle John's Mystery

Hour (recalling Dr. Spooner, who admonished a luckless student: "You hissed my mystery lecture"), though if McCarthy is excited by a topic, his lectures can be awesome.

He has a wonderful talent for nettling his colleagues, often because he expects them to fill in the gaps he is too impatient to attend to as he explains a new idea. "I think he's mellowing on that score," said one of his associates recently. "We had a conversation the other day, and I actually heard him using words like *because* and so *it follows*."

Stephen Coles, a former student of Simon's who eventually went to SRI, tells a story about when he was doing his post-Ph.D. job hunt. He'd been in an enormous hurry to get his thesis finished, and so the slides to illustrate his lecture were faulty—they'd been done under pressure at a late hour, and they showed some bugs still surreptitiously in the code he was using that caused an inconsistency.

So I said to myself, well, I'm not going to rewrite any programs now— the thesis is done and no one would ever see this error; it's only clear to someone who's been an author of the system and worked on it very hard and understands it, and I'll just pass right over it. McCarthy that day at the AI lab was his usual self—reading the newspaper and looking at the ceiling and not paying any attention, dozing off at my presentation like he's totally bored by what I have to say. Other people, of course, are interested and following what I have to say, and so that was good enough—if he doesn't want to pay attention, that's his problem. And at the end of the talk he says, go back to figure blah blah, there's an error there. Nobody else saw it and I was just totally stunned. The guy is phenomenal. He has all these other idiosyncrasies which are hard to overlook, but there's nobody else that I know of in this community who is so sharp at spotting weak spots. And that's really what you need if you're trying to do something original which has never been done before, someone who can challenge, find the flaws, shoot it down.

McCarthy can be enormously provoking. Several people told me about a meeting where the leading researchers in AI had gathered to make a collective presentation insuring the continuation

John McCarthy in high spirits

of funds from the defense department. Not only were all the biggest names in AI there, but several high-level executives from defense department agencies were present as well, and the object of the meeting was to overcome their skepticism. McCarthy had brought along his Polaroid, and shortly after the meeting began, he brought it out and began ostentatiously snapping pictures. Pretty soon he got up out of his seat and walked around the small, crowded meeting room, getting close-ups, fresh angles. After an hour or so, somebody got annoyed enough to ask him why he was doing it. "As a memory aid," McCarthy replied simply, and kept on snapping.

"Anybody could see what was going on," one person at the meeting told me. "Here was supersmart John, but he was in a room full of people who are probably just as supersmart. How else could he distinguish himself except by making a pest of himself?"

Perhaps. McCarthy is also very shy, and shy people sometimes have funny ways of compensating for that. But once he has overcome his shyness, once a listener has John McCarthy for an evening of easy talk, there's no one more fun to be with. He's playful, almost giddy, and the stream of original ideas flows nonstop, most of them fantasies he has for solving the very serious problems of the world with technology, a continuing theme in his life. He is something of an apostle on that subject, believing that technology has been unjustly maligned. He is certain that if used with imagination, technology can in fact solve not only some of the problems its injudicious use creates, but also a great many other problems too. Some arguments along these lines have found their way into print in letters to the editors, but anyone in AI who wants to know what John McCarthy is thinking right now can hook in by computer terminal to the ARPANET, a network of connections among computer installations funded by the Advanced Research Projects Agency of the Department of Defense, and read the latest edition of McCarthy's private newsletter. His faith in technology wasn't shaken in the least by his long involvement with the counterculture during the 1960s: on the contrary, McCarthy was bringing the same message to the Free University of Palo Alto as to the *New York Times*.

In the fall of 1962, McCarthy left MIT for Stanford. Several reasons seem to have caused his move. Partly he was fed up with the politics surrounding the MIT time-sharing project; partly he felt unappreciated at MIT, and Stanford offered him a better position at a higher salary. He hoped to get an artificial-intelligence effort underway at Stanford that would be a tight little group of smart people doing interesting things—a dream rather different from the sprawling project that eventually grew on top of the dry, grassy hills behind Stanford. For what started out as McCarthy and a handful of graduate students working on games, theorem proving, and other such logistically modest efforts was suddenly inflated by scores of people, and the reason for that inflation was robotics.

Robotics, of course, provides the perfect environment for scores of people, in particular graduate students who can bite off thesis-sized chunks. The subject had caught the fancy of the military, which had a lot of money to spend and was willing to spend it on something with such obvious military potential. And it simply cannot be denied that to crack the problem of an intelligent entity that would interact with the real world is as appealing a scientific problem as anyone could dream up.

Bert Raphael, deeply involved in the development of Shakey, the Stanford Research Institute's intelligent mobile robot, has pointed out some of the specifics of robot research. It embraces several aspects of intelligence, such as pattern recognition, problem solving, information representation, and natural-language processing, all of which have continuing research interest. It must be general—an aesthetic canon of science for very sound reasons—and sufficiently rich and open-ended to offer a tremendous range of problems of ever-increasing complexity. That a robot has to cope with the real world puts it in quite a different class from intelligent programs that operate in formal domains, such as mathematics or game playing, for "a problem-solver in a formal domain is essentially done when it has constructed a plan for a solution; nothing can go wrong. A robot in the real world, however, must consider the execution of the plan as a major part of every task. Unexpected occurrences are not unusual, so that the use of sensory feedback and corrective actions are crucial" (Raphael, 1970).

Marvin Minsky explains his interest in robotics somewhat differently. The son of a physician, and married to one as well, he'd become fascinated by the idea of microrobotics for surgery. Why not machines—or, better yet, robotic instruments—that could crawl into arteries and scrape off the accumulated fat deposits, or that could make delicate tissue repair in spots inaccessible to human eyes and fingers? By the time he'd drafted proposals for robotics projects as a consultant at SRI and Bolt Beranek and Newman, he was persuaded to whip one up for MIT, and so he did.

Yet Minsky took, and still seems to take, a lukewarm stand on robotics as a way of doing AI:

You might say that making robots was a sort of hobby which I encouraged but didn't really concern myself with that much, and I always felt that studying the sensory and perceptual systems is not the best way to think about thinking, because the sensory systems are developed in lower animals as well, and come prior to symbolic intelligence. So you can study those things to death and you may only learn about some hard-

ware tricks that were developed over a few million years that don't really tell you how the problem-solving parts of the brain work. We may have looked from the outside like a great deal of robotics work was going on here at MIT, but the things I was most concerned with were the theses like Slagle's and Bobrow's and Raphael's, and such people who were really working on the symbolic problem-solving things.

Thus three large robotics projects got underway in the mid-1960s in the United States—at Stanford University, at Stanford Research Institute (which is nearby, but no longer officially connected with Stanford University, and is known officially now as SRI International), and at MIT. A fourth large project at Edinburgh University, sponsored mainly by the Science Research Council of Great Britain, soon joined in, and considerable exchange of ideas took place among the four centers. Though each project had its own flavor, the general aim was the same—to produce some sort of independent agent that would function in the real world, or at least a somewhat impoverished real world.

With P. J. Hayes of Edinburgh, McCarthy wrote a paper in 1969 that outlined some of the central ideas of robotics. By this time all four robotics projects were well underway, and Hayes and McCarthy were no longer speaking entirely theoretically. They began by pointing out that a computer program capable of acting intelligently in the real world must have some knowledge of that world, and to design such a program requires commitments about what knowledge is and how it's obtained, central issues in philosophy since Greek times. Other points of philosophical debate must also be formalized: the nature of causality and ability, and the nature of intelligence. This is precisely what Minsky and Papert were getting at in a document they wrote in 1971 describing the robot project at MIT, that robotics provides a perfect medium for testing any ideas about the nature of intelligence, for if an idea about intelligence can't be made to work empirically, it probably isn't a very good idea (Minsky and Papert, 1971).

What is a general intelligence, McCarthy and Hayes asked? Turing's idea that a machine should successfully convince a sophisticated observer that it is human for half an hour will do, though such a test has some built-in liabilities[7] and provides necessary but not sufficient conditions for intelligence. Indeed, Bernard Meltzer, also of Edinburgh, wrote a brief essay suggesting that the Turing Test be retired, having done its proper work in the political battle to establish artificial intelligence as a respectable scientific discipline (Meltzer, 1971). A general-purpose intelligence is nearly impossible to specify, Meltzer argued, for even if we could, we would draw only on an arbitrary set of abilities that are a product of the biological and cultural evolution of the human race. Other abilities for performing intelligent tasks exist or can be conceived to exist, including some possessed by other animals, new abilities people might develop in the future, or entirely new ones in the intelligent machines of the future. Meltzer urged his colleagues to concentrate, rather than on a general intelligence, on special or restricted instances of intelligence, for the Turing Test involves an attempt to select only one such set.

It was distinctly possible, Meltzer concluded, that much of the work in robotics would make little contribution to artificial intelligence, but would instead elucidate in a concrete way traditional problems of human epistemology:

> One should not believe that machines which mimic human behavior and to that extent have a rather "general" but limited capacity are necessarily more intelligent, more useful, or more worth research than ones with a lesser range of capacities—or even one capacity only. The

---

[7] And may already have been met. When Daniel Bobrow worked for the Cambridge, Massachusetts, firm of Bolt Beranek and Newman, he reported this in the *ACM SIGART Newsletter* of December 1968:

"One Saturday morning about 9 A.M., a BBN Vice President in charge of selling our Telcomp commercial service arrived at our PDP-1 computer room hoping to be able to use the computer for a demonstration. He found a note on the PDP-1 teletype that read: "TYPE ALL MESSAGES INTO THIS TELETYPE, AND END THEM WITH A PERIOD." This note was left there by one of the programming bums who had just succeeded in making run a LISP version of the DOCTOR program (based on ELIZA as developed by Weizenbaum). However, the VP thought that this note indicated that this teletype was linked with the one in my home, since I often used the computer all night. Therefore, he started typing to "me," but in fact was typing to this version of the DOCTOR program. The following conversation ensued:

VP Do you think that I can use the computer this morning?

TELETYPE Why do you ask?

VP I might be able to make some additional telcomp sales.

TELETYPE Why aren't you sure?

VP My prospective customers have never seen a demonstration of the system.

TELETYPE Does this mean very much to you?

VP Of course it does.

TELETYPE You seem very positive about that.

VP This is ridiculous.

TELETYPE What do you mean this is ridiculous?

VP Is there any reason why I can't use the computer now?

TELETYPE Tell me more about your prospective customers.

VP Please dial me up on 491-1850

"Note that after the last remark the VP did not obey instructions, and left out the period. Therefore, of course, the computer didn't answer him. This so infuriated the VP, who thought I was playing games with him, that he called me up, woke me from a deep sleep, and said:

VP Why are you being so snotty to me?

BOBROW What do you mean why am I being snotty to you?

"The VP angrily read me the dialog that 'we' had been having, and couldn't get any response but laughter from me. It took a while to convince him it really was the computer."

Naturally, there were quibbles about whether Turing's Test had indeed been passed. A short note from one J. Markowitz, also of Bolt Beranek and Newman, was appended to Bobrow's tale:

"Did Dr. Bobrow's story show that a computer had passed the imitation test of Turing? Sure, our vice president failed to detect the presence of a computer in six successive exchanges. But that's not Turing's Test.

"Turing specified that an observer be forced to choose between two (hidden) devices—one known to be a man, the other known to be a machine.

"It is easy to show that Turing's arrangement leads to better performance by the observer, and is thus a stiffer test for machines who would pretend. More importantly, Turing's arrangement minimizes the effect of bias on the observer's part. Clearly, the VP held a strong a priori belief that he would be talking to a man—a severe limitation."

only genius—or near-genius—I ever had close relations with would possibly have failed the Turing Test, as he was practically incapable of carrying on a coherent conversation with his colleagues!

(Meltzer, 1971)

McCarthy and Hayes certainly saw some risk in using human beings as a model for making intelligent robots. For example, we may be mistaken, they declared, in our introspective views of our own mental structure; we may only think we use facts. Moreover, there may be entities that satisfy behaviorist criteria of intelligence but that are not organized in this way. Nevertheless, the construction of intelligent machines as fact manipulators seemed to them the best bet both for creating artificial intelligences and for understanding natural intelligence.

Therefore, they were willing to declare an entity intelligent if it had an adequate model of the world (including the intellectual world of mathematics; understanding of its own goals and other mental processes), if it was clever enough to answer a wide variety of questions on the basis of this model, if it could get additional information from the external world when required, and if it could perform such tasks in the external world as its goals demanded and its physical abilities permitted.

Four kinds of problems are inherent in the construction of an intelligent entity, all having to do with representations) of the world. They are:

1. How to allow the incorporation of specific observations and generalizations from those observations.

2. How to represent data from other than the physical world.

3. How to get knowledge about the world.

4. How to assimilate and express that knowledge internally.

Since these questions have not been solved by philosophers in more than twenty-five hundred years of dispute, artificial-intelligence researchers might well be dismayed about their own hopes for solutions. But McCarthy and Hayes shrug: one may as well begin the journey with a few working assumptions. These are, first, that the physical world exists and already contains some intelligent machines called people; second, that information about this world is obtainable through the senses and is expressible internally; third, that, moreover, our common-sense and scientific views of the world are approximately correct; and, finally, that the best procedure is to use all of human knowledge in trying to construct a computer program that knows. Modern philosophers could take issue with any one of these working assumptions but, as McCarthy and Hayes have pointed out, none offers any scheme precise enough to substitute.

The robots conceived and built at Stanford, MIT, and Edinburgh were hand-and-eye affairs, the eye a "seeing" television camera, the computer brain processing the camera's perceptions and instructing the arm(s) to move accordingly. I can describe these efforts so quickly, and yet the simple acts the robots could perform represent a knot of epistemological problems that no one had ever successfully sorted out before. The painstaking problems of deciding how the image on the retina is converted to a symbol inside the brain, and what form that symbol takes (a full-blown three-dimensional picture? a flexible outline, with only essential details? a token that only stands in for the image?) provided enough problems for a small army of researchers. These questions indeed raised problems in all the aspects of intelligence Raphael had cited, and those problems are still a very long way from being completely solved. But the solutions that were found began to suggest that many problems which seemed at first impossibly nonmechanical—exactly those problems that the most vehement critics of AI have declared will never be solved, such as "understanding" and "meaning"—slowly began to be brought into the domain of ordinary computational processes.

I don't wish to imply that these concepts were clarified only by robotics research. In particular, "meaning" has revealed its mechanisms in a number of different problem domains. But robotics research underscored the intimate, unseverable connections among all these aspects that we cannot see unless we have some knowledge of what we'll see; put another way, we cannot assign meaning unless we have a context in which to assign it. The same principles were to emerge in speech-understanding systems later on. If they are by now psychological commonplaces, we have to remember that robot builders were faced with questions of just how much knowledge was necessary for understanding, couched in precisely which terms. AI researchers call this the knowledge-representation problem. And for some researchers, its centrality still wasn't obvious.

Joel Moses of MIT says,

The word you look for and you hardly ever see in the early AI literature is the word knowledge. They didn't believe you have to know anything, you could always rework it all. And it's a tremendously arrogant person who could believe that you could rework it all on the fly—start with this simple machine and just feed it a few things and all of a sudden you get Einstein's theory of relativity. And nearly everybody bought that view. It took a long time for it to wither away.

Moses sees both the work in robotics and other work in real-world rather than toy problems as a turning point:

In fact, 1967 is the turning point in my mind when there was enough feeling that the old ideas of general principles had to go. I believe we could have worked earlier on issues that involved knowledge in a bigger way. I think the first place we see the old ideas found wanting is in Danny Bobrow's work on STUDENT—here's the first piece of work when someone is giving up on GPS, is trying to solve a problem and see what's really in it. My own work was the same, and in a sense an attack on Slagle's work, which had taken a generalist point of view. I came up with an argument for what I call the primacy of expertise, and at the time I called the other guys the generalists. I was antigeneralist for many years. I think the field of artificial intelligence has been essentially healthy since then, but it took some difficult doing. I think there was a tremendous battle between Papert and Minsky between 1965 and 1970. Minsky's view of my thesis [a system called MACSYMA to aid mathematicians, which is deeply steeped in specific knowledge] was that it wasn't AI. He came to my exam and said, this isn't AI. Papert and he had had a five-hour argument the day before. But the old ideas were dying.

Newell also fought the idea, says Moses, clinging instead to hopes for generalism.

He called my position the big-switch theory, the idea being that you have all these experts working for you and when you have a problem, you decide which expert to call in to solve the problem. That's not AI, you see, but he didn't say that. I think what finally broke Newell's position was [Terry] Winograd. Essentially I think Newell is doing some very good work right now, but it took almost fifteen years. And it took Minsky nearly as long; McCarthy hasn't converted much, and Simon really isn't playing such a major role these days so it's not so critical he change his views. Besides, he always has very interesting positions on things, so it's quite fine. Simple Simon can continue.

In any event, robotics seemed to say very strongly that knowledge—lots of it in depth—was at least as essential as general principles of intelligence. And these results came to stand also as the most convincing denial yet of a difference between mind and body since ancient theologians first made that division and renaissance philosophers calcified it.

That impoverished world I mentioned was usually a world of toy boxes and blocks, moved around here and there by command. After the hand-eye robot at Stanford had proved itself by learning to move blocks around on a table, it graduated to a considerably more complicated task of assembling an automobile water pump from parts scattered randomly on a table. A film exists of the Stanford robot engaged in this task, which runs about four minutes. Even with such brevity, the average viewer is hard put to stay awake. We've seen assembly-line processes accomplish what looks like—what looks like—the same thing, so we remain unimpressed. We have to remind ourselves of how we'd feel watching a beloved four-year-old human child engaged in the same task, sorting out the pieces one by one, "knowing" that the big nut must be used to secure the big screw, and that one part, put on in haste, has to be removed to accommodate a seal that was forgotten in the first attempt. If one deserves awe, so does the other: the human act because at long last we are beginning to understand how really awesome in its parsimony the human brain is, and the robot's act because we are approaching some duplication of that breathtaking parsimony.

Both the MIT robot and the Edinburgh robot (called Freddy) had the same general effect on the naive viewer. Even less naive viewers were often perturbed. In the same essay I quoted earlier, Bernard Meltzer of Edinburgh made some uncharitable cracks about Freddy pushing his blocks around in his toy world, and the waste of good human brainpower to produce nothing more than this feeble creature.

If there was a robot you could feel affection for, it had to be Stanford Research Institute's Shakey. "We worked for a month trying to find a good name for it." says Charles Rosen, who headed the SRI artificial-intelligence group, "ranging from Greek names to whatnot, and then one of us said, "Hey, it shakes like hell and moves around, let's just call it Shakey." Unlike the other robots, Shakey was mobile, and could propel himself from room to room of SRI, evading obstacles and recovering from unforeseen circumstances, such as schoolchildren standing agape at his progress through the halls.

When I saw Shakey he was in retirement in the SRI office of Bert Raphael, one of his designers. Shakey was a sad sight, immobile in a corner. Some sort of mucilage that had kept body and soul together had flowed from his supporting platform, solidified, and looked now for all the world like a hula dancer's skirt, modestly concealing Shakey's wheels. He'd surely seen better days.

Raphael shook his head as he looked over at Shakey, explaining to me the genesis of the project. He himself had received one of the first Ph.D.'s given at MIT in artificial intelligence, had been hired by SRI for this project as the only one who knew LISP and who had had experience with the LISP language and large computers.

Initially our interests were to see what the state of the art was in learning machines and pattern recognition methods [he said], and in symbol manipulation and in modeling, and put them all together and see if we could make the sum greater than the parts. Suppose you have a visual-perception capability that can give information to the problem solver, and a problem solver that can predict what you're likely to be looking at to help the vision system. And the other initial goal was to see how much we could accomplish with limited hardware capabilities. At the time, MIT was working on vision and they were developing or contracting for very high resolution cameras.

MIT intended to exceed the human eye's capabilities, and SRI took the opposite approach: how much could you do knowing something about the environment and having other kinds of information about what the system is supposed to achieve? How effective could they make a simple system by combining the software with their current capabilities in hardware?

The vision capabilities were essential. Nils Nilsson, who was project leader for Shakey sees scene analysis, as it's called, as the one element common to the four robot projects. But each group took a different approach. SRI felt itself somewhere in the middle—its interest was in making sense of the scene, not getting the best images possible. Nilsson believes that Shakey's problem-solving system was probably more sophisticated than those of the others, which seemed to him somewhat ad hoc:

Those of us at SRI were more interested, I believe, in general problem-solving mechanisms for reasoning out the solutions to problems, so I think we were more sophisticated in that regard. We also concentrated a good deal more than the others on the interaction between the plan that was developed by the problem-solving system and the execution of that plan. Other people at other places were interested in such things, but there was never any connection made, I think, in any of their actual robot systems.

Shakey taught its creators some surprising things. Perhaps one of the most important had to do with that elusive property of generality, for Shakey showed that you could not, for example, take a graph-searching algorithm from a chess program and hand-printed-character-recognizing algorithm from a vision program and, having attached them together, expect the robot to understand the world. As Raphael put it, there are serious questions about the interaction between knowledge in different domains.

Another problem had to do with uncertainty in a complicated world—a problem most of us can appreciate.

When a computer chess program says pawn to king-four, it's absolutely assumed that the pawn is now in king-four, and it can go on to think about the next move [says Raphael]. But when we said, Shakey, move forward three feet, the only thing we could be absolutely sure of is that he did not move exactly three feet. He probably would move three feet plus or minus epsilon according to some normal distribution, depending upon the errors in the calibration and slippage in the wheels; but maybe he moves one and a half feet and runs into the wall, or maybe he doesn't move at all because the commands got garbled in transmission, or his batteries are low. So there's an interesting research area that we made some progress on—how to build robust systems, and what kinds of monitoring are needed and how the system has to check whether it accomplishes what it tries to accomplish. We developed ways of using the TV camera and sensory feedback to monitor and update Shakey's own model of the world. We built various ideas of representing information in the robot's mind as in a computer. In a sense, the robot has a model of itself and of its environment. It knows where it thinks it is in the world, and it also knows an expected value of the error in that. Shakey assumed every time it moved that there was some normally distributed error, and that got added into its knowledge of its position. Of course the error kept getting bigger and bigger, and when it got big enough, that would trigger another part of the program, which would come in and say, "Hey, I'm confused enough that I'd better look for some landmark and check where I really am." I think a lot of these considerations came out of this work that are now automatically part of our thoughts when we try to apply computer-control techniques to industrial automation, for instance.

Two major versions of Shakey were produced. The first version, completed in 1969, could manage a combination of abilities in perception and problem solving. Raphael's own informative book on artificial intelligence, called *The Thinking Computer, Mind Inside Matter* (1976), describes how the first version of Shakey worked, and contrasts it with major changes made in the second version. Curiously, the hardware—the robot vehicle itself—was virtually unchanged while nearly everything else was changed drastically. The core memory was nearly tripled, and four major levels of behavior were now programmed, the lowest levels concerned with physical mobility, a third level concerned with planning solutions to problems (a system called STRIPS), which was intended to combine both formal and informal reasoning and also to get around some of the problems GPS had raised, such as how to proceed with a task. The top level was executive, and carried out the plans completed by the second level; it could either rearrange the plans to achieve better results or call for better plans. The lower levels were also capable of detecting and correcting certain kinds of errors without reference to or help from the executive program.

We picked a mobile robot as a project [says Charles Rosen] because we believed that to cope with the ever-changing environment, or an environment that was not fixed, you'd have to solve some elementary problems of intelligence. This would mean a combination of computers, and machineries and sensors to sense the environment, and information about the world that must be stored in the form of models. All of these ideas were very early in the game.

Rosen recalls how they found someone in the defense department who was willing to support the research, though for what Rosen himself considered foolish reasons, namely, that somehow a robot could be developed that could go about surreptitiously gathering information—a mechanical spy, nothing more or less than the original golem. Rosen didn't make that connection, but how can we resist?

"A direct result of the Shakey work is our present work in automation," Rosen says. "We really did get some practical results." Why then was support from ARPA withdrawn?

I think the reasons were more political than technical. We were told later by the ARPA people that there were too many people raising issues of Shakey being a dangerous thing to have, which strikes me as a little silly, when the defense department has some really dangerous machines around like robot aircraft and remote weapons. What Shakey is going to contribute to the weapons system isn't much: the technology there is from very early stuff, on visual perception.

Whether you like it or not, something you develop is going to find its way into a weapons system. It has and it will always. Can't be stopped. Shakey served as a training ground and as a means for putting together a fairly elaborate computer-machine-sensor system. It also pointed out what was needed in a system that was going to solve real-world problems, and then Shakey died. I hope it goes to the Smithsonian with its white blood.

Nils Nilsson suspects that funding from ARPA was terminated because no immediate military application could be seen:

Like the other AI labs, we were very interested in basic questions of mechanisms of intelligence, problem solving. We thought the domain of robots was a good place to pursue those questions. It made us face a lot of real-world problems. But those problems were in the long range, and if ARPA was interested in something they could put in the field within five years, they probably made the right assessment. On the other hand, with our background and what we'd done, we could have transformed the project into something a good deal more applied. I don't know that we would have agreed to, but we could have.

Nilsson sees the legacy of robotics as the integration of visual and problem-solving systems—which might have been done without a robot, but probably wouldn't have been.

Of the four major robotics projects in the 1960s, Shakey probably received the most public attention, perhaps because it was the only mobile robot, and came closest to science fiction notions of robotics. To this Rosen allows,

Well, I was a bit responsible for some of its notoriety. I used to make it very easy for anybody to come and see it demonstrated, and that included a lot of kids, schoolteachers, some of the press, although toward the end we got so burnt with some of the press that we didn't want to show it to them any more.

Rosen clearly meant an article which was published in *Life* magazine and written by Brad Darrach, an article that was more than science fiction: the AI community feels it was victimized in this instance by outright lies. Said Rosen,

That article was very rough and that man didn't do right by us. He came here and all of us spent a great deal of time being very honest and candid with him. Then he didn't present the whole story. He picked the sensational things and left out the others. Like you asked me about Shakey. Some of its principles will be found in weapons of war, but if that's all you say without adding right alongside it that we're doing automation for industry, using many more of the principles from Shakey than the weapons do, then you're not telling the whole story, and I think that's wrong. And that's what he didn't do, that fellow. He didn't tell the whole story. He had a point of view that he wanted to demonstrate and he found the facts. I don't think he fabricated too much. But he didn't tell the whole story, and that's as dishonest as anything.

Bert Raphael, who had spent a lot of time with Darrach, goes further:

I guess the worst part of that article from our point of view was that we didn't imagine he could produce anything that bad. We were completely taken in by his sincerity and interest in what we were doing, and then he went off and wrote this stuff. He wrote it as if he'd seen many things he never saw, and wrote about seeing Shakey going down the hall and hurrying from office to office, when in fact all the time he'd been here we were in the process of changing one computer system to another and never demonstrated anything for him. There were many direct quotes that were imaginary or completely out of context.

Steve Coles adds,

My work is unrecognizable in that Life magazine article, although it was like a nightmare. I really had to strain my imagination to realize that he was talking about what I did.

Marvin Minsky was so exercised that he wrote a long rebuttal and denial of quotations attributed to him, but it was only for members of the AI community, and Darrach's article not only made an enormous, sensational splash when it was published, but has subsequently been anthologized and appears in at least one college textbook where it's taught and read by the credulous.

The Darrach article was one of the negative results of robotics research, a little fantasy that raised blood pressures and choler. The positive results are more varied.

The acquisition and understanding of sensory data has become much clearer since the work on robotics began. In 1971, Minsky and Papert wrote a document describing much of the robotics work underway at MIT and making explicit the human connection in robotics research. The report is concerned specifically with vision, whose mechanisms, despite much effort, have been very difficult to pin down. One reason for this difficulty, says the report, and the thing that makes vision typical of any intelligent act, is that vision is a deeply complex process, drawing on previously known facts and expectations, making analogies, and so forth, as well as being the physical process of light striking the retina. Of course, this idea is not original with artificial intelligence, nor is it a notion discovered belatedly by researchers in that field. Part of the early literature AI researchers claimed as their own was an article by J. Y. Lettvin, H. Maturana, W. S. McCulloch, and W. Pitts called "What the Frog's Eye Tells the Frog's Brain," which makes this point among others (1958).

Minsky and Papert pointed out how our feelings about what we see—for example, the belief that we see everything when we first walk into a room—are not borne out by tests:

> In general, and not just in regard to vision, [they wrote], people are not good at describing mental processes; even when their descriptions seem eloquent, they rarely agree either with one another or with objective performances. The ability to analyze one's own mental processes, evidently, does not arise spontaneously or reliably; instead, suitable concepts for this must be developed or learned, through processes similar to development of scientific theories.
>
> (Minsky and Papert, 1971)

So the ideas about vision that were developed in the course of trying to make a robot eye see suggested new ways to think about thinking in general, and about imagery and vision in particular. Furthermore, these new ways had to pass a test that many traditional notions in psychology and philosophy did not and could not pass: if a theory of vision—or thinking of any sort—was to be taken seriously, one should be able to use it to make a machine that sees, or thinks.

Thus, in the early 1960s, Larry Roberts, then a graduate student at MIT, began intensive work on what is called visual scene analysis. Roberts' work involved analyzing scenes containing polyhedra, a program that matched what it saw with its camera eye to its preformed expectations. Although visual pattern recognition had been an early part of AI, that effort had been limited to two-dimensional patterns, and aimed at merely categorizing patterns rather than describing or "knowing" what the patterns were. Roberts wanted to work with three-dimensional scenes, and was furthermore interested in trying something more sophisticated than the template-match-classify procedure that characterized most visual pattern recognition. His work, and that which followed it, was to lead to some sophisticated techniques of edge-detection, contour-following, and region-finding programs. There even emerged an elegant theory of permissible representation of edges and vertices and their relations to three-dimensional polyhedra—a theory not previously discovered by projective or descriptive geometers.

The intelligent control of effectors—that is, mechanical devices that can make changes in the physical world—increased enormously because of robotics research. Effectors were designed to deal with some of the events that could throw a good solution to a problem off the track, such as initial misinformation, accidental dynamic effects, and such. Again, the most intimate relationship between plan and execution had to exist. Real-world representations inside the computer had to be developed. Finally, when robotics research had got underway in the mid-1960s, the actual physical mechanisms available were generally primitive, and thus robotics improved hardware, in particular hand-arm devices, optical range finders, and special tactile, force, and torque sensors.

But the overwhelming message—not always recognized by those doing the robotics work themselves—was that general principles of intelligence were insufficient. As I pointed out earlier in this chapter, there was considerable resistance to that idea. Edward Feigenbaum and his group at Stanford, who were unconnected with the Stanford robot project but were instead working on a way of assisting chemists to do spectograph analysis, were coming to the same conclusion, but they felt very lonely in that discovery.

Joel Moses, whose thesis had relied on expertise instead of general principles, remembers the frustration of trying to expound that point of view. "Papert almost cried once," Moses remembers. "He said, 'How can you get those guys to listen?' That was 1966, maybe 1968."

But the robots seemed to prove the view beyond a shadow of a doubt. And does the end of robotics work in the early 1970s mean, as Dreyfus predicted, that no robot will ever be generally intelligent? That the robot cannot be intelligent, because it lacks a human body in a human world? The current view is hardly so. Although it is presently too expensive to devise a general-purpose robot for tasks that a special-purpose robot can do more efficiently, Nilsson has written, "It seems reasonable to predict that man's historic fascination with robots, coupled with a new round of advances in vision and reasoning abilities, will lead to a resurgence of interest in general robot systems, perhaps during the late 1970s" (1974).

Chapter Eleven

———

Language, Scenes, Symbols, and Understanding

Human beings have taken to examining our language the way we once examined our spirituality. The flaws in our spirits were what separated us from the angels; the fact of our language is what separates us from the beasts. Or so we've thought. In either case, we're looking for warranties of our humanity, and at least in the instance of language, there's not a small element of self-congratulation about it all.

"Perhaps of all the creations of man language is the most astonishing," crowed Lytton Strachey (1948), just the sort of sentiment you'd expect from a writer.[1] Do musicians and painters carry on so? Strachey might modestly have remembered the old Chinese proverb that a picture is worth a thousand words (poetic if not quite scientific truth—most pictures have more information

---

[1] Readers of this history may be amused by how Strachey continues his essay, called "Words and Poetry":
"Those small articulated sounds, that seem so simple and definite, turn out, the more one examines them, to be the receptacles of subtle mystery and the dispensers of unanticipated power. Each one of them, as we look, shoots up into
　　　A palm with winged imagination in it,
　　And roots that stretch even beneath the grave.
"It really is a case of Frankenstein and his monster. These things that we have made are as alive as we are, and we have become their slaves."
Strachey, a member of the Bloomsbury Group, which also included Virginia Woolf and John Maynard Keynes, was a relative of the late Christopher Strachey, who is mentioned elsewhere in this book.

content than that), doubtless invented by some Chinese sage to put down an uppity poet.

If the capacity for language is part of our human genetic endowment, as is presently supposed, then we're no more or less entitled to brag about it than our naturally curly hair or the wonderful way we smell to a lover. All we can say is that language is an essential property of human beings. But then we owe our biological supremacy to language, P. B. Medawar likes to point out (1977), for it is language that made it possible for us to inaugurate and retain the things we call culture and civilization. Moreover, he goes on, cultural inheritance is Lamarckian: thanks to language, what's learned in one generation may be passed on to the next. The symbol for language, the very word itself, supports Medawar's view: the Indo-European root it derives from means to collect, or gather, and carries with it associations of magic, of healing, of the law; almost an entire culture is bound up in this putting into words.

So we admire ourselves for what we cannot help, and perhaps we should. In the beginning, St. John informs us, was the Word, and the Word was with God, and the Word was God. Language is divine in origin and lends us its spark. In Norse mythology, Odin is supposed to have invented the runic alphabet, and Indra, far away in the Vedas, is credited with inventing articulate speech.

In the Greek word *logos*, thought and word are inextricable from each other. The medieval and rationalist view would later imply that humans invented language to express their thoughts, but we've more or less returned to the Greek view, and believe that human thought and its form of expression developed hand in hand (though Piaget holds that logic precedes and thus shapes language).

It's uncertain how long we've been conscious of this unique role language plays in the human species— obviously a long time, on the evidence of mythology. But another property of ours means a good deal to us as well. We share it with nearly all the other creatures, and still we acknowledge its centrality in our mental processes. Confronted with a thing we understand, we say "I see."

Sight seems to be less emblematic of intelligent behavior than language. Again, this view is supported by human mythology, which gives a place of great honor to those who are blind in the physical sense but seers in the metaphysical sense, and who can of course communicate that vision in words. Minsky's opinion is the same: as we saw in the preceding chapter, he worried that work on vision might teach us only about some hardware tricks that had been developed over millions of years, but nothing about the problem-solving parts of the brain. It happens that he was unduly concerned; we've learned some interesting things about the intelligence required by vision. Again, this requirement is implied in our language when we use vision to mean a grand, inspired scheme, as in Proverbs: "If the leaders have no vision, the people will perish." So I mean to treat vision research along with natural language in this chapter, but it will be somewhat subordinate, and I believe correctly.

"The ability to communicate through language defines us as human beings in a human society," says a popular essay on stuttering, and nobody nowadays would dispute that (Jonas, 1976). From the divine, language has been captured by and emblemizes the human. Where else can it go?

Well, we're having some luck teaching English to chimpanzees, and even more luck teaching it to machines. To accomplish this, we've so far had to resort to a kind of pidgin English—an artificial and highly simplified language that accommodates the needs of both parties, enough, let's say, to do business (the term *pidgin* seems to have come from the Chinese attempts to pronounce the English word *business*). Certainly the early computer languages were pidgins: complexity of expression was meager and they stuck to business; that is, to the very limited domains in which the languages were applied, and often for which they were custom-made.

But what if we could communicate with the computer more easily? By more easily, computer scientists meant using something approximating a natural language, and the closer the better. To be able to speak or communicate by writing to a computer in natural language instead of some arcane computer language would vastly enhance the kinds of scientific work that might get done, in the same way that humans speaking their natural language on the telephone not only eliminated the telegraph operator, but increased the number of wire-carried messages tremendously, and at least amplified, if not caused, some profound social changes.

Also, nearly everyone suspected that a more natural communication with the computer would surely be at the heart of what we mean by understanding, hence, intelligent behavior. If we could figure out how to make a computer "understand" our language, we'd finally know just what understanding and even language is all about. These were two good reasons for studying natural language and attempting to make machines proficient in it, and a number of computer scientists took up the project in the early 1960s.

Other kinds of specialists were also involved in computers and natural language, but these projects had little to do with the understanding of language. For example, it was mainly linguists who first undertook to build automatic translators from one natural language to another, and as it turned out, their approach— stressing syntax rather than meaning—was to come to grief, and serve as a good demonstration of the fact that for any language transaction beyond the most trivial, understanding is essential.[2]

Initial work focused on syntax rather than meaning because linguists then believed that some transformation could be made

---

[2] In content analysis, for example, certain values were assigned to certain key words in diplomatic exchanges. Thus, "We are going to wipe you off the face of the earth" might sum up to a score of 10, while "Come, let us reason together" might sum up to a score of 1. This technique allowed for large numbers of diplomatic texts to be analyzed for their relative bellicosity or peacefulness. Related to this were textual analyses of great writers—did Bacon really write Shakespeare?—which worked on approximately the same principle. The tedious job of putting together concordances and indexes could suddenly be done by machine. I remember consulting one of my own professors in the late 1950s, and after we'd finished our class business, he got to bragging about the concordance he'd just done on the works of one of the Italian greats—Petrarch, it might have been. He was proud and relieved; ten years of grubby scholarship locked into galleys. I gasped, and he was pleased. I ruined the effect, and much of his pleasure too, I'm sure, by explaining that my gasp was dismay, that it all could've been done by computer in a small fraction of the time.

from one language to another by a process of identifying the atomic parts of speech, namely words, and then consulting a dictionary. *The cat is black* easily becomes *le chat est noir*. But the harder stuff made trouble.

It was soon obvious that translation isn't merely transformation, but consists of a process of "world modeling," as Yehoshua Bar-Hillel, the well-known Israeli linguist, put it—the machine must, in some sense, understand the text before it can translate into another language, and it is in reference to the world model that understanding takes place. Thus, such sentences as, "The pen is in the box" and "The box is in the pen," which would utterly confound a mechanical word-by-word translator, are easily translated by a human who carries in his head the model of a world where pens (writing instruments) are to be found in boxes, and boxes (toys, perhaps) are to be found in small enclosed places where infants play. It would be impossible to provide a machine with a world encyclopedia, Bar-Hillel went on, much less the capability to infer new concepts from the facts in such an encyclopedia. Therefore, mechanical translation was impossible; therefore, case closed (Bar-Hillel, 1964).

I've already reported Mortimer Taube's tirades against machine translation, and that a blue-ribbon committee of scientists headed by John Pierce, then of Bell Labs, delivered the coup de grace to funding for machine translation in the mid-1960s. All this activity obscures the fact that some interesting work did get done, and that the translations I've looked at aren't all as hopelessly hilarious as the critics of machine translation claimed. Ten years after the Pierce committee's report, mechanical translators were used for the American-Soviet Apollo-Soyuz project, and a French-English mechanical translator was in use at the European Economic Community Center. But during the decade and a half in which research into automatic translation was at its most feverish, it began to look as if no program with the expertise of a human translator would ever be possible, since translation depended on understanding, and no one knew what that was, much less how to endow a computer with it.

Information-retrieval projects also addressed themselves to natural language, but in a different way. With an information-retrieval program, you'd go to a terminal and ask for all the articles stored in the data bank on the subject of, say, ibexes. By doing a search of its data and matching those data with the request, the machine might tell you more or less what you hoped to find. But the difference between an information-retrieval system and an intelligent question-answering system is like the difference between a library clerk and a reference librarian: one does what it's told and no more; the other understands, and is helpful in unexpected ways. To put it another way, in a mere information-retrieval system, the program makes no changes in the data base, wheras in an intelligent question-answering machine, the program amends the data with new input, draws inferences from the data it has, and so forth.

Why should an intelligent machine take the form of a question-answering device? It's probably because one of our rules of thumb about the nature of intelligence is that if we receive a sensible answer to a question, then we can assume that we've been understood, and understanding is essential to intelligence. Thus our main way of testing understanding in schools is to ask questions and hope for sensible answers.

This assumption also underlies Turing's Test, that durable experiment which proposed that if an interrogator couldn't tell whether he or she was communicating with a human being or a machine, then the machine could be considered intelligent.

In 1959, when most transactions between humans and computers looked to the innocent observer like code from a stock exchange in the netherworld, L. E. S. Green, Edmund Berkeley, and Calvin Gotlieb were inspired to construct something they called The Conversation Machine. With perfect cocktail-party manners, it chatted amiably about the weather. "I do not enjoy rain during July," a person might say to it. "Well, we don't usually have rainy weather in July, so you will probably not be disappointed," the machine would reply politely. Its creators had stored in its memory facts about typical weather, facts about time, and the notion of operators—that is, verbs, specifically change, stop, begin, let up. The Conversation Machine did no syntactical analysis, but it did manage to figure out simple questions and, using a matching process, make sensible answers to them.

What The Conversation Machine showed—as did a similar program called Oracle, done at about the same time by A. V. Phillips as a master's thesis at MIT under John McCarthy—was that if simple statements could be encoded semantically and syntactically, they could be matched to discover how closely they resembled each other. Thus the human's remark was encoded—translated, as it were, by the machine—and matched to coded data assembled in the machine's memory. Depending upon that match, the machine would make a proper response, which it would then retranslate from code to English for purposes of responding to the human.

Two slightly more sophisticated question-answering machines were reported in Feigenbaum and Feldman's collection *Computers and Thought* (1963). One answered questions in ordinary (though limited) English about the month, day, place, and teams and scores for each baseball game in the American League for one year. As cocktail conversation, this is a cut above the weather, but there were several restrictions even so, and the Baseball program was stymied by semantic ambiguities, and furthermore would return its answers in outline form. It might understand some limited English, but couldn't speak it, rather like a foreign tourist who can read the *New York Daily News*, but is tongue-tied when asking for breakfast. The program's developers, a group at Lincoln Laboratories under the direction of Bert F. Green, wrote that the problem of ambiguity remained unsolved, and that probably a way of allowing the computer to query the questioner would be the most powerful means of resolving ambiguities.

We are back to pens in boxes and boxes in pens (bull pens? press boxes? pity the poor machine now!). And we are back to how we know the world, and how we bring this knowledge to the task of understanding language. In fact, we might amend Warren McCulloch's lovely question—what is a number that a man might know it, and a man that he might know a number?—to ask, What is language, and what are humans that they know it?

A program contemporary with Baseball did, in fact, build an internal model of its universe and answer questions based on inferences it made from that model. This was SAD SAM, the work of Robert K. Lindsay, who had been a student of Newell and Simon at Carnegie. SAD SAM answered questions about kinship relations in a particular family; its information was given to it in English sentences (unlike Baseball's, which was stored ahead of time), and it slowly built an internal model of those relationships. In this sense it understood: it acquired knowledge, it made inferences, it paraphrased. But even in this very simple universe, some formidable problems arose having to do not only with denotation of words and concepts, but with connotation and implication as well:

> Knowing more than one is told is a characteristic of human performance which is present in most behaviors which are called intelligent [Lindsay wrote]. We have argued that this characteristic is necessary for machines which are to solve the real problems of information retrieval, language translation, and problem solving. And furthermore, we must find more efficient ways to store implications if we are to develop intelligent machines with finite memory capacities, that is, if we are to develop intelligent machines.
>
> (Feigenbaum and Feldman, 1963)

We understand more than we are told. That elusive idea seemed right and baffling at the same time. Chomsky's ideas of deep structures, that is, universal built-in ways of understanding language, might account for some of the answers, and the discovery of those deep structures would be helpful. But was there more? Surely there was. What was it? Words are units of meaning; we all know that. But faced with automatic translation, a word-for-word translation had turned out to be impossible. We seemed to be at a point where Hubert Dreyfus could make his assertion (though he hadn't yet done so) that humans derive their intelligence from living in human bodies in a real world; moreover, it seemed true that the language we learn shapes our perceptions of that world.

Well, maybe we know more than we're told because we can see too.

The perception problem was being tackled in pattern recognition and other kinds of visual understanding, as we saw in the preceding chapter. In some sense, the language and vision problems were parallel, because they both dealt with knowledge and its representation, with learning concepts, and with the specification of rules for recognition. And all of these elements have to do with learning, which is not only making analogies and generalizing from examples, but also includes the construction of new knowledge by *transforming* old knowledge.

Very little of these relationships was obvious to researchers in the early 1960s. Various efforts were made, some even combining language and scene analysis, such as Robert F. Simmons's program that generated natural-language descriptions of line drawings: "the dog is beside and to the right of the boy," "the boy is to the left of and taller than the dog." Simmons and his coworker D. Londe showed that geometric inferences could be made and then expressed in natural language by the computer, though they hoped that some of its principles might be extended from picture to nongraphic examples too.

With another team, Simmons helped build a system that tried to answer questions from an encyclopedia. They set themselves the problem of designing a program that would accept natural English questions and search a large text to discover the most acceptable sentence, paragraph, or article as an answer. It was a multiple-level program that consulted an index and a synonym dictionary, used logic, and assigned scores to possible answers, which it compared with the question posed. But the researchers found that human intervention was essential to resolve ambiguities, both syntactic and semantic, once again, a real human who operated in the real world.

In 1963, Thomas Evans, one of Minsky's students at MIT, constructed a program that would solve the kind of geometric-analogy problems that appear in intelligence tests. In these questions line drawings are supposed to be compared: figure A is to figure B as figure C is to which of the following figures? Evans developed a system that decomposed the figures into primitive elements and made descriptions of them. Another part of the system then compared the objects or descriptions and made relations between them. Finally, a higher level part of the system assessed the similarities and differences between the descriptions, and chose the best candidate. Not only did the program perform well, doing college-level work, but, as Minsky was later to point out, it suggested that this kind of description comparison appears to be of enormous importance, not just because it shows that such concepts as analogy could be handled by machine, but because through some of the techniques Evans used other problem-solving systems might be freed from their bonds of specialization.

A year later, in 1964, another of Minsky's students, Bert Raphael, who was later to work on the robot Shakey, published his doctoral dissertation on his Semantic Information Retriever, or SIR. Like SAD SAM, SIR too "understood" because it could accumulate facts and then make deductions about them. And like the Specific Question Answerer, a program developed by Fischer Black at Bolt Beranek and Newman, it understood some aspects of the meanings of words, though it did not have a very general way of combining them. SIR was mainly concerned with the organization of logical connections among facts rather than the problems of grammar and language. The Specific Question Answerer, for its part, followed the suggestion of John McCarthy's hypothetical Advice Taker, and could be reprogrammed by the question asker with additional information.

While Raphael was a student at MIT, still another of Minsky's students, Daniel Bobrow, did his doctoral dissertation by developing a natural-language machine, one that accepted ordinary high school algebra problems and transformed them into equations that could be solved arithmetically. Minsky says that Bobrow's program, called STUDENT, is a demonstration par excellence of the power of using meaning to solve linguistic problems:

By knowing in advance that a collection of sentences must describe some algebraic relations between some things, the program is able to make good guesses as to what are the relations and what are the things. The result is that, within certain limits, it can do the really hard part of high school algebra: "setting-up" the equations from informal verbal statements.

(Minsky, 1968)

The Project MAC laboratory[3] in the mid-1960s was fermenting with ideas. This was due largely to Minsky himself, whose relationship with his students is close and supportive, imbuing them—at least the ones who succeed—with self-confidence that borders on arrogance about their accomplishments. Once we were speaking of his role as thesis advisor, and he compared it to gardening:

Yeah, grappling at the weeds is what I do. We don't know very much about what makes the actual flowers work, but it's easy enough to get rid of the weeds. Now, in the case of the very good theses, it seems all the work is done by the student, and the effect I have is on viewpoint occasionally; that is, the main effect is when they're picking the problem and working on it and we just talk about it. I'm never happy with conversations with my students because I don't feel that anything crystallized, but usually after a while something good happens.

If Minsky was gardener in the 1960s, he's more like wizard in the 1970s. He spends less time at the AI lab, for his own interests have changed—"he's paid his dues," says one colleague, "and now he can go do what interests him"—and those interests happen to be the composition of serious music. But every now and then he sweeps through the lab, swift, intense, and perhaps he'll stop and talk. In no time at all, students have gathered to hear him, to exchange ideas.

Minsky's diction is as precise as a trained actor's, his knowledge nearly universal. He shares with the rest of the founding fathers of AI an omnivorous appetite for experience and knowl-

---

[3] Project MAC was the name given to a large assortment of computer research projects at MIT involving Man And Computers, or Machine-Aided Cognition. The acronym was deliberately ambiguous because no one knew which direction the research would take.

edge—of music, medicine, science fiction, history, engineering, mathematics, politics, futurism, fantasy. Which is not to say he can't be cutting. Watching Minsky slice up a colleague whose ideas seem ill-considered to him is as much terror as sport, and though the voice is well-modulated, the hands are fidgeting and tense, a sign of his emotional involvement. But toward those less well able to defend themselves from his staggering intellect, he is gentle, which may be why his students' feelings for him verge on adoration, and the only complaints are occasionally about his lack of attention to organized lectures, deadlines, and other worldly matters.

Minsky's gentleness is without condescension. Once when we were speaking of an early worker in AI who had some spectacular success and then disappeared from the field, he cautioned me about dismissing it all as a fluke. "Funny things can happen in a person's life to undermine his confidence, having nothing to do with his intelligence. We just don't know why some people don't go on, what life experiences they have that seem to stop their work." His personal loyalty is deep. His friends are not only from graduate school days but also from grammar school; he even talks every now and then to his fourth-grade teacher at the Fieldston School in New York, a man who encouraged his interest in science (as Minsky's father, a physician, also did).

And it was from the work that was going on in his laboratory in the early to middle 1960s that Minsky began to realize that the whole problem of "learning"—which had baffled psychologists and everybody else who wanted to think about intelligence—was, as he puts it, a nonproblem:

It took us a long time to realize—and people elsewhere still haven't—that in a sense, once you have the right kind of descriptions and mechanisms, learning isn't really important. It's important to find out how something might be done at all, and once you understand that, you might be able to see quite easily how a task can be learned—and that's a very deep change in viewpoint. The problem in learning now as we see it is how do you decide what you want to have in your memory? It depends on having good descriptions, and then finding the difference between the descriptions, and saying, that's the new thing. The differ-

ence isn't between the things, it's the difference between their descriptions. And that took a very long time for people to appreciate.

The answer to the problem of induction was again description, as it was to the problem of understanding. This conclusion would be tested not only in applications of certain computer programs, but also in Seymour Papert's laboratory for teaching schoolchildren, considered more fully in the next chapter. That description, or representations of knowledge, as it came to be called, was central to learning and understanding was not obvious until the early 1970s. It would take many approaches to understand understanding.

In his survey of natural language work up to the mid-1960s, Robert Simmons detected a number of similarities among the question-answering machines. Data had to be carefully organized. Language had to be both syntactically and semantically analyzed. Matching was the key operation allowing a program to determine whether a particular answer was possible or appropriate to the question being asked. Steady, even rapid, progress was being made, Simmons concluded, though the effort was only five years along. The most difficult questions had defined themselves: How does one characterize the meaning of a sentence? How are ambiguous interpretations, both syntactic and semantic, to be dealt with? How are inferences to be made without exhaustive searches? How are partial answers, widely separated in the text, to be combined? To what extent can we or should we translate from English into formal languages? Can these studies be attacked from a theoretical point of view, or do they yield best to the empirical approach of building large systems as test vehicles?

Nonsense, scoffed Vincent Giuliano, in a comment appended to Simmons's paper. Giuliano, then on the research staff at Bell Labs, wrote, "The paper might lead a casual reader to believe that considerable progress is being made—I tend instead to see evidence mainly of motion, with little real evidence of progress."

He pointed out what he called some brutal facts. The data the programs operated with were highly restricted, almost trivial. Beyond those areas, only the foggiest sort of understanding of semantics existed, and research in the area was likely to take many years to make real progress. There was still the problem of ambiguity, now resolved by human intervention, and nobody understood the relation between meaning and logical formalism. And, anyway, how do we decide what is relevant, which logical formalism cannot tell us anything about? Giuliano was skeptical as to whether any general principles had emerged from this work. The progression Simmons had cited of syntactic, semantic, and then logical analysis reminded Giuliano unpleasantly of the notion widespread among workers in machine translation that such translation should proceed through the stages of lexical, syntactic, and semantic processing, which hadn't been achieved and didn't seem likely to be.

> In summary [wrote Giuliano], my reaction is that in a rush to demonstrate that question-answering can be done by computer, sight has too often been lost of the fact that much is yet to be learned about language, and that a demonstration can only be as good as the knowledge of language that goes into it. The existence of procedures of alchemy does not create a science—theories are needed which lead to testable hypotheses, and artifacts of computer usage are likely to be of utility only insofar as they are based on such theories or hypotheses.
>
> (Simmons, 1965)

Simmons made a cheerful reply. Yes, question-answering machines were in their infancy: reader beware. But he took issue with Giuliano's version of how science is done (a version Mortimer Taube had also advanced in his early attack on AI). We eagerly apply what little theory is available from linguistics and logic, Simmons wrote, but theory often lags far behind model building and sometimes derives from it. We are not alchemists in search of an elixir of life or a philosopher's stone; we are scientists accumulating knowledge by the toughest kind of experimentation—that of building small, very complex models and testing their limits.

So Hubert Dreyfus's metaphor of alchemy was already in the pages of the *Communications of the ACM*, one of the main journals of computer science, and the dispute between those who would rush ahead and try things out and those who would wait until theory supported models was before us again. "If we wait til the physiologists get around to give us a theory of mind, we'll be waiting forever," Herb Simon had grumbled at the same kinds of criticisms of his own early work in artificial intelligence. The conflict seems to be a constant in this field, as it is in most of interesting science.

Four years later, in 1969, Simmons undertook another survey of natural-language answering systems. By this time we knew that words were not the atomic units of meaning after all; that feeble idea had come to grief most embarrassingly in machine translation. But other notions—or paradigms, as Simmons called them, borrowing Thomas Kuhn's term—were abroad, and these new paradigms had infused the issue of question-answering machines with renewed vigor. Not only had the systems builders learned from their earlier efforts, but they were aided enormously by the advent of higher level programming languages, just as other artificial-intelligence efforts such as game playing had been aided. Along with new ideas and better languages in which to express them, there was time-sharing, allowing an experimenter to fiddle with his system on the spot, bit by bit, instead of running it at a distance and waiting painfully for the total result, forced to hunt around in haystacks of code for a needle of error.

Taking advantage of these new languages and the interactive capabilities of a time-shared computer system, a restless young engineer named Joseph Weizenbaum produced a system he called ELIZA (for, like the famous Miss Doolittle, it could be taught to speak increasingly well). Weizenbaum had been working for General Electric, and had been responsible for the integration of software (programming) and hardware (the things you can put your hands on in a computer) for the Bank of America's ERMA project, one of the first large-scale computer data-processing efforts for banking, which have changed the way we do our everyday financial affairs in some profound ways. At loose ends after the successful completion of the ERMA system, Weizenbaum got interested in language. Ed Feigenbaum introduced him to a Stanford colleague named Kenneth Colby, a psychiatrist who had grown disenchanted with ordinary one-on-one psychotherapy and turned to computers as a possible way of gaining new insights into neurotic behavior, perhaps even producing new modes of therapy. The collaboration and friendship between Weizenbaum and Colby that began then would eventually be wrecked by highly differing views of good science and ethical behavior. But at this time, Weizenbaum was demonstrably helpful to Colby in the design of the DOCTOR program. How helpful—and whether the credit was properly shared—is part of their later dispute.

In 1963, Weizenbaum went to MIT, and it happened that as a faculty member, he was supplied with a computer terminal at home. It was a marvelous toy, and to have some fun, he designed a program that would answer such questions as, Is this April? Is today Thursday? It was a short, tricky program, based on sleight of hand, and it led Weizenbaum to ask himself some very serious questions about mystification and the computer that would later become the catalyst for a full-length book. I will deal with those issues later on; for the present I want to follow another set of questions it raised in Weizenbaum's mind, related as well to his work with Colby. If you could do a simple question-answering machine, why not a complicated one? How different would complexity make such a machine? Could you seem to have complex responses based on simple rules?

Question-answering machines were in the air. Bobrow was at MIT working on STUDENT; Raphael was working on what would be SIR; the Baseball program was a Cambridge-area product. To add impetus, Weizenbaum drove into work many a morning with his neighbor Victor Yngve, who had developed the COMIT language, for pattern matching. If you were going to play around with matching patterns, why not the patterns in English words and sentences?

ELIZA was the result. ELIZA was intended to simulate—or caricature, as Weizenbaum himself suggests—the conversation between a Rogerian psychoanalyst and a patient, with the machine in the role of analyst. There were a number of reasons for that choice. Partly it had to do with the illusions of mutual understanding that Weizenbaum senses human beings entertain. He explains:

What I mean here is the cocktail party conversation. Someone says something to you that you really don't fully understand, but because of the context and lots of other things, you are in fact able to give a response which appears appropriate, and in fact the conversation continues for quite a long time. We do it all the time, not only at cocktail parties. Indeed, I think it's a very necessary mechanism, because we can't, even in serious discussion, probe to the limit of possible understanding. I might say to you, "Well, this is rather like the quantum mechanical something or other," and you'll say, "I understand." Well, maybe you don't understand. Maybe you don't know anything about quantum mechanics except in the most rudimentary terms. You don't really quite see how it's like that. But you say, "Okay, in this context and for the purposes of this conversation, I can say I understand." You might even dig as deep as you can into your shallow pool of quantum-mechanical knowledge and say, "Oh, you mean it's the Heisenberg Principle," and that's really all you know. And I say, "Yes, wonderful—you really do understand," and all that sort of thing. That's necessary. It's not cheating.

And when Weizenbaum was looking for a context where he could carry on that sort of illusion, he needed one where ignorance would not destroy the illusion of understanding:

For example [he goes on], in the psychiatric interview the psychiatrist says, tell me about the fishing fleet in San Francisco. One doesn't say, "Look, he's a smart man—how come he doesn't know about the fishing fleet in San Francisco?" What he really wants to hear is what the patient has to say about it. Whereas if a policeman and a citizen were conversing, and the policeman said, "Tell me about traffic lights," that would destroy the illusion completely.

Illusions, yes—this notion informed Weizenbaum's choice, along with his own admitted fascination with the psychoanalytical process, his acquaintance with Colby and Colby's work, and more work yet by people from Massachusetts General Hospital who were curious about the things ELIZA might have to tell them about two-person conversations.

Once the program was up and running, Weizenbaum was worried about how to present it to the world. Though it did indeed simulate the conversation between a psychotherapist and a patient, Weizenbaum was convinced that it might be misunderstood as giving some insight into therapy, into madness. As it turned out, he was right to be alarmed.

ELIZA is seductive, and its protocols are wonderfully funny to read—that is, as long as you remember that the conversation is taking place between a computer and a human being, and not between two humans. But that distinction is very hard to maintain, at least for the human involved.

The first time I saw ELIZA in action was at the Stanford Computation Center, where a visitor from the Soviet Union was being shown around. He was terribly polite, as foreign visitors always are, even though we could detect the strain of three weeks away from home, in a foreign culture, and most trying of all, in a foreign language. He sat down in front of a computer console on the second floor of Polya Hall, and was connected with Kenneth Colby's version of ELIZA, called DOCTOR, the psychotherapist. I no longer have the precise protocol, but it went very much like this:

DOCTOR Good afternoon. Please type in your name.

[The visitor did so, and the doctor continued with some polite preliminaries.]

DOCTOR What brought you here to see me today?

VISITOR Oh, nothing much. I'm feeling a bit tired, that's all.

DOCTOR Why do you think you're feeling a bit tired?

VISITOR Well, I've been traveling a lot, and away from home.

DOCTOR Tell me about your family.

And it was here that the conversation suddenly became intimate. Our visitor, an internationally respected computer scientist, who knew very well that there was a computer at the other end of the line and not a human being, nevertheless began to disclose his worries about his wife, his children, his distance—both geographical and emotional—from them. We watched in painful embarrassment, trying hard not to look, yet mesmerized all the same. Something about that impartial machine had evoked a response from the visitor that the norms of polite human conduct forbade. If a sophisticated computer scientist could be lured into participating in such a conversation so that he became nearly oblivious to the spectators about him, what effect might such a conversation machine have on a less sophisticated person? It was just such possibilities (and they were to happen: frantic people who telephoned Weizenbaum and pleaded with him for just a little time with ELIZA in order to straighten themselves out) that worried Weizenbaum and made him seek advice about publishing.

First, Weizenbaum entitled his paper "ELIZA—A Computer Program for the Study of Natural Language Communication between Man and Machine," thereby squelching anybody's idea that his work had been about psychotherapy. On the advice of his colleague Robert Fano, who was then head of electrical engineering at MIT, he elected to publish in the *Communications of the ACM*, a computing journal, rather than other places the paper might logically have appeared, such as the psychiatric journals. He was dismayed to discover that in the publications lag that takes place in nearly all professional journals, somebody else had jumped the gun. Kenneth Colby was to publish a short note in the *Journal of Nervous and Mental Diseases* stressing the therapeutic aspects of the program, or his version of it, which Weizenbaum had helped him set up and get going. Weizenbaum insisted that an addendum be inserted to the effect that his own purposes were different from Colby's, which Colby did.

What additionally irked Weizenbaum, and helped accelerate the split between him and Colby, was the feeling that Colby had seized ELIZA and made it his own, under the name of DOCTOR, without giving due credit to Weizenbaum. The AI community, well aware of the split between Weizenbaum and Colby, generally believes this to be its cause, and nobody laughs. You needn't take sides to see that the assignment of credit in science is not a small matter. A scientist's only currency qua scientist is the sum of his ideas. If someone else gets credit for them, the scientist is robbed, not in the way we might be of money, but rather in the way we might be robbed of our good name. To have one's good name filched is to be poor indeed; in science, it's a scientist's identity as scientist, nothing less.

But the main disagreement, Weizenbaum strongly insists, is his fundamental belief that the program is of no therapeutic significance, whereas Colby maintained very strongly that it could be. This is not just a matter of two scientists disagreeing on a scientific issue. It speaks to a fundamental view of what machines are, what humans are, what psychotherapy is, and what intelligent machines might be and do. Weizenbaum's view of these matters is the theme of his *Computer Power and Human Reason*, which would appear ten years after ELIZA.

ELIZA was an impressive program, and although its mechanisms were relatively simple, it gave the illusion of deep semantic analysis. Unlike a lot of its contemporaries, ELIZA generated natural English responses to the natural English statements it received. When something puzzled it—when no logical match was possible—ELIZA could fall back on "I see," or "That's very interesting," or "Go on," even as humans do in the same situation. The clever use of the psychotherapeutic situation lent this technique even more credence than it has in ordinary conversation. Thus ELIZA was to language processing what pattern recognition was to visual scene analysis. That is, neither relied on understanding in any real sense. Rather, a process of matching and classification took place which worked pretty well in certain limited instances—the psychiatric interview, or the recognition of carefully constructed lettering—but which failed in harder tasks, such as genuine conversation, or cursive script, or a complicated landscape.

In question-answering machines, then, a thousand flowers bloomed during the mid and late 1960s. Simmons's survey mentions most of them. Yet the questions Giuliano had raised, and those Simmons himself had raised at the end of his first survey, were not yet answered. The domains of discourse were still highly restricted, and the problems of transforming from natural to formal languages hadn't been resolved. Nevertheless, Simmons again claimed that significant progress had been made. Syntactic processing, at least, was well understood; semantic analysis worked in some well-defined instances. But the problems were still gargantuan, and Simmons wondered if these simple models were just too simple, whether models that could handle the large grammars, semantic systems, and dictionaries with tens of thousands of entries that ordinary humans manage so handily might not have to be constructed quite differently. He looked hopefully at one of the new paradigms advanced by such linguists as Noam Chomsky, which suggested that every use of natural language refers to an underlying structure of concepts or data, and that meaning is a set of operations upon this underlying structure. These operations consist of making and breaking connections, finding equivalences, and so forth, all of which can finally be expressed in a formal, as opposed to a natural, language, the formal language standing as an agent operating upon natural-language processes. Translation between a pair of natural languages, such as English to Russian, Simmons went on to say, can be seen as a special case of this model, along with question-answering machines, conversational machines, and even, *mirabile dictu*, story-writing systems.

But Simmons, a good scientist, didn't claim eternal truth for this paradigm. He expected it to be gradually replaced by a finer, more accurate model, though it would stand as a good guide into the second decade of language work before it became obsolete.

Is that all there is? This crown of human achievement, over which we've been congratulating ourselves, is nothing more than a data structure with connections and operators? It does seem a bit skimpy. But then what are crown jewels, but a bit of carbon and stuff, heated and pressed, then later chipped and polished?

For some uses, such a description of crown jewels is appropriate. For some uses, such a description of language is too.

They were—we are—onto something else, which we all suspect. It is that language is a unique expression of human thought, a window onto the mind that nothing else provides. In the act of exchanging words with one another, we are displaying human thought at its most intricate. This casual thing we do so naturally, the conversation that every normal human being learns without tears or strain, calls into play a variety of faculties that contain the key to human cognition. In the act of speaking to one another, we are analyzing, reasoning, adjusting our internal model of the world, and expressing those phenomena in symbols that are more or less accessible to our fellow humans so that they too can analyze, reason, modify, and in turn express. Strachey was right: language is a most astonishing thing.

But even if the sentence-by-sentence analysis by computer program in the late 1960s was far better than the word-by-word analysis it replaced, it was still insufficient to the ambiguities of natural language. Take, for example, the sentence, "Sandra hit it." We only understand its meaning because we already know what "it" is. Thus we grasp at once whether Sandra has taken a baseball bat and squashed a mosquito, or whether she's made an astute guess and won a game of Twenty Questions after three tries. It's unlikely, under the circumstances, that we'll mis-hear the final letter of "hit" and assume that Sandra is hiding something, though the same string of sounds in another context would give us exactly that notion. We understand in *context*, which means not only that we mutually understand the rules of our native language, but also that we've agreed upon a topic of discourse, that we talk about it in a specific setting, and that we have some knowledge about the world and each other's ideas.

Thus, context, a shared world view based on mutually agreed upon facts, and a means of organizing this material for easy modification and access—all of these elements would come to replace the principles that had guided artificial-intelligence research up until now. Instead of searching for a few general and uniform principles of intelligent behavior, AI researchers were beginning to suspect—reluctantly, for it violated the scientific canon of parsimony—that intelligence might very well be based on the ability to use large amounts of diverse knowledge in different ways.

Meanwhile, pattern recognition as mere classification, whether of visual or linguistic symbols, had moved away from AI and become a discipline of its own. AI still pursued the big question: how does a system—human or machine—understand? In this sense, vision is an extremely difficult problem, and much of it remains to be solved. The same is true of language understanding. There are some successes along the way, but theories are tentative.

Physiological studies of vision in animals suggest that the computation involved in seeing is enormous. As Patrick Winston, still another former student of Minsky's and now a professor himself at MIT has put it,

> Knowing what the primary cells do does not determine how they do it or what they are to do next. Consequently, there can be no sensible effort at present to make a computer simulate the visual machinery of biological seeing machines. Instead, the effort must be, as in dealing with other dimensions of computer intelligence, to make a computer be a seeing machine, exploiting hints from all quarters. We must study the issues inherent in the problem because the hardware is too inadequately understood to be copied.
>
> (Winston, 1977)

Early vision research in the 1960s had concentrated on the blocks world, the cubes and pyramids that robot arms moved around on command. To understand such a simple universe seemed, as Winston says, like a summer's project. It wasn't. The problems were very hard indeed, and depended not only on understanding image processing, but also on the predispositions of the observer, how that system—human or robot—understood limits, and how it represented its knowledge internally.

Here too were a thousand flowers, growing in the rich soil of the robotics projects. I've already mentioned Evans's work in geometric analogies, and the primacy it gave to description—the system must be able to make up little descriptions of what it sees in order to understand well enough to make analogies. The later work of David Waltz at MIT would show that one great simplification in scene interpretation is the knowledge of how edges come together in the real world. A few thousand ways are physically possible, and it's this knowledge, and not deep reasoning power, which allows a seeing system to analyze shapes and identify them (Waltz, 1972).

If intelligence—understanding both what one is told and what one can see for oneself—turned out to be the use of specific kinds of knowledge in a given context, then this hypothesis needed further tests, and one such test was made in an MIT doctoral thesis done by Terry Winograd, now a professor at Stanford.

Just as college students have despaired of separating style from meaning when they were asked to analyze poems and stories, Winograd took as his working assumption that semantic and syntactic knowledge could not be separated from facts, that meaning is embodied in procedural structures, and that language is a way of activating such procedures within the hearer. Instead of the kinds of representation of knowledge other language programs had used, such as patterns to be matched (ELIZA), knowledge in SHRDLU, Winograd's whimsical name for his program, was embodied as a piece of executable computer code. The belief was that basic computer programming methods, such as procedures, iteration, and recursion, are also basic to human cognition. Thus, reasoning, semantic analysis and parsing were functionally integrated; they occurred simultaneously instead of being accomplished sequentially. This simultaneous attack on several fronts saved SHRDLU from the embarrassing dead ends toward which semantic ambiguities had always led programs in the past, because at each stage of the understanding process, each bit of expertise made a contribution.

Winograd believed it was impossible to isolate one aspect of language from another, or a person's linguistic knowledge from his or her use of other knowledge. But he also knew that, given the present state of computer science, it was impossible as well to give a computer program all the knowledge a human being brings to a conversation. Part of his solution to this problem was to pick only a small part of the world to talk about, but to give his program a deep kind of knowledge of that little world. The little world chosen was the blocks world, where a simple arm could manipulate blocks on a table top. To place the discussion on its proper level of scientific abstraction, no real robot arm moved wooden blocks about on a real table top. Instead, the whole world was abstracted and displayed as line drawings on a television screen—blocks, arm, table top; even the dialogue appeared as subtitles to the display. This expedient also did away with the mechanical engineering problems that had bedeviled the designers of real-world robots.

In the course of the dialogue, the arm could be asked to manipulate the blocks, doing such things as building stacks and taking things out of a box. It could also be questioned about the current state of blocks on the table, about events that had already gone on in the discussion, and, to a limited extent, about its reasoning. "The conversation goes on within a dynamic framework," Winograd wrote (1973), "one in which the computer is an active participant, doing things to change his toy world, and discussing them."

"Pick up a big red block," the human participant types onto the teletype. "OK," the computer answers back, and a linear "arm" moves across the television screen, picks up a green block on top of the red one, finds a space on the table top to store the green block, then picks up the big red block. "Grasp the pyramid," says the human. "I don't understand which pyramid you mean," the machine replies, for there are three pyramids in the scene, and the program knows that phrases beginning with the article "the" are intended to refer to a specific object the speaker has in mind. "Find a block that is taller than the one you are holding and put it into the box," the human says, and the system must figure out what is meant by "one" and "it," as well as using deduction to decide which block it is holding. It inquires for clarification: "By 'it' I assume you mean the block that is taller than the one I am holding." There are two possible referents for "it": the one selected, and "the one you are holding." The program has heuristics, rules of thumb, for making such choices. And so it goes.

The human can ask abstract questions, such as, "Can the table pick up blocks?" which the program answers in the negative, for it knows as a fact that picking up must be done by animate objects, and the table is not animate. Or perhaps the human would like some history: "Had you touched any pyramid before you put the green one on the little cube?" "Yes," the program replies, "the green one," for it has deduced from its memory which event is being referred to, and can relate other events to it in time. Something new comes up. The human says "Does a steeple—" "Sorry," the program interrupts, "I don't know the word steeple." "A steeple is a stack that contains two green cubes and a pyramid," the human explains, and that definition is added to the program's knowledge in a form compatible with all the words it already knows. "I understand," it says, and in a limited but real sense it does.

Along with the functional integration of procedures, another key feature of Winograd's program was that its internal model of the world was symbolic. Symbolic representation of meaning, which seems to be the way humans handle large amounts of data, means that some kind of token stands for the object to be represented, and that token can be simple and abstract indeed. Winograd rejected the idea that atomic units of meaning existed. Instead, he assumed that humans categorized their experience along lines which are relevant to thought processes they will use, and that categorization is generally neither consistent, nor parsimonious, nor complete.

Clearly, formal techniques of logic and mathematics are not easily applicable to such holistic models, as the critics of AI have been pointing out for years. But the fault was probably with logic and mathematics, Winograd declared. AI approaches to modeling cognitive processes can provide formalism without the limitations of mathematical or logical formalism. This involves a computer notion of *procedure* instead of proof, and the viability of this approach stands or falls on how well it provides a model of what we mean by understanding.

Well, what do we mean by understanding? Simon argues that understanding is a relation among three elements: a system, one or more bodies of knowledge, and a set of tasks the system is expected to perform (1977). He expects that future work in AI will concentrate on constructing systems that understand by enriching both the bodies of knowledge available to the systems and the procedures for using that knowledge in the performance of wider and wider ranges of tasks. "Knowledge without appropriate procedure for its use is dumb, and procedure without suitable knowledge is blind."

This declaration seems so broad as to be downright unarguable. But Simon speaks out of the experience of trying to construct just such systems, embodying knowledge, procedures for using that knowledge, and the accomplishment of tasks. How is that knowledge to be acquired, represented, and made accessible for later use? What procedures are to be used?

"Much remains to be done," Winograd had written, "in understanding how to write computer programs in which a number of concurrent processes are working in a coordinated fashion without being under the primary hierarchical control of one of them. A language model able to implement this sort of 'hetararchy' found in biological systems (like the coordination between different systems of an organism) will be much closer to a valid psychological theory" (Winograd, 1973).

This hetararchy Winograd suggested was coming up.[4] It would appear in a speech-understanding program developed at Carnegie

---

[4] In fact, it had already come up a long time ago. Just after the Dartmouth Conference, Oliver Selfridge got to thinking about some of the problems in his pattern-recognition system, and hit upon the idea of "demons" as a way of controlling decisions a program might have to make on the basis of a lot of information. His description of how little demons might work—"all of whom shout the answers in concert to a decision-making demon"—gave a name to his Pandemonium paradigm, and predated by some fifteen years the hetararchical organization of programs, composed of sets of experts (or demons) that presented their advice to some central control system to weigh before it acted. Selfridge was explicit that Pandemonium, as an assembly of quasi-independent modules, could be modified as the need arose without altering the entire program, a notion which also informed the design of the speech-understanding systems. A number of problems kept Pandemonium from success in the late 1950s, among them the primitive state of computer languages and the relatively limited amounts of memory available in the computer. But once it could be implemented, the idea was to prove highly fruitful (Selfridge, 1959).

Mellon some five years after SHRDLU appeared. SHRDLU itself was retired for the time being until some of its harder problems could be solved. Some scientists even argued that SHRDLU was essentially unextendable, and that a better way would be to put knowledge into the control structure instead of the procedures. Regardless, SHRDLU had suggested something that designers of other kinds of artificial intelligences, namely intelligent assistants, had simultaneously arrived at in their attempts to build programs that performed at the level of human experts in a given field. It was that humans operated successfully not by using powerful, underlying general rules, but rather by using a large amount of detailed knowledge, organized in special ways. This was the factor that distinguished the specialist from the amateur. And this might also explain why we have brilliant mathematicians who play dull chess, and gifted writers who can barely carry on a conversation. How this special knowledge was to be represented—organized, controlled, acquired, and modified— would preoccupy AI researchers in the 1970s.

Meanwhile, SHRDLU was improved by Gerald Sussman to contain another feature of general interest. When an instruction could not be carried out—a tower of blocks could not be built, say, because one of the blocks was somehow hidden or too big or whatever—an early type of AI program would have returned to the original state and tried a whole new strategy for construction. But the new version of SHRDLU, called HACKER, had the ability to debug—that is, to examine the procedure it had just undertaken and to identify the small flaw that had thrown it off the track. Thus it learned from its mistakes. It came to know that subgoals on the way toward a goal may sometimes conflict, and it attempted to reorder those subgoals so that the most pressing goal would be achieved first. HACKER exhibited expertise about debugging and repair, coupled with the ability to examine its own problem-solving goals and actions so that it was able to supply this debugging expertise to its own reasoning. In short, it had self-consciousness.

In describing HACKER, Ira Goldstein and Seymour Papert say this:

> We emphasize the importance of debugging skill because a vital element of intelligence is knowing how to handle a vast variety of situations. We cannot do this by knowing all about these situations because each real world situation is different. What we need is (1) knowing how to tell which old situation is sufficiently like the present one, and (2) knowing ways to adapt—"debug"—the old procedure for handling the old problem to a new procedure that can deal with the new situation.

Thus, they go on, such subjects as analogy, similarity, metaphor, are at the heart of the new formalisms. In the older methods of AI—linguistics, logic, and psychology—these issues were embarrassing, hard-to-explain phenomena. In the new debugging technology, they became concrete, manipulable types of knowledge (Goldstein and Papert, 1976).

SHRDLU and its offspring were by no means the only approach to language understanding in the early 1970s. Roger Schank and his students at the Stanford AI lab took a computational-linguistics approach to the problem, which assumed that if language were represented in a sufficiently deep way, any two sentences with the same meaning would, at this deep level, take the same form, a view similar to Chomsky's. It follows, of course, that identical, or even similar sentences with different meanings would register that difference once they were represented in that deep way, thereby eliminating ambiguity as a problem. Schank and his colleagues suggested that concepts, or primitive units of meaning, could be identified and were related to one another by dependencies. Such concepts and their dependencies could be represented in graphlike schemes.[5] He illustrated his points with

---

[5] It's in the nature of new scientific fields to have muddy or redundant nomenclature. Schank's "concepts" aren't quite the same as Winograd's, and a whole group of names exists for other kinds of data structures. Even experts are hard-pressed to tell the difference between scripts and frames and chunks. Maturity will bring order and more generally agreed-upon terms. Meanwhile, AI at least has the virtue of using comfortable, everyday terms that, in proper AI fashion, summon up in the observer's mind useful parcels of prior experience, allowing us to begin understanding what the notion in question is all about.

some memorable paradigms: "I hit the boy with the girl with long hair with a hammer with vengeance," or "The old man's glasses were filled with sherry."

Schank subsequently moved to Yale and began working with Robert Abelson on integrating these concepts into larger contexts, which they call scripts. The script idea says that for most situations humans find themselves in, there is a script of action which they expect to follow. We all know how to behave in restaurants, or we learn pretty quickly. The script idea is similar to the frame idea, which was developed by Marvin Minsky, inspired by some of the work on computer vision and then generalized to language and other tasks requiring intelligence. Scripts, or frames, assume that few situations are really new. Most have enough in common with previously encountered situations that the main features can be preanalyzed and stored for future use. The knowledge is highly particular, not general, and the script or frame serves as a skeleton upon which we hang our own rich associations.

Insofar as language is concerned, we have been dealing all along here with understanding *written* natural language—a string of symbols punctuated by stops and clearly defined spaces between each word, sentence, paragraph and section, as in the material you've been reading here. But it's an altogether different order of difficulty to understand human speech, whose elements aren't so clearly demarcated, and which requires a sense of context, as students of elementary French will remember from their odious *dictées*. Sounds and words are not one-for-one, and the pauses between words, the stops, vary not only from speaker to speaker, but with the same speaker on different occasions. In this sense, speech understanding is similar to image understanding, and a look at one speech-understanding project will illustrate what some of the difficulties have been, and how, in part at least, they've been solved.

Systems have been developed that *recognize* discontinuous spoken words—that is, word by word with clear pauses in between— with as high as 99 percent accuracy, but they have not been extendable into more sophisticated systems, and they aren't very useful for most applications. Even connected-speech recognizers were getting underway in the 1960s, although their accuracy was lower than the isolated-word recognizers, and their vocabularies were severely limited.

But what about a system that *understands* continuous speech, even as humans do? Is such a beast possible? What would it look like? What, in this context, does understanding mean, anyway, as distinct from merely recognizing? How soon before we conduct conversations with our computers, as the astronauts did with HAL in the film *2001*?

In 1971, a group of artificial-intelligence researchers met together at the direction of ARPA and studied the properties a working speech-understanding system might have. The group, chaired by Allen Newell, decided that any such system must combine semantics and syntactics, and because so little work had been done, the most fruitful way to proceed was with hard empirical investigations, in other words, to build a system that would understand speech and see what made it tick. This plan echoes the words of Newell and Simon in their Dartmouth proposal fifteen years earlier: "The present need is for a large population of concrete systems that are completely understood and thereby provide a base for induction."

The reasons for building speech-understanding systems were clear. Humans, the study group pointed out, come equipped to communicate in multiple ways: through spoken natural language, written natural language, body gestures, pushing buttons, making checks in boxes, and so forth. Each of these has its advantages and disadvantages, but the advantages of speech are striking. It's fast—substantially faster than writing, at least to transmit, if not to receive. It can be used when the hands must be free, or are otherwise occupied, and you can walk around while you're doing it. In small groups anyone can be the speaker, and all others not only receive the information, but are aware that others have received it too; also, speaker-listener roles can be switched in seconds, though this also depends on visual cues.

Speech is preferred whenever the spontaneous generation of information occurs between humans—in legislatures, conferences, social gatherings, the courts, and the marketplace. Everywhere are human subsystems—known as secretaries, stenographers, typists, and so on—to take dictation so that the generator of words can speak and others who cost less can produce the written documents. We spoke before we wrote, and normal humans manage the skill without hesitation.

But I found one notion from the speech-study group's report the most interesting:

> Perhaps as important as the rate is that in spontaneous communication with speech the human appears not to be speech limited, but rather thought limited, whereas with writing the opposite is true. That is, a person knows what he wants to communicate faster than he can write it, but not faster than he can say it. Even when saying predigested material, our speech apparatus is never used at close to capacity, at least as we currently know how to measure such capacities.
>
> (Newell, 1973c)

In any event, if the advantages of developing a speech-understanding system were clear, for some reason addressing the abuses of such a system weren't considered part of the group's responsibility, though in these more sensitive times, we'd expect to find at least a short paragraph on the dangers of the proposed project. It's the dangers that have worried people lately—the startling opportunity a truly sensitive speech-understanding apparatus would provide for, say, monitoring telephone conversations on an enormous scale. Alexander Solzhenitsyn's remarkable *First Circle*, a novel about scientists and their ethical responsibility for the applications of their work, was published in the United States well before the speech-understanding group's 1971 report. The machine in *First Circle* merely recognizes; it doesn't understand. But that's sufficient to send at least one man to prison. Of course, it can be argued that if human ingenuity can result in a way to monitor, human ingenuity can result in ways to jam. In principle, technology offers no more opportunities for abuse, nor protection from it, than we've ever had. Our only remedy against ourselves and our machines is our mutually agreed-upon laws, and thus it has ever been. But we're right to beware.

In the 1950s, the Bell System had hoped to develop a system that would allow us to speak telephone numbers into a receiver instead of dialing, but that effort was a failure. In the 1960s, some simple speech-recognition systems had been constructed, among them one by P. Vicens and Raj Reddy as graduate students at Stanford. Reddy was to become a professor at Carnegie Mellon and a member of the 1971 study group, and his early effort was for its time relatively sophisticated, but on any absolute scale still at the baby-talk level.

The 1971 study group recommended a five-year plan that was neither hopelessly ambitious nor sure-fire modest: at the end of five years a system would exist that could accept continuous speech from many cooperative speakers of the general American dialect in a quiet room over a good quality microphone, allowing slight tuning of the system per speaker but requiring only natural adaptation by the user. The vocabulary would be one thousand words with a highly artificial syntax in the context of such a task as data management or computer status. The system would have a simple psychological model of the user, and it would provide "graceful interaction" tolerating less than 10 percent semantic error, and, a rate as close as possible to that of normal human discourse. The emphasis was on understanding, as opposed to faithful recognition. Thus, if the system could correctly guess—infer, induce, whatever—what the user wanted, its inability to determine exactly what the user might have said was not to be held against it. This latter characteristic was probably its most human-like one.

In the next five years, several attempts to meet these specifications got underway around the country. And in 1976, the Carnegie Mellon project came in on target. It was not only on time, it not only met the specifications, but it was within budget—altogether a rare specimen in AI.

The supervisor of the project was an energetic young scientist, the same Raj Reddy who had originally shown up at Stanford in the early 1960s as a graduate student to study numerical analysis, the solution of large numerical problems by computer. "But it didn't take long to decide that in fact the problems of artificial intelligence were both exciting and challenging. And it never occurred to me," he adds with a brilliant smile, "that it could not be done. That may be exactly what's needed for anybody who wants to go into this field, namely, blind optimism with no reasonable basis for it."

One day as we were talking he mused on the best kind of researcher for artificial intelligence, and it describes him and his own temperament very well:

If you're very calculating, very careful, a systematic and rational person who weighs every little research project you take on to see whether it will succeed, and you pick only those things which are sure things, then you obviously can't go into artificial intelligence. We were interviewing a potential faculty member here, and he said, "I can never think about problems that take me more than two years to do." But in AI you have lifetime problems—things that take twenty, thirty years. You've got to be prepared to find out that you've spent your whole life on the wrong aspects of the right problems, perhaps—or all your time using the wrong approach or the wrong attack.

"Doesn't that make you uneasy?" I asked. He answered thoughtfully,

No. Basically it's the same old story I had in India. It will make you nervous if you think you're so special that you have to leave a major impact on this world. We work towards that, but must be realistic and say we might not succeed. Maybe less than 1 percent of the people succeed. But if you don't work on problems that can make a major impact, then for sure you never will.

Reddy is talking about ambition, which led him from his rural village in India—barely changed in hundreds of years—to the frontier of one of the most sophisticated technologies in the world. He still slips easily from one world to the other, as easily as he slips from Western to Eastern clothing, and both lives are essential to him.

As Reddy describes his relationship to John McCarthy during his graduate school days, it was more like the relationship of spiritual novice to spiritual master than student to professor. McCarthy thought his speech project a good idea but wouldn't be involved on a day-to-day basis; he made a few germane suggestions to Reddy and left him alone. "It didn't bother me," Reddy says, "because I was quite happy to go do what I wanted to do. But some others wanted to work with him who needed a lot more help, and they didn't get it because John doesn't operate that way."

Reddy soon saw the speech problem as a central one of artificial intelligence, as the balance between very large numbers of facts and much smaller numbers of general techniques for processing those facts. For example, he and his group went to work on the problem of how many different kinds of knowledge a person uses to decode an utterance, working on the hypothesis that many different and virtually unrelated kinds of knowledge and techniques are called into play. Thus they tested different kinds of knowledge simultaneously, assuming that if one kind was eliminated, then the system ought still to run, but not with the same accuracy or speed. If the system was unaffected by such an elimination, then that element was judged not to be essential to the speech-decoding problem. In short, the project addressed big questions: how to build systems, how to make them all cooperate, how to organize control structures, what the role of knowledge is, what happens if knowledge is absent, and similar questions, the very ones being tussled with in image understanding. Everybody was up against the biggest problem yet: what is *understanding*? It was epistemology in the most exigent sense.

Reddy and his colleagues took a behaviorist view and dreamed up six different ways of identifying that understanding had taken place. Some ways were straightforward, such as giving the right answers to questions, paraphrasing a paragraph, or drawing inferences from it, or translating it from one language to another, and even being able to predict what a person might say next—all these are aspects of understanding, and they work at different levels. There is, as we've always suspected, deep understanding and less deep understanding. How deep, everyone wondered, did understanding have to be in order to be useful?

In addition to levels of understanding, the group identified kinds of knowledge: semantic, pragmatic, syntactic, lexical, phonemic, phonetic, and so on. The Hearsay System, as it came to be called, had five subsystems corresponding to these kinds of knowledge. There were still more kinds of knowledge: task-dependent, conversation-dependent, speaker-dependent, and analysis-dependent; each of these interacted with the other five kinds of knowledge. It is believed that the design of Hearsay will be useful as a solution for the general problem of knowledge-based systems in artificial intelligence, and not merely for its present use in understanding speech.

Hearsay is designed around a so-called blackboard that allows each knowledge source to take a look at hypotheses generated by all the other knowledge sources about what an utterance might be and mean, and then say yes, no, or maybe, based on the kinds of knowledge each of them has at the moment. The knowledge sources are independent—that is, anonymous to each other—and are treated uniformly by the system controlling the blackboard. This independence of knowledge sources allows each one to be modified or replaced with relative ease; indeed, hearsay can continue to function with the absence of one or more of them, although its speed and accuracy are impaired, and some sources must remain for the game to be played. Some of the knowledge sources are high level, some of them low, but all can be linked—bottom up or top down, simple or complex, depending upon what seems appropriate for the task at hand (and other considerations, such as cost). The systems work simultaneously and asynchronously, roughly analogous, says Reddy, to a group of people attempting to solve a jigsaw puzzle, with each person working on a different part of the puzzle but each modifying his or her strategies based on the progress being made by the others.

But if Hearsay proved that connected-speech recognition is possible by computer, basic questions exist for which we have only vague answers. How essential is understanding to recognition of speech? Will understanding turn out to be so much more costly than mere recognition that recognition systems supersede understanding systems, or will such an economy turn out to be only a short-run phenomenon? Speech-understanding programs are presently very complex, large, unwieldy, slow, and in Reddy's view, contain too many ad hoc decisions. If significant progress has taken place—if the roles of knowledge are better understood, and error and ambiguity manageable—big problems remain.

Maybe it can't be done at all. Some researchers, such as Leon Harmon, who worked very early at Bell Labs on the problem of speech recognition, doubt that it can. Reddy's declaration, that he's prepared to accept the fact that he may have worked for a good part of his life on the wrong, or the intractable, seems suddenly a bit melancholy.

But we needn't feel too sorry yet. Hearsay, like the entire natural-language effort and the image-understanding effort, illuminated some significant aspects of intelligent behavior. These essentials emerged elsewhere too: in the efforts to build intelligent assistants, for example, or in chess programs. To summarize briefly, we seem to be discovering that large amounts of specialized knowledge couched in procedural rather than declarative terms ("when this happens do that" rather than "all men are mortal") are essential to intelligent behavior, and, furthermore, many rather than a few of these parcels of specialized knowledge come into play in a given situation. How such parcels are organized and controlled is of prime importance.

Knowledge, then, is mainly dynamic and not static. The symbols that stand for knowledge are entities with a functional property. Symbols can be created; they lead to information; they can be reordered, deleted, and replaced. All this is seen explicitly in computer programs, but also seems to describe human information processing too. Understanding is the application—efficient, appropriate, sometimes unexpected—of this procedural information to a situation, the recognition of similarities to old situations and dissimilarities in new ones, and the ability to choose between the doing of small repairs, or debugging, and changing the whole system.

This way of defining knowledge and understanding is altogether surprising, and even sounds vacuous until we see such a description of intelligent behavior made concrete in a computer system. Then such a theory seems intuitively right (and so it might, since so much of it has come out of introspection) and shares some ideas with epistemological philosophies of the past, including, of all things, phenomenology. These ideas include the situation; the notion of a dynamic, flexible, and often idiosyncratic response to the situation; contingency; and a sense of purpose or goals. Indeed, Minsky holds that most precomputer phenomenologists were philosophers in search of an information-processing vocabulary, and would probably have been in the thick of artificial-intelligence research if the computer had existed to give them its rich possibilities for metaphor and modeling.

More hard questions remain. How do we and computers learn in really novel situations, if our major learning takes place by comparing the present with the past and by calling up and debugging old routines to make them fit new situations? One partial answer is that culture, in the form of teachers of various kinds, helps us do that—which accords with our long-time suspicion that nothing under the sun is really new, at least to the species, even if it is to the individual. And how do we acquire those old routines in the first place? Are they wired in? If what keeps us from going down the wrong path (in the case of ambiguities, say) is some warning mechanism triggered by experience or previously acquired knowledge, why do the warning mechanisms (known in AI as demons or sentries) work better for some intelligent agents than for others? Where does intermediate knowledge, which is neither procedural nor declarative, such as "knowing" a tune, fit in? Can we learn to be more intelligent? Can we organize knowledge more efficiently in our human heads?

We don't yet know.

As the 1970s drew to a close, knowledge representation was perhaps the most hotly debated topic in artificial intelligence. At the 1977 International Joint Artificial Intelligence Conference, a panel with representatives from the entire spectrum of opinions, ranging from the most formal to the most contingent, drew shouts and cheers from the nearly one thousand scientists present, acting as if they were watching a football game. As I sat among them, amused by the noise, I thought how much at odds with the stereotype of the cool, disinterested scientist this demonstration was. More important, what a marvelous and accommodating structure science has, for sooner or later the issue would be resolved on the basis of the best choice—maybe a mode of knowledge representation which hadn't even yet been dreamed up—and the partisanship would disappear, or more accurately, find its expression in the next big issue.

Chapter Twelve

⌒

Applied Artificial Intelligence

The speculative nature of science—at least the most interesting sorts of science—has laid it open to endless mockery, and artificial intelligence is no exception. Far ahead of Mortimer Taube, Hubert Dreyfus, or Sir James Lighthill, there was Jonathan Swift, holder of the original patent. In 1727, Swift has Gulliver report from the land of Laputa on a thinking machine, a dotty burlesque of Ramon Lull's *Ars Magna*, as it happens, and described by its inventor this way:

> a Project for improving speculative Knowledge by practical and mechanical Operations. But the World would soon be sensible of its Usefulness; and he flattered himself, that a more noble exalted Thought never sprang in any other Man's Head. Every one knew how laborious the usual Method is of attaining to Arts and Sciences; whereas by his Contrivance, the most ignorant Person at a reasonable Charge, and with a little bodily Labour, may write Books in Philosophy, Poetry, Politicks, Law, Mathematicks, and Theology, without the least Assistance from Genius or Study.

Swift meant us to laugh, and we have been doing so ever since.

Scientists themselves have often led the way in such mockeries. An amusing little paper appeared in 1961 spoofing the Perceptron. It was called "The Chaostron: An Important Advance in Learning Machines" (Cadwallader-Cohen et al., 1961), and it spins nonsense upon nonsense, and ends by thanking the only genuine scientist named in the piece (and therefore rumored to be its true author) for "manually simulating the 704 simulating STRETCH simulating Chaostron, to complete run 133 after the budget funds ran out."

For any number of reasons, most scientists outside the field who are acquainted with artificial intelligence at all assume that the Laputa contrivance and the Chaostron are the last word in AI research. Though some of this ignorance can be traced to a resistance to the whole idea of artificial intelligence, which I've commented upon elsewhere in this book, much of it is also ignorance of a more common variety. It's simply a nuisance to keep abreast of matters outside your own field. Not, of course, that ignorance ever kept anyone from confident assertion on foreign matters, but few of us have Dr. Johnson's moral courage to confess, once found out. The following account, therefore, means to make impure the ignorance of all those who still declare that nothing has happened in artificial intelligence, nothing can, and nothing ever will.

The two examples of applied AI to be described are useful instances of the way this field seems most likely to affect us in the immediate future. First is an intelligent assistant, working in a narrow but difficult task domain and helping a human expert to do some of the taxing but essential parts of a particular job. This new version of the sorcerer's apprentice is called DENDRAL, and behaves as a chemist's assistant in interpreting the data from mass spectrography, working at the intellectual level of a chemistry Ph.D. DENDRAL was the first such intelligent assistant to be designed and put into use, but many others have followed. They include a mathematician's assistant called MACSYMA, designed by Joel Moses and his group at MIT, which now works faster than humans in manipulating algebraic expressions involving constants, variables and functions; the Sussman and Stallman program for understanding electronic circuits, also at MIT; and the burgeoning number of medical-diagnosis programs at Rutgers, Stanford, the University of Pittsburgh, and elsewhere, which specialize in internal medicine, bacterial infections, pulmonary-function diagnoses, and other medical specialties (Feigenbaum *et al.*, 1971; Moses, 1971; Sussman and

Stallman, 1975; Pople, 1977; Feigenbaum, 1977; Amarel *et al.*, 1977). A program is even underway to provide intelligent advice to a molecular geneticist on the planning of experiments involving the manipulation of DNA. In general, at this writing the medical programs are still in the demonstration stages, and are not yet in use by physicians on actual cases, which is why I'll have no more to say about them. But the chemist's assistant and the mathematician's assistant are in practical, daily use by scientists all over the country who can take advantage of such high-powered help.

The second project I want to describe is a somewhat different application of artificial intelligence. In the effort to make machines think, some insights have been gained into how humans think, and therefore learn. Though these insights are very far from complete, they suggest some ways the process of educating schoolchildren might be altered to make it more effective and pleasant for everyone concerned—child, parent, and teacher.

So first to the chemist's laboratory, filled with extensive unwieldy apparatus for fantastical purposes even Dr. Faust might have paused at. Or maybe he least of all. This being a history, it takes some background to get there.

When Herb Simon strolled into a classroom at Carnegie Tech in January of 1956 and announced that he and Allen Newell had invented a thinking machine over the Christmas holidays, he won hearts and minds, but none so thoroughly as those of a certain undergraduate who'd been given special permission to take the course, normally restricted to graduate students. He was a low-key redhead named Edward Feigenbaum, who remembers a boyhood of happy monthly trips to Hayden Planetarium from his suburban New Jersey home. He also remembers scouring the Carnegie catalog for unusual courses that would rescue him from being "just an engineer," and finding an odd lot in the Graduate School of Industrial Administration, the business school of all places, which is how he happened to be listening to Simon's extraordinary announcement.

Feigenbaum was to become a graduate student in GSIA just when the pregnant notion was first introduced that an orderly process, such as a decision-making process in a firm or an individual's decision-making process, could be modeled on a computer. This idea was then so novel that the paper he wrote on it with two faculty members, James March and Richard Cyert (now president of Carnegie Mellon University, né Carnegie Tech) was turned down by the *American Economic Review* as being too extreme.

"*Behavioral Science*, which was a pretty far out journal at that time, accepted it," Feigenbaum now remembers, "and one year later, the *American Economic Review* published an entire simulation issue; we were only a little early! But it was all part of the general bag of tricks, methodological innovations, that were taking place at Carnegie at that time in GSIA." He recalls the atmosphere then:

It was a spectacular intellectual environment. It wasn't just a question of a high level of innovation being tolerated. A high level of innovation was absolutely necessary for your survival there. It was an explosive intellectual atmosphere, driven by people who were pushing along so fast that you could hardly keep up. You were bewildered at the process of just keeping up with their intellectual steam, Simon and Newell being prime examples. And it was quite clear that the most significant things in computer science were taking shape there right under our noses—compilers, languages, the IPLs.

Feigenbaum worked on several versions of the IPLs, the information-processing languages, and even went to RAND during the summer of 1957 to write a version for the IBM 704, considered a monster of a machine by standards of the day.

He was typical of what we might call the second generation of artificial-intelligence researchers. Unlike the previous generation, the second had come of intellectual age just at the moment when the computer was beginning to propagate, and took to it as the instrument of choice—and to the notion of artificial intelligence—with hardly a second thought. McCarthy and Minsky had surrounded themselves with a small group at MIT, which included Daniel Bobrow, Bert Raphael, and James Slagle, all to go on to distinguished scientific careers; and at Carnegie, Newell and Simon had a handful of disciples in the Graduate School of Industrial Administration which included Feigenbaum, Julian Feldman, Robert K. Lindsay, Fred Tonge, and others.

When Julian Feldman received his degree he was hired at Berkeley by the business school, where he continued with his own research in simulating a classical psychology problem of choice under uncertainty. Feigenbaum, after spending a year abroad as a Fulbright fellow at the National Physical Laboratory outside London, where he had a memorable friendship with a young and somewhat eccentric South African scholar named Seymour Papert, joined Feldman at Berkeley.

There was a certain amount of skepticism among the older faculty members at Berkeley, especially those who thought that the business of a business school was teaching accounting and marketing and insurance, just the fuddy-duddy courses that Simon and his colleagues had decided to avoid in their own business school. Then the two new kids received a grant from the Carnegie Corporation of New York for work over a three-year period in the amount of $70,000. In the late 1950s, that was a staggering amount for a grant, and professors who were used to $2,000 and $3,000 grants—enough, say, to pay a research assistant and a part-time secretary—were appropriately staggered. Feigenbaum and Feldman were somewhat protected from the outrage by being associated with a group called the Center for Research in Management Science, made up of the few others in the business school of like mind, so Feigenbaum at least, doesn't remember any special pain.

The Berkeley business school was quite a benign place [he recalls] because of the management-science group having its own little bailiwick, and no one bothered us much. There was a little bit of flak when it came to tenure time, about why should we give tenure to someone who is so bizarre. And that had an important impact. When you're in the middle of a tenure flap, even if you're happy, you tend to look around for alternate places. I'd say if there were reasons for my being unhappy at Berkeley, they had to do with psychologists and electrical engineers, not business school people. The business school people were quite tolerant. They didn't know what I was doing, but at least they thought it was kind of sexy. Psychologists were impenetrable. I was associated with the Center for Human Learning. As a member I'd go to all the luncheons and all the seminars; I'd talk their ear off about information-processing theories, about EPAM and GPS and all the other things that were being developed, and the people who have now adopted all the right phraseology—by 1972, 1973 they're talking about search retrieval, data bases, all the right language—at that time were absolutely impenetrable. They'd listen; they'd nod, how nice; and they'd go off and do their own silly experiments. Anyway, by that time I'd just gotten tired of talking to psychologists. There was just no headway to be made. They didn't really appreciate what was going on. There were a handful of psychologists in the world who did, but they didn't happen to be at Berkeley.

Feigenbaum's first very big successful piece of research, for which he'd received his Ph.D., was in cognitive psychology. It was something called EPAM, which stands for Elementary Perceiver and Memorizer (and which. Herb Simon notes, is also a nod to Epaminondas, for along with his work on the Logic Theorist and IPL and GPS and other matters, Simon was, characteristically, teaching himself Greek, and took a fancy to the Theban general and statesman of that name). EPAM addressed another classical problem in cognitive psychology, the rote learning of nonsense syllables. Says Feigenbaum,

Herb had an idea about how one might make progress in this particular area—it can't really be all that hard, and look at these results; you see how the literature kind of lays itself out in relatively simple terms; we can model that, can't we? Sure we can—and off we went to the 650 to do the initial EPAM simulation.

Feigenbaum and Simon saw themselves doing pure behavioral-science research. They had in mind to model certain classical phenomena of rote learning that were well known in the literature. This was somewhat different from the two-fold motivation of the Logic Theorist effort, and even the General Problem

Solver, where not only did a strong thread of the thinking go into understanding and modeling human problem-solving processes, but also some motivation existed to construct smart problem-solving devices. There had been at least as much excitement about the fact that the Logic Theorist could prove some reasonably difficult theorems as there was because it proved them with mechanisms similar to ones that humans could conceivably use, Feigenbaum recalls. "But the EPAM payoff was a good explanation of data, of psychological data. Could we come up with something that was cogent, clear, fairly simple, realized by a program, whose structure at least psychologists could understand, and that could explain a wide range of phenomena?"

The answer was yes, and for many years EPAM was probably the prime exemplar among the information-processing psychology models that interested psychologists outside, because it addressed problems of classical concern to them. Yet EPAM provided a mechanism that was useful elsewhere too, called a discrimination net, used for the recognition and flexible storing of objects in an associative way. Feigenbaum says, "I think the adaptivity of the EPAM net had particular impact, the fact that it could grow over time to incorporate new stimulus objects that needed to be recognized in a fairly efficient manner." The basic EPAM structure is still being used. For example, Simon has been doing work on chess perception that relies strongly on EPAM nets for encoding, for chunking patterns on chess boards. So, although EPAM wasn't a complicated program, it was a seminal work in the field in that it showed the way, it provided a rather clear and simple information-processing structure; the patterns were easy to understand and gave rise to complex behavior and interesting explanations of phenomena that were well understood experimentally.

At Berkeley, Feigenbaum and Feldman tried to interest bright young graduate students in this odd effort called artificial intelligence—to recruit a third generation, so to speak. The two began by teaching together a course in computer modeling of thought. Feigenbaum says,

In order to get it into the curriculum in the business school, it being a rather bizarre subject for business schools, we enlisted the aid of Herb Simon and his Social Science Research Council Committee on Computer Simulation of Cognitive Processes. They gave us a few thousand dollars to buy our way into the curriculum by giving this money to the dean of the business school to let us teach this course. That's the way the course got established. Then we began to teach it once, maybe twice a year to Berkeley students. It attracted a wide variety of students from different parts of the university, ranging from things like neurophysiology all the way through economics, physics, math, business; almost anyone you could imagine dropped into that course.

Well, it turned out to be extraordinarily difficult to teach the course because the relevant sources were spread all over the map in odd journals, obscure places—an example being a paper by Newel on GPS which appeared in some obscure German proceedings because Al happened to give the paper in Germany. How would one get hold of all this? There seemed to be no single reference work that summarized the state of the art for the student, gave him some interpretive remarks along the way, that put a structure on these papers, gave him an adequate bibliography and an overview.

The answer obviously was to put together a book, and Feigenbaum and Feldman did so.[1] They selected articles not only to inform students; they also felt that scientists in other fields, as well as the general intelligent reader, should know that such an effort was underway. Thus the most readable, and not necessarily the most technically competent reports, were chosen; also, only working programs were included, not speculations about what might be done. "Plus we made sure we included ourselves and all our friends," Julian Feldman grins.

---

[1] An English major who had been working her way through college by typing afternoons in the business administration school, and watching the clash between the old ways and the new with some amusement, was invited to spend the nine months between graduation and law school in the autumn doing odd jobs for *Computers and Thought*. "Sure," I said, "what's the book about?" "Artificial intelligence," replied Ed Feigenbaum. "What's that?" I said. He told me. "Sounds like fun," I said. It was. And changed my life.

Edward Feigenbaum
(Stanford University)

In his role as consulting editor for Prentice-Hall, Herb Simon recommended against publishing this collection, now called *Computers and Thought*. An interesting idea, but no market for it, he declared. Simon was mistaken: McGraw-Hill published the book in 1963, and it not only sold remarkably well from the beginning, but sustained its sales over a period of a decade, extraordinary for a scientific book. It was translated into most of the European languages, and Japanese and Russian (though that edition was carefully edited by Soviet censors to suppress deviant social thought).

In some sense the book was to be a Debrett's Peerage: if you were in it, you were really in. Contributing authors included Minsky, Selfridge, Gelernter, Newell, Shaw and Simon, Samuel, and many others. Because it was the only book of its kind for many years, a newcomer to the field could easily have believed that these scientists were the only ones doing work in the field—it had become the canonical collection. It deliberately excluded work on the Perceptron and other neural-net research, on the assumption that those kinds of projects were somehow different, were getting enough attention on their own, and, says Julian Feldman, were frankly too speculative. "Our selections were substantive things—programs which really worked, did what they said they would."

I once asked Feigenbaum if he thought this collection had later come to shape the field. Perhaps, he allowed, and continued,

AI is hard to define anyway; most of the time it ends up being viewed as those things which the AI people are doing. That's what this collection was. So by virtue of its existence, it sort of set the definition of the field—a student might get the view that what AI was all about was what was in COMPUTERS AND THOUGHT. *Time has tended to weed out certain things, and time has done some injustices, but basically the shape of the field as it left the 1960s was the shape reflected in the book. I don't want to assert that the book did it. What I'd like to assert is that* COMPUTERS AND THOUGHT *was a prescient view of where things were and where they were going. If there was a hot idea that would have shaped the field in some other way, the fact that the book existed wouldn't have stopped that development.*

Feldman doesn't think the book did much shaping. It helped AI gain a little respectability, he believes, making it easier for funding agencies to justify their support. Both men also saw the book as an antidote to some of the overselling they thought AI was getting by people who loved to talk but didn't do much.

The political situation at Berkeley, involving a dispute about where computer science should be, whether connected with electrical engineering or an entity on its own in the college of letters and sciences, had begun to heat up. Long before it was settled, both Feigenbaum and Feldman departed, Feigenbaum to Stanford, where John McCarthy hoped he would become a collaborator, and Feldman to the University of California campus at Irvine, where he was to be dean of social sciences. Their departure left at least one bewildered soul. Feigenbaum had invited Bert Raphael to join them at Berkeley when he got his Ph.D. from MIT. Raphael arrived on the Berkeley campus to find himself utterly unknown and unexpected, and after some hair-raising times with Berkeley's fossilized payroll office, he finally fled to the Stanford Research Institute, where he would soon go to work on the robot Shakey.

With his move down the peninsula to Stanford, Feigenbaum discovered he was tired of EPAM. He was tired of hassling with psychologists and trying to get his ideas accepted. His EPAM phase ended altogether in 1965, and he describes himself now as almost a spectator, who reads "with interest" Herb Simon's continuing work on EPAM. What intrigued him instead was a problem that had more of an artificial-intelligence than psychological flavor. By that I mean how closely the task mimicked human methods began to be less interesting than how well the task was performed, regardless of method. In the end, those two concepts were less distant than had originally been believed, but I will come to that in a moment.

I began to get interested in a set of problems that it seemed to me hadn't been well explored by earlier AI work, namely tasks of empirical induction [Feigenbaum says]. Given a set of data elements, construct the hypothesis that purports to explain that set of data. And I viewed empirical induction as first of all being prototypic of scientific behavior. So I was interested in modeling what scientists were doing in their problem solving, particularly in what they were doing in empirical-induction activity. Secondly, I was interested in induction per se, and scientists were the professional inducers in society, so maybe they could make more explicit than others what they were in fact doing during induction, since they did it routinely and that was their job. And third, I just thought it was an interesting rock to turn over. I didn't know what kind of mechanisms one would find if one tried to construct performances like that. I didn't know if there would be different ones, or if it would turn out that mechanisms already exhibited in other programs could just be employed to do empirical induction tasks. So I began a search for the right problem to study empirical induction. That search lasted a couple of years.

Feigenbaum was meanwhile involved in administration at Stanford, and as head of the computing center, brought in the first time-sharing system, a task with more difficulties than anyone had imagined. He also discovered that his work habits and McCarthy's were at odds, and though they remained personally friendly, each went his own way professionally. In McCarthy's case the move was literal: he set up shop a few miles from the main campus in a modernistic building that had fallen to Stanford from one of the periodic fiscal shakeouts in peninsula electronics, and there began his large-scale robotics project. Feigenbaum re-

mained on the campus proper, and once he'd unburdened himself of computing matters, he picked up his research again.

The precise problem he was to use as a task environment was
first suggested to him by his colleague at Stanford, Joshua
Lederberg, a Nobel laureate in genetics, whom Feigenbaum had
already met before moving to Stanford. Lederberg was to become
a full-fledged investigator in the project, teaching himself computing and AI as he went along. The task environment was the
analysis of data from a mass spectrograph.

*It was a problem which had all the elements of classical empirical
induction [Feigenbaum says]. Here's an array of data that comes from
a physical instrument, the mass spectrograph. Here's a set of primitive
constructs out of which to compose a hypothesis about what organic
compound is being analyzed. Here's a legal-move generator for generating all possible hypotheses. The problem is to find good ones out of
the set of all possible ones, since in the general case, you don't want to
generate all possible hypotheses. How do you find the good ones? And
how do you employ knowledge of the world of chemistry, mass spec-
trometry, to constrain the set of alternatives, steering away from large
sets of unfruitful ones? That was the framework.*

They named their system DENDRAL. In many ways it was
similar to the chess problem, and documents written by researchers on the project sometimes pointed out the similarities. Consequently, Feigenbaum and his team paid some attention to the
kinds of work going on in chess and it was often very suggestive:

*We began developing a performance program that would accept this
kind of physical data, and produce adequate hypotheses to explain the
data, adequate in the sense that a chemist would say, yes, that's a good
hypothesis, or in fact that it was the right answer in test cases. That led
not only to a very high performance program, one of the first, but also to
more abstract issues of scientific methodology, philosophy of science.*

DENDRAL was a precursor to the next generation of AI programs called knowledge-based systems. For Feigenbaum and his
colleagues soon discovered that human chemists who had been
doing this analysis carried around enormous amounts of special-

ized knowledge in their heads, and it was simply impossible to do the job without having that specialized knowledge. So another difficult task was to sit down with those human chemists, watch them work, ask them questions about how they made decisions—questions they weren't always able to answer in ways that would fit comfortably into a computer program—and then figure out some way of representing that knowledge, those rules of thumb, in a large data base.

As suggested earlier, AI workers in their role of knowledge engineers often make a distinction between methods and data, or between what they call procedural knowledge and declarative knowledge. But in practice the distinction is blurred. "In DENDRAL," Feigenbaum says, "we looked not only for the relatively hard [declarative] chemical knowledge about valences and stability and mass spectral processes, but also for the relatively soft [procedural] knowledge: how a particular scientist makes a particular kind of decision when he's not really sure, when there's a variety of evidence, a lot of ambiguities. How does he select?" The two kinds of knowledge, hard and soft, declarative and procedural, are intimately tied to one another.

The DENDRAL project sounds somewhat contradictory. If performance was suddenly more important than humanlike behavior, why were human experts so assiduously consulted? Feigenbaum answers,

Even though we set out to build a device whose behavior didn't have to match human behavior in detail, we ourselves didn't have any particularly good ideas about better ways of doing it, in ways different from the way, roughly speaking, human chemists do the same job. We were out to milk all we could from the human expert.

He paused, then brought up a significant point:

We've come to the view lately, in asking ourselves why our program turns out to be so much like humans, and did it have to be that way—we've come lately to the conclusion that it would've been impossible for us to construct a program as complex as DENDRAL if the mechanisms in the program were vastly different from human prob-

lem-solving mechanisms, because what was going on inside the program would have been virtually incomprehensible to us and the chemists. We would've had no way to relate to the processes going on inside, to add to them, to modify them, unless they were pretty much like what our human chemists were using.

The project team was to discover some interesting things about expert chemist behavior, which they could soon generalize. First, it became clear that the acquisition of specialized knowledge was a bottleneck in the design and building of intelligent agents. The knowledge inside an expert's head is largely heuristic knowledge, experiential and uncertain—mostly good guesses in lieu of facts and rigor. Much of it is private to the expert, not because he's unwilling to share publicly how he performs, but because he's unable. He knows more than he's aware of knowing, something the early language-machine designers recognized. But if a second party, or even the expert himself, undertakes patient observation of that expert in the act of doing what he does best, the knowledge can be teased out and made explicit.

Thus has been born a new creature, the knowledge engineer. This engineer works intensively with the expert to acquire the specific knowledge the expert has and to organize it for use by a program. Of his favorite knowledge engineer, Feigenbaum has written,

> Simultaneously she's matching the tools of the AI workbench to the task at hand—program organizations, methods of symbolic inference, techniques for the structuring of symbolic information and the like. If the tool fits, or nearly fits, she uses it. If not, necessity mothers AI invention, and a new tool gets created. She builds the early versions of the intelligent agent, guided always by her intent that the program eventually achieve expert levels of performance in the task. She refines or reconceptualizes the system as the increasing amount of acquired knowledge causes the AI tool to "break" or slow down intolerably, She also refines the human interface to the intelligent agent with several aims: to make the system appear "comfortable" to the human user in his linguistic transactions with it; to make the system's inference processes understandable to the user; and to make the assis-

tance controllable by the user when, in the context of a real problem, he has an insight that previously was not elicited and therefore not incorporated.

(Feigenbaum, 1977)

This painstaking method, first used in DENDRAL, has subsequently been applied to a variety of areas, especially medical diagnoses.

The DENDRAL project is almost the antithesis of earlier AI programs, say, the General Problem Solver. The hope with GPS had been that methods by which problems were solved were independent of the content of the problem itself. DENDRAL seemed to say that specialized knowledge—a lot of it—was essential for solving any really significant problems.

For a while [Feigenbaum recollects], we were regarded at arm's length by the rest of the AI world. I think they thought DENDRAL was something a little dirty to touch because it had to do with chemistry, though people were pretty generous about oohs and ahs because it was performing like a Ph.D. in chemistry. Still, DENDRAL seemed like a large collection of specialized facts about chemistry, with very little artificial intelligence in it. Now the DENDRAL view is the au courant view, so people beat a path to the DENDRAL door trying to find out what kinds of knowledge representations we used, how we extracted the expertise from the experts, what do we know, and so on. I feel quite good about that. I did write a paper which appeared in 1971 on the question of generality versus specificity in problem solvers, and came down strongly on the side of specificity of knowledge for high performance in problem solvers. And I suggested that generality was powerful not at the performance level but at higher levels at which information is extracted from nature or from texts or from some other source that got translated into specific pieces of knowledge applicable at the performance level. It was the viewpoint that was reflected in the so-called Meta-DENDRAL program, which is an attempt to extract from nature by more general processes the specific pieces of knowledge needed by the DENDRAL program in doing its empirical induction.

Edward Feigenbaum simulating a well-known war hero.

Meta-DENDRAL aims to automate knowledge gathering, which must otherwise be painstakingly evoked from human experts, whether by face-to-face interviews, or from texts they have written.

As Feigenbaum says, the reaction to DENDRAL in the AI community was mixed. Some members viewed it as so specialized a piece of knowledge engineering that it wasn't worth studying in detail. In part this was the fault of Feigenbaum and his group, who addressed their published papers mainly to chemists. When it came time to talk about the AI implications, they found themselves more anxious to go on with the research than to sit down and figure out in some orderly, general way what they had to say to their colleagues in AI.

Some of those colleagues, however, were nosy enough to push in for themselves and see what was happening, and they were very impressed. DENDRAL also had the advantage of working, and working very well indeed, so that it has become a sort of model program AI researchers can point to when they're asked what useful work has been done lately in the field. And then there's a third group, who Feigenbaum laughs about:

They're the folks who think that DENDRAL is an absolutely marvel-ous and super—in the same sense that Athena was absolutely marvel-ous and super, or Zeus with his thunderbolts was absolutely marvel-ous and super. For them it's a piece of mythology. They've heard about it. They know it exists. But they've never read a paper on it. I ran into that abroad where people had heard of DENDRAL, what a magnifi-cent program it was, but they admitted they'd never made their way through a single DENDRAL paper. In fact, my suspicion is growing that Allen Newell is the only one in the world who has ever made his way through an entire DENDRAL paper.

Aside from the fact that DENDRAL has provided a highly efficient and helpful tool for working chemists, which is no small thing in itself, it has provided something more. The framework of DENDRAL has provided a base with sufficient complexity and detail to make some headway on the problem of automatic knowledge acquisition. DENDRAL suggests ways to extract the regularities from nature and put them into programs by com-puter program, rather than by handcrafting the knowledge from the head of an expert.

The automatic knowledge-acquisition problem [explains Feigenbaum] is seen by virtually everybody to be absolutely critical to AI's progress—and so Meta-DENDRAL is right in the critical path. Now Meta-DENDRAL was in the critical path a long time ago, and it was the kind of activity that Lederberg and I were talking about at the time we starting working on a heuristic DENDRAL. But we just didn't have the chutzpah to try the theory-formation task at that time. We just couldn't see the lay of the land. We didn't quite understand what target we were shooting at. But having spent many years building a system which now does a very creditable job of problem solving in a domain, we now in effect have a target to shoot for. We now have the knowledge framed that a Meta-DENDRAL needs to infer automati-cally. It turns out we framed it by hand, but at least we know the answer to the problem. It's like having to do a test problem in which there are answers at the back of the book. We now have the answers at the back of the book. We can tell whether a Meta-DENDRAL is on the track on not, and even how to frame it.

Applied artificial intelligence is viewed by many as a major and important task, and DENDRAL is one of the key examples. It may not have started out as applied AI, but it has ended up that way, and instead of being a spectacular curiosity, it has become the source of inspiration and ideas about how to structure knowledge in specific task domains. Adapting its principles, other such programs have begun to appear, as we've seen. Though the details differ from program to program, some common themes are shared.

First is what AI researches call "generation and test," a technique first made explicit in Newell, Shaw, and Simon's General Problem Solver, in which the intelligent agent makes a plausible guess, and then tests to see how well that guess fits all the circumstances.

Next is a theme of situation-leads-to-action rules. Each of these programs has a specific set of such rules in the form of *if* this is the case (and this, and this, and also this) *then* do that (and that, and that, and that).

A third theme is one of specific knowledge—the amount of detailed expertise that plays a crucial role in organizing and limiting to sensible proportions the search for answers. This knowledge base must also be flexible enough to be modified at will.

Another central organizing principle of these systems is that their line of reasoning be explicit, and this comprehensible to a human expert. This is necessary not only for debugging and extending the knowledge base, but also for judging, by the expert, whether the ultimate action suggested by the artificial intelligence—a medical diagnosis or a decision to bomb a hamlet—is consonant with common sense or human intelligence.

Finally, each of these programs effectively must coordinate many sources of knowledge, just as we've seen in the speech- and image-understanding programs. This is a formidable problem. A medical diagnosis is made not only on the basis of known facts about the behavior of microorganisms, but also on the basis of the individual patient's situation, the kinds of remedies available, and tolerable, and so forth. The decision to bomb a hamlet is not only a matter of military tactics, but also of human ethics, the strategy of species survival.

DENDRAL, in any case, has passed from the hands of the knowledge engineers to the hands of the chemists, where it's in everyday use by Stanford chemists and their colleagues at universities and industrial laboratories elsewhere. A commercial computer network allows access by other users, and the British government is currently supporting the transfer of DENDRAL to Edinburgh, where it will be used by workers in the chemical industry of Great Britain.

Few of us have jobs to do that require the expertise of a mass-spectrograph chemist. We'd be glad to have someone—something—to do the vacuuming or the laundry. We might even be glad to have some sort of built-in nanny who reminds us about our calorie intake for the day or whether we've had enough roughage. If AI applications aren't quite that homely yet, an interesting project has been underway in Cambridge-area schools. There, schoolchildren have been getting applied AI, and they seem to like it.

What these schoolchildren have not been getting is the kind of education that computers were first associated with. They are not sitting at televisionlike terminals being drilled in arithmetic or spelling, the terminal patiently correcting errors—being concerned more with declarative than procedural knowledge, so to speak. Instead, the LOGO project has been conceived as a means for applying what artificial-intelligence researchers have discovered about human thinking and problem-solving processes in the course of trying to get computers to think and solve problems. It's concerned with teaching children to think—or, as Seymour Papert, the principal investigator on the project once put it, LOGO aims to teach children to be mathematicians instead of teaching them about mathematics. And not only mathematicians—musicians, physicists, engineers, story tellers, even teachers themselves.

The chief of this project is that young South African mathematician Ed Feigenbaum had encountered one afternoon in his London rooming house, where he overheard Papert using the only telephone in the place and realized from the conversation that Papert too was associated with the National Physical Laboratory, and that they were both interested in information-processing psychology.

Papert had taken a somewhat different route to arrive at Great Britain's National Physical Laboratory at Teddington. A man of legendary personal charisma, Papert grew up in South Africa, where as a high school student he became passionately interested in logic, and was permitted to attend the logic seminars at the university at Johannesburg. And there, he remembers, great arguments ensued about whether logic could be formalized. He believed that it could, so built a little logic machine that worked out syllogisms. "It was the first computer I ever saw," he says. "It was an actual physical machine. It had lights and buttons to push, and you could say to it, 'All men are mortal....'" He had no notion that other such machines might exist, and only discovered they did when he showed his machine around. Later, he read Shannon's thesis and discovered that Shannon had used Boolean logic to explain switching circuits.

Papert studied philosophy as an undergraduate, but it soon seemed very arid to him, and he took up mathematics instead. But he was always preoccupied with trying to understand how symbolic thinking came about in people. This led him at last to work with Jean Piaget in Geneva. Says Papert,

I think Piaget is not so widely recognized as he should be as a precursor to the kind of thinking that we take for granted now to be symbolic cognitive psychology. He really was trying to do the sort of thing AI is, that is, trying to formalize some other kind of thinking than the highly logical, so-called correct thinking that was the only sort of thing that had been formalized before by logicians. I think he ran into troubles because he lacked the appropriate sort of model to that kind of formalism, and computers were going to give that. But I think he does a lot to prepare the way.

Piaget did not want to think about computers, and Papert did, which is how he came to be at Teddington and to meet Feigenbaum. It was just about this time that Papert and Minsky also met, both attending a symposium in London and appearing with papers that proved virtually the identical theorem in different ways, a theorem addressing Perceptron-like devices. I suppose this was the kernel of their later book. In any case, Papert was immediately invited to the United States by Warren McCulloch, but his political activities in South Africa held up an American visa[2] until the mid-1960s, when one day he walked into Minsky's MIT office, and they sat down immediately to work on a problem together.

If Papert didn't see eye to eye with Piaget on every aspect of the issue, he was nevertheless profoundly influenced. Along with Piaget, and for that matter with Dewey and Montessori too, Papert believed that children learn by doing, and *by thinking about what they do*.

That children think about thinking is probably one of the most controversial parts of Papert's program. He writes:

> It is usually considered good practice to give people instruction in their occupational activities. Now, the occupational activities of children are learning, thinking, playing and the like. Yet we tell them nothing about those things. Instead we tell them about numbers, grammar and the French Revolution; somehow hoping that from this disorder the really important things will emerge all by themselves. And they sometimes do. But the alienation-dropout-drug complex is certainly not less frequent.
>
> (Papert, 1971)

But why don't we teach children about thinking and learning? The objections to it are usually two: we know so little about cognitive psychology that we surely don't want to teach such half-baked theories in our schools; and teaching children to be self-conscious about their learning will surely spoil the whole process. Like the millipede who was so flummoxed by the question of how she did it that she couldn't take another step, the argument goes, children who begin to think about thinking won't be able to go on. But as Papert points out, children themselves are constantly engaged in inventing theories about everything, including them-selves, their schools, and their teachers. So the choice is not be-tween half-baked theories or cognitive innocence. Either we must give the child "the best ideas we can muster about cognitive pro-cesses, or we leave him at the mercy of the theories he invents or picks up in the gutter. The question is: who can do better, the child or us?"

Both Piaget and Papert himself have made extensive queries into what children believe about learning and why they believe it. Among common theories that children hold (and not a few adults) is that learning consists of "getting it"—it being mathematics, let's say—in a flash, all at once, ready-made. This model has two states: I get it, and I don't get it. Children who believe in this theory of learning lack, and even resist, a model that allows un-derstanding to occur gradually, through a process of additions, refinements, debugging, and so on. Papert writes, "These children's way of thinking about learning is clearly disastrously antithetical to learning any concept that cannot be acquired in one bite" (Papert, 1971).

Another theory is one of faculties. Papert notes that most children seem to have and extensively use an elaborate classifica-tion of mental abilities: "he's a brain," "he's a retard," "he's dumb," "I'm not mathematically minded," and so on. Papert doesn't add but might as well that these models are carried on into adult life, sometimes with catastrophic results. A child who considers him-self dumb has lost all incentive for learning, for adapting to new situations, for shaping his world with any sort of will of his own. A child who is considered a brain might very well have a different fate, a belief that schoolbook intelligence will solve every prob-lem of life, and that no other solutions are worth considering.

My own experience with mathematics in school cannot be very different from that of most people who label themselves "not mathematically minded." My story doesn't come from the im-poverished inner city nor even the Dark Ages. On the contrary, I was educated in the well-heeled and generally progressive public schools of northern California during the 1940s and 1950s. Thus:

I struggled through twenty sums a night, I learned so-called short division, then was taught the "theory"—that is, long division—and promptly forgot how to do the much more convenient short division; I arrived in high school bewildered and bored by it all, knowing only that I had to master it—get it—if I wanted to go on to university. When I did badly in an algebra placement test, I was put in a remedial class that tried to teach me in ways that had already failed to teach me. When 1 finally struggled into plane geometry, it was my splendid luck to sit next to somebody who was very good and who, each morning, shyly passed me his finished proof so I could fill in the blanks on my own homework paper (if you're reading these lines now, Mr. Kuramoto, please accept my much belated but deep thanks). It offended my honor to cheat this way on homework, but I justified it by imagining that the whole thing was inessential anyway except as a means of getting into college, an obstacle course I had to negotiate by hook or by crook. Nobody—certainly not my mathematics teachers, not my parents, not my friends—ever contradicted that belief.

There were patches of blue. At some point I found out I could balance my checkbook with precision and without tears, though by using methods of arithmetic I would be embarrassed to confess,

We are to thinking as the Victorians were to sex [Seymour Papert likes to say]. We all know that we have all these horrible moments of confusion when we begin a new project, that nothing looks clear and everything looks awful, that we work our way out using all sorts of odd little rules of thumb, by going down blind alleys and coming back again, and so on, but since everyone else seems to be thinking logically, or at least they claim they do, then we figure we must be the only ones in the world with such murky thought processes. We disclaim them, and make believe we think in logical, orderly ways, all the time knowing very well that we don't. And the worst offenders here are teachers, who present crisp, clean batches of knowledge to their students, and look as if they themselves learned that knowledge in a crisp, clean way. It didn't happen that way, but the teachers don't admit it, and the students groan inwardly, feeling so hopelessly dumb.

In graduate school I was forced to read a book by Alfred North Whitehead which proposed that mathematics was a language, and could be learned as a language. Why hadn't anyone told me that before? I had all kinds of language skills I could've put to use. And then, as I began research for this book and looked at the kinds of methods the Logic Theorist embodied, I was stunned. Reasoning backward? Nobody, not once, had ever told me that reasoning backward was a useful way of proving a theorem. If it occurred to me at all, I must have guessed it was cheating on a par with copying from my neighbor, and if you must do something unethical, copying was a lot easier.

But, says Papert, merely talking to children about the bad theories they hold regarding learning is almost certainly inadequate. A child's intellectual growth must be rooted in experience. In the late 1960s, therefore, Papert proposed to create an environment in which children would become highly involved in experiences rich for the growth of intuitions and concepts for dealing with thinking, learning, playing, and so on. He wanted an environment where technology was used not in the form of machines for processing children, as they had been in computer-aided instruction and even television, but as something the children themselves would learn to manipulate, to extend, to apply to projects, thereby gaining a greater and more articulate mastery of the world, a sense of the power of applied knowledge, and a self-confidently realistic image of themselves as intellectual agents. All the children were to be their own knowledge engineers.

Though a computer would be central to this particular environment, teaching children to be mathematicians and musicians, Papert was able to point to other such painlessly acquired skills. Children acquire language, for example, easily and painlessly. That is, all normal children learn to talk. But children could learn speech either because language is not really learned but is essentially innate, as Chomsky's theory suggests, or because a great amount of time is spent on language learning irrespective of the conditions of learning. However, Papert conjectures that speech results from neither innate differences nor quantity of learning so much as from the learning *process*. That is, nobody, least of all the child, expects fully formed paragraphs to emerge at eighteen months. Learning occurs gradually, by accretion. It is encouraged and richly rewarded by the much deeper participation in the environment it gives to a child, including but not limited to a lot more power— "Cold, mommy" gets a sweater a lot faster than a general wail of discontent. In any case, the question of innateness or quantity versus learning conditions eventually will be settled by experimenting with building totally different learning environments.

Another of Papert's noncomputer models is a favorite not only because it's natural and has a large degree of painless success. He likes it because amateurs and experts, children and adults, are all connected in a highly gratifying social structure. It's an organization in Rio de Janeiro called the Samba School. Papert says,

> I suspect that were such a thing to exist in the United States, it would not use the word school in its name. It would be more likely to describe itself as a "club," for although it is a school in the sense that people do learn there, it is not a school in that learning is no more the primary reason for participation in the Samba School than it is for membership in a baseball team or for playing any game.
>
> (Papert, 1976)

Here is Papert's description of the Samba School:

> If you dropped in at a Samba School on a typical Saturday night, you would take it for a dance hall. The dominant activity is dancing, with the expected accompaniment of drinking, talking and observing the scene. From time to time the dancing stops and someone sings a lyric or makes a short speech over a very loud P.A. system. You would soon begin to realize that there is more continuity, social cohesion and long term common purpose than amongst transient or even regular dancers in a typical American dance hall. The point is that the Samba School has another purpose than the fun of the particular evening. The purpose is related to the famous Carnival which will dominate Rio at Mardi Gras and at which each Samba School will take on a segment of the more than twenty-four hour long procession of street dancing. This segment will be an elaborately prepared, decorated and choreographed presentation of a story, typically a folk tale rewritten with lyrics, music and dance newly composed during the previous

year. Some of the many ways in which this can be taken as a success model for education are too complex to discuss briefly here as a mere example. So I shall select just one, the simplest and most visible of its school-like aspects: its function as a dance school.

From this point of view a very remarkable aspect of the Samba School is the presence in one place of people engaged in a common activity—dancing—at all levels of competence from beginning children who seem scarcely yet able to talk, to super-stars who would not be put to shame by the soloists of dance companies anywhere in the world. The fact of being together would in itself be "educational" for the beginners; but what is more deeply so is the degree of interaction between dancers of different levels of competence. From time to time a dancer will gather a group of others to work together on some technical aspect; the life of the group might be ten minutes or half an hour, its average age five or twenty-five, its mode of operation might be highly didactic or more simply a chance to interact with a more advanced dancer. The details are not important: what counts is the weaving of education into the larger, richer cultural-social experiences of the Samba School.

So we have as our problem: to transfer the positive features of the Samba School into the context of learning traditional "school material"—let's say mathematics or grammar. Can we solve it?

(Papert, 1976)

Samba, *si*; but mathematics? Everyone knows dancing is fun. Few people believe mathematics is. Too bad for us. Papert goes on to remind us what mathematics is like in real-life schools, and how we dance. We dance for fun. We dance with each other. But for schoolchildren, mathematics is often a lonely, impersonal experience of manipulating symbols in accordance with rules learned by rote:

> If you denatured dance by reducing it to rules to be learned and operated in the same alienated way... it would not be more fun than denatured mathematics. On the other hand, mathematicians find that math is fun. Does mathematics have to be denatured to be done by kids?
>
> (Papert, 1976)

Well, yes, he says, it does. And the reason it does is the consequence of the technology traditionally employed in schools, the technology of paper and pencil. It's static, and to be contrasted with the dynamic technology of computers. If you can do mathematics with a dynamic technology instead of with a static one, then perhaps you can do real mathematics instead of denatured math, and thereby open the possibility of a Samba School effect. And in the experimental classrooms, that is precisely what happened. Groups formed and dissolved around given problems; innovators were admired and copied; children taught children, and were also taught—in several different ways—by adults. It must have been fun for everyone. One graduate student's report says he regularly taught his official number of children and an equal number of unofficial pupils who wouldn't be shooed away.

Thus artificial intelligence is being applied at the LOGO project in two ways. The first I have already mentioned—that in trying to teach computers how to do things, AI researchers have had to be much more careful about defining how humans do things, discovering and naming tricks, processes, rules of thumb, which may not lend themselves to the formalism of a mathematical proof, but which we (and now computers) nevertheless use in order to think. These heuristics are embedded in the idea of *procedure*, by which children have already learned to instruct a computer to do things. That, of course, is the second application, intimately tied to the first, and having to do with the power of computer technology itself. The computer at LOGO is not the exclusive property of authority—the schools, the teachers, the government. In the best sense, it belongs to the children themselves. They have access to it physically and intellectually; they make it work for them. And it does.

LOGO, as Papert himself once said,

is a grander vision of an educational system in which technology is used not in the form of machines for processing children but as something the child himself will learn to manipulate, to extend, to apply to projects, thereby gaining a greater and more articulate mastery of the world, a sense of the power of applied knowledge and a self-confidently realistic image of himself as an intellectual agent.

(Papert, 1971)

I'm reminded here that Simone de Beauvoir wrote in *The Prime of Life* that the only end capable of justifying human undertakings is liberty. "An activity is good when you aim to conquer those positions of privilege, both for yourself and for others: to set freedom free." It's an attitude which is central to the LOGO Project, and one to keep in mind as we come to examine Joseph Weizenbaum's quite opposite view of the machine in the next chapter.

In appearance, the LOGO project at MIT (it has satellite locations in the public schools) isn't entirely depressing, but any visitor used to other kinds of model educational projects, or even used to most middle-class public schools, will find a surprising lack of amenities. No carpets, no brightly colored pictures, no black (or brown or green) boards. Some of these are budgetary constraints, as it happens; the LOGO project hopes one day to have a more pleasant environment where, for instance, children can take naps between projects. But presently, a lot of not especially appetizing hunks of machinery are sitting around on the bare tile floors, and wires dangle from the ceiling in what seems haphazard fashion. The whole place looks unfinished, and in several very deep senses, it is. The project needs human beings—usually schoolchildren—to bring it to life, and its workers consider it unfinished in the larger sense that by no means are all problems solved, all methods debugged to everyone's satisfaction.

There are several computer terminals, one at hand with a television screen attached. The visitor sits down, about to get her first lesson in Turtle Geometry, a new form of nondenatured mathematics invented by the LOGO group to take advantage of the dynamic nature of the computer, and to allow youngsters to participate in the fun of doing, even inventing, mathematics.

What the visitor does isn't what she'd call inventing mathematics. But it is fun. First she instructs the turtle—in this case, a white dot on the TV screen—to move four units to the left. It does so, leaving a neat white line in its track. Plucking up her courage, she adds a few more units, and then a few more. She decides to have the turtle make a square, and moves the turtle up the same number of units she's already moved it left. And then she moves it to the right the same number of units, and then she shifts the figure so it becomes a diamond, and then she—then she remembers she's here on business, and stops.

"Uh, how do the children react to this?" she asks Cynthia Solomon, the staff member who's showing the visitor around. Solomon laughs. "How do you like it?" "Oh," the visitor admits, "I'm really enjoying myself here and almost got carried away, but the children?" They react the same way, Cynthia Solomon assures her. The visitor feels free to get carried away and does so, making squares and triangles and spirals and other such figures to her heart's content. Can she square the circle?

There were no schoolchildren present when I was there, which is probably just as well. Adults have long since learned to be embarrassed when they make wrong guesses publicly. Some of us would even find the Samba School painful, having invested so much *amour propre* in our dignity and in being correct under all circumstances. Too bad for us. We close ourselves off from learning except under the most private circumstances, missing not only the joy of mutual human activity, but the acceleration it lends to individual learning. This is partially what Papert is talking about when he says that for mathematicians, mathematics is fun. They often do it together, pulling each other out of holes, making their whole contribution greater than the sum of its parts. Committees have a bad reputation, but the collaboration of a few like minds in science (and more often than is supposed in the arts) can be dazzling.[3]

Anyway, with some collaboration from Cynthia Solomon, I learned a few things that afternoon. For one thing, I learned some

---

[3] Papert once was speaking of scientific partnerships, which often produce rich work in congenial surroundings. We may not even need our partner's physical presence, he said, for we've internalized a model of him so well that we can imagine what he'll say about a given idea (which is not necessarily what he does say when asked), keeping the wildest ideas in check. More important, a partner amplifies the other's work, an important part of thinking that is little studied. A good partnership is one where a balance exists between checks and amplifications, keeping out the hopeless ideas but encouraging the fruitful ones, and this turns out to be a major problem in designing artificial-intelligence programs. "Are the collaborative mores of our intellectual culture an important intellectual discovery, one that's been embedded in the society? That's plausible. I think it's an important part of understanding the way knowledge happens."

basic principles of programming a computer. I made it work for me, and when it didn't, I knew it at once, and could pretty soon figure out why. If I'd sat there longer, I'd have learned some short-cuts that would've speeded my learning, allowed me to do more complicated things more quickly, such as achieving more complicated geometric figures by decomposing those figures into subsections, learning to design a subsection and then multiplying it all over the screen. Or, as one fifth-grader did, I might've had the courage to wade in and change a procedure I'd already learned and thus discover a whole new geometrical figure, called a squiral, since it was a spiral set of squares, or a squared spiral, or, if you like, a figure bounded by parallel lines.

More important, I was getting a concrete introduction to the idea of a procedure. When it didn't work, I could see why it didn't—which step had thrown me off—and fiddle with that until it did.

The name Turtle Geometry comes from the fact that in addition to the cathode ray tubes, or TV screens, there are small mechanisms at the LOGO project with humped backs that rather resemble turtles. They crawl over that bare tile floor, attached by those haphazard wires to the computer and manipulated by a child at the terminal. The turtles have pens in their innards, and with the command PENDOWN, can execute on paper (or on the floor) any design the child commands. PENUP signals the end.

As the figures the children wish the turtles to make become more complex (expressing, say, planetary movements) children receive instructions from their LOGO teachers that go something like this: if you cannot solve a problem as it stands, try simplifying it; if you cannot find a complete solution, find a partial one. No doubt, Papert writes, everyone gives similar advice. The difference is that in this context the advice is concrete enough to be followed by children who seem quite impervious to the usual math. There are no wrong answers, only improperly debugged ones. In Turtle Geometry, the location of the bug in the procedure is usually easy to spot and can be remedied. The bug has probably shown us the unanticipated side effects of an instruction.

The LOGO turtles are the historical descendants of Grey Walter's early homeostatic beasts of the late 1940s and early 1950s, thereby adding a satisfying sense of poetic completeness—a high priority in the LOGO project that can be seen in the careful efforts to cross boundaries separating traditional school subjects and make learning a holistic experience for the child. The turtles also represent another priority, to make computers a part of instead of alien to our human culture: "Thus we are fundamentally concerned with creating a kind of computer presence which is compatible with our culture, one which is resonant with people's sense of who they are and who they want to be" (Papert, 1976),

The turtles are not present merely to motivate children, a sort of gadgetry to disguise the real business of learning. On the contrary, they are an authentic means of allowing children to have personal power. They do things for children that children want them to do. The turtle encourages the beginner to anthropomorphize, though in this instance for good reasons, not bad; the LOGO language puts the child in charge. This personal power is what Joseph Weizenbaum (see Chapter 13) sees as an evil thing in his description of hackers. But Papert and his group view it as a good thing, an intellectual power extended to those who have traditionally been the most powerless, namely, children. The effect of the turtles is all the more vivid in children with learning difficulties, not only those who have been badly educated in the past and now have psychological problems with learning, but also children with severe physical handicaps, such as cerebral palsy. Manipulating the turtle is an exercise in setting freedom free.

The LOGO project has as its overall goal the design of new learning environments. This means building them, experimenting with them, and thereby refining their theory. As in all empirical sciences, these three efforts are intimately associated. But an easy way of comprehending LOGO is to examine each effort as if it were separate.

Thus the physical environment has a number of educational devices. Among them are the computer-controlled mechanical turtles. Others are the televisionlike computer displays, which allow students to work on mathematics and other things—for example, designing their own teaching programs, which requires them not only to understand thoroughly whatever they wish other students to learn from the program (for how else could they write a precise procedure?), but also to anticipate misunderstandings a novice might have, difficulties that might be encountered. These difficulties in turn require the designers of the teaching programs to teach themselves, to understand more deeply and significantly, what it is they want to teach. There are devices that allow students to compose music and hear it at once, where they not only examine their notions of aesthetics but learn something about the physics of sound. Special terminals have been designed for very young children; instead of a typewriter keyboard they have a small series of buttons that correspond to computer commands. Special-purpose computer systems have been designed, including a personal student computer whose design is being directed by Marvin Minsky, who has always shared Papert's view that computing belongs to everyone.

LOGO also specializes in something called "bridge activities," which connect the computer experience and the familiar informal experiences of children. With the learning of such skills having been cast into "people procedures," analogous to "computer procedures," children have been taught quickly and painlessly to walk on stilts, ride a unicycle (in the interests of poetic completeness, I should repeat that's a skill for which Claude Shannon has some notoriety in the Cambridge area), and to juggle.

Once at the Boston airport, as it happened, I was talking with Seymour Papert about these bridge activities. "Oh," I said innocently, "can you juggle, Seymour?" In answer, he jumped up, swept the ashtrays from nearby tables, and treated me and the travelers of Boston to the astonishing sight of an MIT professor juggling glass ashtrays to beat the band.

Apart from the fun involved, these bridge activities have other implications. Papert quotes Jerome S. Bruner, a distinguished psychologist of learning who has written in his *Towards a Theory of Instruction* that he finds words and diagrams "impotent" in getting a child to ride a bike. This is very much a piece of the same attitude that relies on tacit knowledge—knowledge that cannot be expressed, but that all (or most, or some) humans "know." Adherents to this view say that there are just some things (or a lot of things) where we may as well forget about giving good instructions.

But the evidence of the LOGO lab is contrary. Papert writes that Bruner shows at best that some words and diagrams are impotent:

> Now in our laboratory we have studied how people balance bicycles and more complicated devices such as unicycles and circus balls. There is nothing complex or mysterious or undescribable about these processes. We can describe them in a non-impotent way, provided that a suitable descriptive system has been set up in advance. Key components of the descriptive system rest on concepts like: the idea of a "first order" or "linear" theory in which control variables can be assumed to act independently; or the idea of feedback.
>
> (Papert, 1971)

Here then is the theory, the human as information processor, made deeply, serenely aware of our own processes, and put in the position of being able to modify them as it suits our personal tastes. The intuitive becomes explicit, and all the more precise and useful to our own purposes for that explicitness.

Juggling is fun. But so then are mathematics, physics, biology, music, games, and language: the theory expands, is experimented with, on fifth graders, seventh graders, first graders, college undergraduates, teachers, the average, the below-average, the above-average. Each finds something to engage and stretch his or her intellectual capacity.

Though much has been done with LOGO, very much more remains to be done. It's hard work—though fun—for everyone, because it isn't a mere manipulation of familiar curriculum, say sums illustrated by prettier pictures, arranged horizontally instead of vertically, or called by different names. LOGO represents a fundamental change in educational concepts and reflects in miniature the fundamental changes the computer will make in society as a whole. The usual analogy is a comparison between pre- and post printing-press societies. But that comparison misleads in its simplicity. Add transportation, electronic communication, and the germ theory of diseases, and you might begin to sense the shape of the revolution to come.

LOGO and DENDRAL are two quite different but working examples of the application of artificial intelligence and its principles to real-life situations. I could have described other applied programs that work, but none that has been so thoroughly accepted by persons outside the field. And that points to a problem that few have addressed, the problem of transfer. If a program exists, say, for diagnosing bacterial infections in humans and prescribing cures, one that on the whole works with better success than human physicians, why aren't patients and physicians using it? Our easy guess might be that any intelligent program that replaces professionals at what they do and get well-paid for, or even a program that is an intelligent aid, is going to meet mighty resistance. But the facts are that no resistance has been recorded because no one has had the resources to attempt a large-scale transfer from the laboratory into the field. The LOGO case especially would require intensive retraining of teachers as well as purchase of equipment (which luckily gets cheaper all the time, since the cost of computers is rapidly diminishing). Large amounts of money would be necessary to make that transfer, which surely could only happen over a long period of time.

People in AI concede that they have not done much of a job alerting or training engineers who could get to work applying even what is already possible. Until now, the rewards in the forms of admiration and contracts have gone to pure research. Given the newness and volatility, or at least tentative nature, of results, this is perhaps appropriate. But the time is coming shortly when these transfers to real applications will be made, and the impact of artificial intelligence will seem sudden, profound, and inevitable. Changes which are sudden, profound, and seem inevitable are often viewed as deeply threatening to us, and this is the theme of the next chapter. But to paraphrase Warren McCulloch, let us say they need not be, and then proceed serenely.

Part Five

The Tensions of Choice

We dance around in a ring and suppose,
But the Secret sits in the middle and knows
— *Robert Frost*

Unmitigated seriousness is always out of place in human
affairs. Let not the unwary reader think me flippant for
saying so; it was Plato, in his solemn old age, who said it.
— *George Santayana*

Part
Five

The Tensions of Choice

Chapter Thirteen

Can a Made-Up Mind Be Moral?

To this point we have been concerned with whether a machine can think. Now we turn to another question, suggested by the ambivalence—one might even say terror—evoked in us by the first: *Should* a machine think?

To answer that, we have to make some guesses about how such thinking machines might behave, and in particular, how they would affect our lives. If such machines are to be our slaves, we can anticipate some splended benefits, but what about intelligent machines to whom we would be slaves? Is that a real possibility? Would they treat us benevolently? On the other hand, is it sensible to ask whether, if they are our slaves, we owe them benevolence? Most important, if at last we come to have machines who think, will we have to adjust our own view of ourselves radically, just as we did when Copernicus told us we weren't at the center of the universe, and Freud told us we weren't the altogether rational creatures we'd assumed?

Mary Shelley's *Frankenstein* expresses our eternal suspicion of what might happen if we don't treat our intelligent artifacts with good will: They'll turn on us and wreak gross destruction. Misery, after all, is what made the monster a fiend. But, then, Frankenstein's nameless monster is a human caricature, doing for us what Coyote does for the Western Apaches: on him we project our best and worst, our contradictory nature, which longs for goodness and finds badness so tempting. A later writer, Isaac Asimov, suggests that artificial intelligences are alien, though they have many humanoid features; it is for this reason that his race of robots is equipped with built-in safeguards, the Three Laws of Robotics mentioned earlier (which are Talmudic and not robotic, John McCarthy once observed). They deprive the robots of the ability to do any physical harm to their human creators. About other kinds of injury, the question remains.

The precise nature of the alien is troublesome. As I've already pointed out, it's the powerful who define that nature, excluding whom or what they will by rules having more to do with convenience than logic. It sometimes seems that more nonsense has been perpetrated on the human race in the name of keeping up moral standards than by out and out villainy. But we must still struggle with these issues. For that we have to consult other kinds of experts[1] and examine other examples.

The experts are those who've been immersed in a given field, whether biology, hitchhiking, or artificial intelligence; the examples are ideas and technologies that have had deep and surprising impacts upon us. We can't put undue faith in such experts and examples, but neither can we totally trust our own instincts toward the good. Before we make up our minds, say, to hang a rascal, we do well to consult an expert in legal procedure who can remind us of the long-range consequences of capital punishment, which are complex and surprising. So it goes with artificial intelligence. The decision to have it or not is ours, likewise what it will look like and the uses it will be put to.

We're about to enter into a moral debate then, consulting experts on both sides. Consultation won't get any of us off the hook; it's just more fruitful than making up our minds ahead of time. We'll do well to keep a firm hold on our skepticism and our sense of history—not to mention our sense of humor—as we plunge ahead.

---

[1] I'll argue in the next chapter that predicting the social impact of new technology or new ideas is one of the skills human beings are worst at.

In the early 1970s, a debate exploded publicly about the morality of intelligent machines and the effects they might have on our view of ourselves and our human rights. "Let the ideas speak for themselves," more than one scientist told me, "and never mind the people involved." Alas, it isn't quite that simple. There are compelling reasons, I think, for putting this debate into context, for examining the credentials of the interested parties.

The first reason is that not inquiring into the background would violate the whole truth as I see it. If, as Jacob Bronowski often said, there's no absolute knowledge, we still ought to try and capture truth as fully as we can. The second reason is that a lot of special pleading, private agendas hidden within, has tried to pass as disinterested truth in science. So it does everywhere else in human affairs, but nonscientists have difficulty detecting it in the statements of scientists. This isn't only because the substance of science is arcane. It is also because scientists themselves have nurtured a semimyth that humans are detached from fact; they use this notion to protect themselves from personal accountability for the material they select to present as fact. There's nothing sinister in such a selection, or in having private motives: every project we humans undertake springs from a wide variety of urges. We volunteer for a political campaign not only because we have ideas about the issues involved but because we long to meet others of like mind, and even harbor some political ambitions ourselves. We write books about topics far afield from our specialty because it's fun to learn new things, because we discover virgin territory we can stake out as our own, because we might make mountains of cash, and because the newspapers flatter us for capsule opinions.

The following account has some obvious goals. There are issues to be aired, and advocates who speak for many sides. But it is also intended to welcome scientists cordially back into the human race.

In the spring of 1976, a book appeared called *Computer Power and Human Reason*, by Joseph Weizenbaum. It was a detailed attack on artificial intelligence, with some aspects echoing earlier attacks, and other aspects quite distinctly original. Among its distinctive features was the fact that it came from one of the major centers of AI, Massachusetts Institute of Technology, and that it was written by a man who had made an early major contribution to the field, the conversational program called ELIZA, described in Chapter 11. In his book, Weizenbaum is the first to raise the question of morality explicitly.

The book deserves consideration from us all, and deserves it at several levels. First is its main premise, that there are domains where computers ought not to intrude, whether or not it's feasible for them to do so. Previous critics have spent most of their energy quarreling that computers were *unable* to do certain human things, and that on these grounds, AI deserves contempt. But as that niche in the ecology of science is getting more difficult to hold on to, Weizenbaum has presented a new space to fill. Though he uses much space to debunk AI, he still maintains that his major concern is the moral, not the scientific, issue. There are, he says unequivocally, domains where computers ought not to intrude. These domains are where such intrusion represents an attack on life itself, where the effects can easily be seen to be irreversible and the side effects are not entirely foreseeable, and where a computer system is proposed as a substitute for a human function that involves interpersonal respect, understanding, and love.

Weizenbaum's book also deserves attention for its second premise, that most of the work done in the field is not science but technique, and is done not by scientists but by compulsive programmers, known around computer installations as hackers, who have been carried to extremes by their megalomania.

A third premise, which Weizenbaum might place first, is that we've embraced the machine metaphor as a description of ourselves and our institutions much too readily, that in this embrace we're in acute danger of yielding what is essentially human—our dignity, our love, our trust—to ideas and artifacts that don't deserve it and that may destroy us once and for all. Weizenbaum fears that we're already slaves to our machines, the real and the metaphorical ones, and that we value ourselves the less for yielding our autonomy. This devaluation has already led to inhuman results.

Since Weizenbaum's book is itself a human artifact appearing in a human society, it represents not only the overt, extrascientific purposes of its author, but perhaps some covert ones as well. Certainly it affects his audience extrascientifically, and as outsiders attempting to judge his opinions, we must examine some of those extrascientific circumstances.

The opening sentences are significant:

> This book is only nominally about computers. In an important sense, the computer is used here merely as a vehicle for moving certain ideas that are much more important than computers... a major point of this book is precisely that we, all of us, have made the world too much into a computer, and that this remaking of the world in the image of the computer started long before there were any electronic computers.

Weizenbaum is more or less true to these opening words, because he illustrates one of his points—that we've adopted the machine metaphor for ourselves far too readily, and to our diminishment, even our ultimate demise as human beings—with events that took place long before the advent of the computer, most notably the Jewish holocaust of World War II, which Weizenbaum himself barely escaped.

Weizenbaum then goes on to review his own involvement in AI, describing the three great shocks he received in the mid-1960s from the reception to ELIZA: First, that some practicing psychoanalysts seriously believed that the DOCTOR program (as ELIZA was known when it was carrying on in Rogerian therapist mode) could grow into a nearly completely automatic form of psychotherapy; second, that people who conversed with DOCTOR got quickly and deeply involved with it, almost regardless of their sophistication about computers; and finally, that many of his fellow scientists made exaggerated claims for ELIZA, especially as a clue to understanding natural language.

If specialists could misinterpret and exaggerate the importance of scientific accomplishments; if they could quickly and deeply involve themselves with something they knew very well was only a computer program, confessing to and consulting it as if it were a human therapist, what assessments were being made by the general public? This question bothered Weizenbaum deeply, because if public conceptions of science were misguided, he reasoned, then public decisions about science and technology were likely to be equally misguided.

Thus, he reports, he asked himself even more questions. What is it about the computer that's brought the view of man as machine up to a new level of plausibility? Well, only with computers have machines been able to perform even a modest set of human intellectual functions, and the dividing line between human and machine intelligence is unclear. If man is nothing but a clockwork, then such a reality and its consequences must be divined and contemplated. If humans normally show strong attachment to and involvement with machines—their cars, their violins—it's hardly surprising that a machine that extends human intelligence will demand greater attachment and involvement. But why would humans willingly cede this autonomy? Is it possible that we're bringing to our views of thinking machines far less skepticism than we ought; that we are like children at a magic show? Why are we so convinced that machines can be, or ought to be, or finally must be autonomous, even social machines such as governments or welfare systems?

Weizenbaum's own milieu at MIT convinced him all the more, he says, that some sort of atrophy of the human spirit was taking place, an atrophy that trusted only "science" to interpret reality. He did believe that this atrophy was recognized at least by some, though they were unable to find other ways of looking at the world. The computer had only recently begun to be mentioned as part of a much larger problem, but now a full-scale debate about the computer was developing. Weizenbaum writes, "The contestants on one side are those who, briefly stated, believe computers can, should, and will do everything, and on the other side those who, like myself, believe there are limits to what computers ought to be put to do."

Our autonomy and corresponding responsibility as human beings, Weizenbaum goes on, is a central issue to all religious systems, and we have always striven for principles that could organize and give sense and meaning to existence. But the substitution of scientific principles for religious or ethical principles is one that Weizenbaum deplores, for he has a persuasive mistrust of "scientific facts"—a mistrust in that sense shared by most thoughtful scientists, it's only fair to add—and he sees that with the ascent of science as the one true way, other ways of apprehending the world and the human condition, in particular the arts, have lost their function and have degenerated when they exist at all, to mere entertainment.

In short, the world is presently a bleak place, and likely to get worse.

Weizenbaum then reminds us that tools are not mere adjuncts to human development, but themselves shape our understanding of the world and ourselves in it, influencing the human imagination in unexpected ways. If this aspect has some arguable points, the chapter where it's discussed seems to me one of the best in the book, raising questions of objectivity versus subjectivity, and, another of Weizenbaum's themes, that computing systems have in fact not revolutionized the world as we know it, but instead have been used to shore up existing systems, such as government welfare, capitalist banking, and so forth, which would have collapsed of their own weight without the computer.

> Yes [he writes], the computer did arrive "just in time." But in time for what? In time to save—and save very nearly intact, indeed to entrench and stabilize—social and political structures that otherwise might have been either radically renovated or allowed to totter under the demands that were sure to be made on them. The computer, then, was used to conserve America's social and political institutions. It buttressed them and immunized them, at least temporarily, against enormous pressures for change.

Like many arguments in this chapter, it's a chewy one, but goes without being amplified or addressed again.

After two chapters on the workings of the computer, there follows a very curious chapter describing that compulsive programmer, the hacker. This creature, working twenty-four hours a day over a hot terminal, oblivious to everything in the world but the universe he is creating, fed on megalomania and superstition, a nut, a pathetic soul, convinced that his computer universe will obey him unconditionally and shape its reality to suit his own warped one, seems an odd detour from the main theme for Weizenbaum to make. The idle reader can substitute artist, say, for hacker, and hardly falsify a sentence. We all know that such folks exist (some of us even become that way from time to time) and while it might be better for them and their immediate families if they were, ah, better-adjusted, it's from such obsessive and badly balanced persons that much of humanity's treasure has come: Nobody would have thought living with Beethoven was a picnic.

But if we think that Weizenbaum merely wants to tell us slaves to hot computer terminals have joined slaves to pianofortes and typewriters and paintbrushes, I believe this misreads the book. Weizenbaum is onto something different. That something different is the implication—not said in so many words, but what else is the reader to infer?—that most of the workers in artificial intelligence are merely glorified hackers, that their work is technique and not science, that their obsessions are ignoble. And a good part of the rest of the book is devoted to just that very point.

He excoriates his colleagues in AI for their tunnel vision, evidenced by the claims made in the past (decades past, to be sure) and the paucity of results. In a chapter on the computer and natural language, where Weizenbaum's credentials are his pioneering ELIZA program, he berates Allen Newell, Herbert Simon, Roger Schank, and Terry Winograd for mistaking "empty heuristic slogans" for general theories of language understanding. Here Weizenbaum asks two questions which, he feels, go to the heart of the question about whether there is any essential difference between a human being and a machine.

First, are the conceptual bases that underlie linguistic understanding entirely formalizable, even in principle, as Schank suggests and as most workers in AI believe? Second, are there ideas that, as I suggested "no machines will ever understand because they relate to objectives that are inappropriate for machines"?

These two questions are inextricably linked, he goes on to say, because if the whole of a human experience and the belief structure to which it gives rise cannot be formalized, then there are indeed appropriately human objectives that are inappropriate for machines. Weizenbaum is finally on the side of those who say it can't be done:

> Sometimes when my children were still little, my wife and I would stand over them as they lay sleeping in their beds. We spoke to each other in silence, rehearsing a scene as old as mankind itself. It is as Ionesco told his journal: "Not everything is unsayable in words, only the living truth."

His next chapter on artificial intelligence in general is somewhat confusing. On stage is a production entitled Should Computers Think? From the wings we hear a whisper: It's okay, we know they really can't. If as nonspecialists we haven't been alerted by some of the questionable assumptions in Weizenbaum's earlier chapters, we certainly notice them now. For example, this passage:

> Questions like "Can a computer have original ideas? Can it compose a metaphor or a symphony or a poem?" keep cropping up. It is as if the folk wisdom knows the distinction between computer thought and the kind of thought people ordinarily engage in. The artificial intelligentsia, of course, do not believe there need be any distinction. They smile and answer "unproven."

Well, what the folk wisdom "knows" is enough to make even the most committed humanist shudder a little. The folk wisdom "knew" the earth was flat and women had no function in life beyond making babies and keeping men happy. To use one of Weizenbaum's favorite examples, the folk wisdom was easily persuaded that Jews were subhuman and deserved what they got. "Even calculating reason," he says later, "compels the belief that we must stand in awe of the mysterious spectacle that is the whole man." But why is awe only possible with mystery? Can't we be awed by the known as well? Or even more? A Japanese maple outside my study window turns from green to flame. That I know the chemical causes of that change enriches my delight in the autumn spectacle, as does my knowledge of that dwarf tree's relationship to New England, and to the quite different culture that honored the cultivation of such a tree. In other words, to explain doesn't explain away; and there are many ways of viewing phenomena, each appropriate for certain occasions. I will return to this point again.

In a chapter called "Incomprehensible Programs," Weizenbaum nods politely at two programs that have taken on "highly technical functions" very successfully because they are based solidly on deep theories—as distinguished from most other artificial-intelligence work, we're left to infer. Weizenbaum ignores the complaint that nobody knew how shallow most theories of intelligence were until scientists tried to make them work in the form of computer programs. It doesn't seem quite fair to blame AI for the disarray in which it found cognitive psychology early on. But mainly the chapter is an attack on behavioral scientists, hackers, and bureaucrats, with the faint but unmistakable leitmotif that the attentive reader has been noticing all along, that things were once upon a time all right. And when the reader perceives that this curious yearning for a long-gone Eden is one of Weizenbaum's deeply felt personal themes, more of the book falls into perspective. We may be left wondering just which premachine paradise it was where all humans enjoyed dignity and autonomy and self-respect, but in Weizenbaum's assumptions we see we've fallen from that paradise and grievously, an old familiar theme.

Computer Power and Human Reason hardly surprised anybody in artificial intelligence. In 1972, Weizenbaum had published an article in *Science* called "On the Impact of Computers on Society," which addressed some of these issues, and at the 1973 summer meeting of the International Joint Conference on Artificial Intelligence at Stanford, he repeated them publicly and got into an exchange from the platform with Kenneth Colby. This was too bad, because it lent credence to the impression that Weizenbaum's moral outrage was fired more by his feud with Colby over scientific credit than by deeply felt horror. Whatever the merits to that belief, it doesn't change the significance of the issues.

Colby was clearest, as it happened, in expressing some of the reasons why he hesitated to reply to the attack on him in *Computer Power and Human Reason*:

> When attacked on moral grounds, the working scientist is faced with problems of justification and reply. If he believes the moral criticism to be justified, he yields and changes his work. If not, he continues working but has to decide whether he should take the time and effort to reply. Is the controversy worth engaging in? Assuming a moral critic is serious and sincere, then he deserves a reply if his charges carry any weight with the public concerned about the issue. Silence in the face of criticism may be interpreted as spinelessness or guilt or humble resignation may disguise a moral pygmy. The decision to reply itself is a moral one since it is a question of value and responsibility how a scientist chooses to best spend his powers, time, talents, funds, and resources. If one is silent, an interested public may begin to believe the unanswered charges are correct. If one responds, the moral critic may be delighted to see he has gotten a rise out of his target. He takes the response as evidence that his efforts have not gone unsung.
>
> (Colby, 1976)

In the same memo, Colby draws a distinction between moral criticism and scientific criticism:

> Scientific criticism is communal. The scientific community values criticism as a way to correct errors, to stimulate new ideas, and to offer encouragement. Personal attacks and name-calling are unacceptable since they are detrimental to inquiry.

And then he says:

> When a moralist tries to assert a law-and-order authority over scientists, he opens himself to a number of moral countercharges, particularly those involving free inquiry, free thought, and free speech. Over the past four centuries the scientific community has come to mistrust suppressions of inquiry, not only because they protect the status quo but because so often the finger-wagging moralist has turned out him-

self to be morally confused, piously self-serving, and irresponsibly blind to the consequences of his own oppressive actions.

In Chapter 11 I referred to the priority dispute between Weizenbaum and Colby, which Weizenbaum still maintains was not enough to propel him into public dispute. If the issue of credit for ELIZA/DOCTOR is annoying to him, what really troubles him is what he calls the con job, that DOCTOR is represented as having the potential to help psychiatric patients.

In *Computer Power and Human Reason*, Weizenbaum quotes Colby as saying that further work is necessary on the psychiatric program before it will be ready for clinical use. If the method proves beneficial, Colby said, then it would provide a therapeutic tool that could be made widely available to mental hospitals and psychiatric centers suffering a shortage of therapists. Because of the time sharing capabilities of modern and future computers, several hundred patients an hour could be handled by a computer system designed for the purpose. The human therapists, involved in the design and operation of this system, would not be replaced, but would become much more efficient, since their efforts would no longer be limited to the one-to-one patient-therapist ratio as now exists.

Weizenbaum comments:

> I had thought it essential, as a prerequisite to the very possibility that one person might help another to cope with his emotional problems, that the helper himself participate in the other's experience of those problems, and in large part by way of his own empathetic recognition of them, himself come to understand them.

If there are many techniques a therapist uses, he or she is still a human being engaged in a human process, a healer healing, not, as Colby and his colleagues would have it, an information processor and decision maker with a set of decision rules which are closely linked to short-range and long-range goals.

Colby's reply was tart:

> That the function of psychotherapy is to dispense respect, understanding, and love is one of those characterizations from the layman

which takes clinicians aback. It seems to confuse psychotherapy with, for example, marriage. To confound professional working relationships with affectionate marriage relationships reveals a fundamental misunderstanding of what psychotherapy is all about.

(Colby, 1976)

Later he says,

There are great difficulties in programming a computer system to participate in therapeutic dialogs. But even if it could be achieved and even if it helped people, it ought not to be done at all, according to our critic. Why not? He offers no reasons: he seems confident that his word is enough. Presumably he believes it is better to let people suffer than have them helped by a computer.

In fact, work on programs that might provide therapy has been stopped because of the difficulties of writing programs which understand natural language. Colby says,

Until this problem is solved to a greater degree, there is little prospect of a computer system participating in complex therapeutic dialogs as humans now do. However, as a clinical realist, I believe the idea of transmitting useful therapeutic information by a computer system will in time be explored because the human need for such information will always outstrip the manpower able to provide it in direct one-to-one relations. The patient suffering from mental distress deserves every chance he can get. To not explore the use of the best tools and instruments available is immoral since it violates a basic principle of the helping professions which are devoted to the relief of suffering of everyone, not just a privileged few.

I'm not old enough to remember the outrage over the suggestion that lunatics were sick instead of wicked. However, I am old enough to remember the cries of alarm when drug therapy was suggested as a substitute for lengthy psychotherapy in some cases. My stars. Never mind that drug therapy really relieved a person's distress, didn't take as long by far, and didn't cost as much by a long shot. Therapeutic drugs wouldn't unearth the root problem, said the critics, they would only mask the symptoms. Implicit in this argument was a view which held that serenity was only to be achieved through prolonged suffering, a moral debit and credit system that had very little to do with the realities of mental illness. Drug therapy turned out to be no panacea, but it also revealed that a great many disturbances which we'd attributed to bad relations with parents were in fact largely biochemical, and relieving a biochemical distress helped ease the bad relations between parents and child remarkably. The lesson in this, as Colby is the first to admit, is that we know less than we wish to about human mental functioning, and we'd better be careful about dogma. Nowhere does Colby say that a computerized system should be used even if it can't be shown to be beneficial. That makes him something of an oddity in the helping professions right there.

Some AI researchers chose the forum of a book review to answer Weizenbaum's charges. One of the briskest and funniest came from John McCarthy, who'd already extracted an apology from Weizen baum for having misquoted McCarthy in the book (McCarthy, 1976).

McCarthy points out some fuzziness in the book's text, but makes an attempt to summarize its main points nonetheless. He replies to these systematically: If artificial intelligence hasn't yet been achieved, there is no reason to believe it cannot; a hundred years passed between Mendel and the discovery of the genetic code for proteins, and it's reasonable to expect that human intelligence is at least that complex. Though Weizenbaum, like Dreyfus, argues that the sensory experience of a human being is necessary to human intelligence, would they or anybody else deny full intelligence to the deaf, to the blind, or to paraplegics? The question about whether language and the concepts that underlie it are formalizable confuses formalization with simulation; the former may be impossible, but the latter need not be. If Weizenbaum says there are tasks a computer should not be programmed to do, McCarthy replies that what shouldn't be done shouldn't be done at all—by anyone or anything. Weizenbaum's arguments against some computer tasks remind McCarthy of Renaissance-era religious objections to dissecting the human body, though the Vatican has since come around to the notion that regarding the human body as a machine for scientific or medical purposes is quite compatible with regarding it as the temple of the soul. And so it goes, with McCarthy worried that "when moralizing is both vehement and vague, it invites authoritarian abuse either by existing authority or by new political movements."

Weizenbaum answered this review, as indeed he has answered a great many of them, pouring his major energy since the book was published into the debate—and thereby surely accumulating enough material as a gloss on the original text to comprise a whole new book. There's little he can say to McCarthy, he finally admits, because he and McCarthy are speaking from such hopelessly irreconcilable points of view.

But what strikes me as moral earnestness in the book is often sanctimoniousness in the schoolyard you-saids and I-did-nots that have characterized the replies to the replies. On the whole, the responsibility for this tone is Weizenbaum's. His book is ambiguous, and often makes odd assumptions—say, that the Freudian model of human behavior is obviously preferable to the information-processing model, or that we have ceded to the machines a personal autonomy we once actually had, or that people in artificial intelligence are obsessed with a single-minded view of humans that admits only one way of looking at them.

Weizenbaum's eagerness to remain engaged in the dispute instead of allowing his book to speak for itself pushes outsiders to examine additional motives he might have had for writing in the first place. And so we look, and find a tangled business indeed. If Weizenbaum and Colby have bad feelings between them, Weizenbaum can honestly claim that his moral judgments about Colby's work have not interfered with his scientific judgments. Asked to sit on a panel a few years ago which was deciding whether to continue funding for Colby's work, Weizenbaum said yes to that continuation. The computer metaphor is a powerful one, Weizenbaum reiterates, and ought not to be thrown out entirely, and Colby was then the only trained psychiatrist who was working with computers, in this way. Anyway, no such feud exists between Weizenbaum and, say, Herbert Simon, though Weizenbaum is angry enough with Simon to misread his work pretty badly, and to omit his name altogether from his account of the EPAM research; Weizenbaum writes approvingly of EPAM, but gives the impression that it was Edward Feigenbaum's alone, though in fact it was done equally by Simon and has since been extended by him.

A view in the community prevails that Weizenbaum has taken this debate up because he can no longer do science. "He fell out of the field ten years ago, and hasn't done a damn thing since ELIZA," one MIT colleague said contemptuously. An old friend of Weizenbaum's shook his head sadly. "I have the impression that if Joe could do science, he wouldn't be doing this. When I gave a talk about my AI work a couple of years ago at MIT, it was Joe who came up to me—and I can't tell you the feeling he said this with—that he'd give his right arm to have done what I'd done. MIT is an incredibly competitive place, and regardless of whether you have tenure, the pressure to produce is terrific. Joe hasn't produced science, so he's got to do something. I wish he hadn't chosen this."

Weizenbaum, of course, might very well retort that he's glad his moral concerns have eclipsed his science, and it would be a good thing if it happened to more scientists. On the other hand, for workers in AI to take the view that what they're doing is a good thing on the whole for society, with more benefits than drawbacks, is not going to surprise anyone. It would be astonishing if they didn't feel this way. Regardless of motives and ego investments, what finally matters are the issues, and perhaps people who are less involved can shed some light.

Reviewers from outside artificial intelligence found much to like in the book. One was Daniel McCracken, himself a writer of several textbooks on computer programming, and the chairman of the Committee on Computers and Public Policy of the Association for Computing Machinery, the largest professional society of computer experts. McCracken's review appeared in *Datamation*, a widely read periodical in computing (McCracken, 1976). Though McCracken had what he called some minor quibbles with Weizenbaum, he was generous in his praise for the book and his hopes that it would reach a wide audience, from those in the artificial intelligentsia (whom he described as already working on rebuttals, the sound of typing drowned out by the sounds of gnashing teeth) to general readers, who were invited to read it as members of the Book-of-the-Month Club. He affirmed Weizenbaum's view that there are deep and fundamental differences between human beings and computers that will never disappear no matter how many new tricks we learn about using computers.

Another highly favorable review appeared in the London *Times Literary Supplement*, written by the British cyberneticist N. S. Sutherland, who concluded that "Weizenbaum writes with passion and conviction and if the arguments he adduces are not always compelling, he raises important issues that are too often ignored. He repeatedly and correctly insists that computers lack wisdom, but if computers are put to ill use, it is because we, not they, lack wisdom." And then Sutherland ruefully adds, "He appears to be in danger of projecting our own vices on to our artifacts, but on the crucial problem of how we are to find the wisdom to use them sensibly, he has nothing to say" (Sutherland, 1976).

Most other reviews from the AI community exhibited less gnashing of teeth than McCracken, at least, might have supposed. Some, such as Russell Taylor at the Stanford AI Lab, didn't think it was good at all that Weizenbaum was content to rest some of his major arguments with the declaration that certain ideas of artificial intelligence were obscene. Unlike McCracken, who did not share all Weizenbaum's judgments as to what was obscene but suggested that this was quibbling, Taylor wrote,

> I do not believe that these are minor quibbles. On the contrary, it is this characteristic of Weizenbaum's argument that constitutes the major weakness of his book. It is certainly true that computers (or anything else) should not be applied for immoral ends. But it is exactly in situations where such value judgments must be made that clear reasoning is most necessary...
>
> (Taylor, 1976)

Weizenbaum responded that he was unable to give reasons as to why he found obscene the idea of disembodying the brain and eyes of a cat, keeping them artificially alive, and using them as visual receptors of a computing system. If simple human decency allows nothing to be taken for granted, he said, then he retreated before the onslaught of his critics. Some things he did take for granted, he went on, and one of these was that a machine of the sort here in question would appear a monster from which he would turn with disgust and revulsion:

I have in mind also the teaching urged on us by some leaders of the AI community that there is nothing unique about the human species, that in fact, the embrace of the illusion of human uniqueness amounts to a kind of species prejudice and is unworthy of enlightened intellectuals. If we find nothing abhorrent in the use of artificially sustained, disembodied animal brains as computer components, and if there is nothing that uniquely distinguishes the human species from animal species, then—need I spell out where that idea leads?

Well, where? Neither logic nor jurisprudence would admit such reasoning, and human wickedness doesn't wait around for it.

Joshua Lederberg, whose DENDRAL program was one of the two Weizenbaum treated approvingly in his book, was asked to review the book for the *New York Times* (which paid him, but for inscrutable reasons of its own never used his review). The Computer Science Department of Stanford University published this review instead (1976), with two others by Stanford AI workers, Bruce Buchanan and John McCarthy, which had appeared elsewhere.

Lederberg's review is restrained, and touches only gently upon what he considers the logical lapses, admitting that some readers will be attracted by the "lyrical anti-technology slogans, which the author's technical reputation will make the more persuasive." Lederberg says that though Weizenbaum makes a conscientious effort to distinguish his assertions of faith from the scientific consensus, the nonspecialist reader will still have to look closely to be sure. For example, Lederberg the geneticist warns that what Weizenbaum takes as proven reality about the physiology and organization of the brain hemispheres is really as yet speculation. He agrees with Weizenbaum that the world-knowledge underlying human understanding (compassion and judgment) needs the life-long experience of having been human—in a word, of having shared love. It is unlikely and undesirable that machines be offered that privilege, he writes, and then many realms will remain uniquely human. Indeed, he adds, we must make equally sure that the fellow-creatures to whom we confide our trust for ethical and aesthetic leadership justify this confidence on the same grounds. And those who deify the machines deserve the human sacrifices that may result from neglecting the human responsibility for moral decisions (Lederberg, 1976).

Where Lederberg radically parts ways with Weizenbaum is at Weizenbaum's insistence that the human brain is fundamentally unknowable. Scientists can contribute more by trying to find out what can be learned about our own nature and putting it to human good than by arguing what may or may not be knowable, says Lederberg, and reminds us that the same kind of mysticism about the core of the cell's reproductive capability being unknowable in chemical terms bears much of the onus for long delays in our understanding of the structure of DNA. "After the fact, this proved to be remarkably simple." And all the more awesome, he might have added.

Lederberg points out Weizenbaum's oscillation between disparagement of AI and his dismay in case it should come to pass: "That policy-makers, the public, and computer scientists alike should take a more critical and pragmatic view of the field than the zealots of twenty years ago may be granted; many well-informed people within the field clearly do, without having reacted as strongly as Weizenbaum" (Lederberg, 1976).

> The abuses [Lederberg goes on] might be either ideological or technological. If human intelligence were more successfully mirrored in the machine, will that not justify treating human beings as if they were MERE machines? [Weizenbaum's] position on this is colored by the experience of Nazi Germany; but the argument is confused. The most savage tyrannies that I can find in history, including Nazism, had no doubt about a unique *élan vital*—just that one folk or

credo had more than an equal share. People who are philosophically concerned about the mechanistic basis of life are also overawed by its complexity, and too concerned about learning more about it to occupy themselves with holy wars. They are the least likely to be sacrificing either people or machines on the grounds of ideological conviction.[2]

If Weizenbaum points to very real concerns about machines that could interpret speech—they could be used as large-scale wire tappers—he overlooks the possible benefits that could relieve millions of office workers from "the mindless tasks of transcribing the words of others, and free them for more creative responsibilities. Both of these contingencies lay heavy burdens on the adaptability of our social institutions, and it is important that we be alerted to them."

Bruce Buchanan, who was also a major investigator on the DENDRAL project, reviewed the book for *Pharos* magazine, and while he concluded that the main issues of the book are important for everyone, he is impatient with Weizenbaum's antiintellectualism, or more precisely, his antirationalism:

> The guidelines that he gives are certainly incomplete for research on energy, communication, transportation—and almost anything interesting enough to be applied in the next century would have unforeseeable side effects or could be used to assault life. They are offered as expressions of his own subjective criteria, and perhaps because they are subjective they cannot be expressed adequately in the language of the brain's left hemisphere (as he reminds us in another context). Such guidelines, even when precise, also fail to admit the value of research aimed at defining the limits of what computers can do by working on programs at the boundaries between men and machines.
>
> (Buchanan, in Lederberg, 1976)

---

[2] Carl Sagan, who wrote enthusiastically about Weizenbaum's DOCTOR program in *The Dragons of Eden* (1977) is one of these. Speaking of the enormous numbers of possible brain states in a human being, which far exceeds the total number of elementary particles in the universe, he says, "From this perspective, each human being is truly rare and different and the sanctity of individual human lives is a plausible ethical consequence."

Like Lederberg, Buchanan sees a symbiotic relationship between humans and machines as the most realistic and desirable outcome of AI.

Gentle reader, we have consulted the experts. What are we non-specialists to make of these charges and countercharges? I feel a bit like Omar Khayyám, that self-described stitcher of the tents of science:

> Myself when young did eagerly frequent
> Doctor and Saint, and heard great argument
> About it and about: but evermore
> came out by the same door where in I went.

The facts, so far as I can see them, are these. Artificial intelligence is the latest manifestation of an enduring human impulse to create artifacts that will imitate our essential human property of intelligence. In the midst of such an act, we participate in two related and equally enduring impulses: the one to know ourselves as deeply and in as much detail as possible; the other to render that knowledge in a way that is accessible to our fellow creatures—always the hope of the biological and behavioral sciences, philosophy, and in some ways, art. Thus *artificial intelligence* strikes me as a happy phrase for what this field seems to be about.

Artificial intelligence as science has been less *apparently* successful than its founding fathers hoped, but its progress in the twenty or so years of its existence has been enough to confound its critics—it performs better and better, slowly breaking down each barrier the critics declared could never be surmounted. Whether it will surmount them all remains to be seen. Each step toward a fully realized intelligent artifact suggests that it can be done, but it hasn't been yet. Each of these steps, however, forces us to refocus our view of ourselves, and at such points people who ponder the human condition have the right, the responsibility, to consider our proper place in the universe, our possibilities and limitations, our individual and collective purposes.

Joseph Weizenbaum's book, his cry from the heart, is as representative an artifact of this history as any chess machine or robot. It rests deep in the heart of what I call the Hebraic tradition: saying no instead of yes, restraining instead of freeing, yearning for a past grace instead of an unknown future.

Weizenbaum is deeply committed to his beliefs. This is neither unusual nor bad: scientists must believe in the work they do as writers believe in their words. The observer's task is to examine the nature and the substance of that attachment: we need not quarrel with the commitment itself.

Are there domains where computers, even if they somehow can, ought not to intrude? To reiterate, Weizenbaum suggests three such places: where computers would represent an attack on life itself; where the effects can easily be seen to be irreversible and the side effects are not entirely foreseeable; and where it is proposed to substitute a computer system for a human function that involves interpersonal respect, understanding, and love.

But life itself is an attack on life, when it comes to that. Women still die in childbirth, and we eat living plants and animals to sustain ourselves. We consider it good to go on having children (and making childbirth as safe as we possibly can), and we try to be good stewards of other forms of life, though the meaning of "good" has been highly variable throughout our history. We can expect our notion of good to continue to change, so we must be wary of blanket proscriptions.

Computers ought not be introduced where the effects can easily be seen to be irreversible and the side effects are not entirely foreseeable. I've met no one in computing who disagrees with this, though Buchanan makes the reasonable point that almost anything interesting will have these properties. This may then be an argument for preserving the status quo, but few people are altogether in love with things as they are. John McCarthy argues that intelligent machines will be helpful in these kinds of predictions, able to foresee by simulation effects and events we ourselves might not (McCarthy, 1976).

No computer system should be substituted for a human function that involves interpersonal respect, understanding, and love. Weizenbaum's examples here are two, a psychoanalyst and a judge. I have already reported Colby's response on the subject of psychotherapy. On judges, John McCarthy says in his review, "The second quotation from me is the rhetorical question, 'What do judges know that we cannot tell a computer?' I'll stand on that if we make it 'eventually tell' and especially if we require that it be something that one human can reliably teach another." Lederberg agrees that no computer system should be substituted for such human functions, but goes on to make the point that we'd better be equally sure about those humans in whom we invest respect, understanding, and love.

This brings us to the notion of artificial intelligence as alien, as not having shared our human experience, and being therefore deficient. When Weizenbaum and I were discussing this idea—and indeed he makes the point in his book—he again said how difficult it is, perhaps impossible, for somebody from one culture to act as judge in another. I laughed, and said that I probably spoke for most women, minorities, and others excluded from power in my own culture when I said I'd rather take my chances with an impartial computer. Our human experience is a fact of our existence, but how much we share with one another is limited not only by gender and class and color of skin, but also by temperament: one person's irresponsibility is another's high spirits. McCarthy also reminds us that we know next to nothing about what constitutes the nature of human experience or understanding; when we know more about them we will be able to reason about whether a machine can have a simulated or vicarious experience normally confined to humans, and we will be able to define whether a machine understands something or not.

Is what the workers in artificial intelligence do science or mere technique? This issue is impossible for a nonscientist to judge. It is even very difficult for specialists to judge, and will finally be decided by history. But there are some indicators. The Turing Award, considered the most prestigious in computing, is made yearly for significant contributions to computer science by a committee composed of computer scientists representing the Association for Computing Machinery. The membership of this committee, which changes yearly and represents the spectrum of computing research, scrutinizes the work of researchers throughout computer science and votes on their selection. Among the early winners of the Turing Award are John McCarthy, Marvin Minsky, and Allen Newell and Herbert Simon. This list represents nearly a third of all the Turing Awards given. By the same sort of peer review, all the people Weizenbaum complains about have received funding from government and private agencies. Herbert Simon has won a Nobel Prize for work he considers very much a piece of his artificial-intelligence research. Neither prize winning nor receipt of funds is positive proof: only time will give us that.

Have we embraced the machine metaphor too readily? I suppose what is meant here is that many more people feel unable to control their lives than in the past, that a political and personal impotence has seized a significant portion of the population, that things are in the saddle and ride mankind (a nineteenth-century sentiment, it might be good to remember).

Well, history is how you read it. My morning newspaper is full of the yellers and the yelled-at, each making more than just a joyful noise, which is doubtless exhilarating to the aggrieved and scary to the entrenched. This is no country for old men, and this is not new in human affairs.

Have large numbers of people adopted the machine metaphor to describe themselves? Weizenbaum offers us workers Studs Terkel interviewed in *Working* who complain that others seem to regard them as machines but that they themselves certainly don't; hence their grievance. This attitude is exactly the opposite of Weizenbaum's point. No, I fail to see that we have assumed the machine metaphor wholesale for ourselves as individuals, although we do feel legitimate anger at systems in society that operate as if humans can be regarded that way. This anger often expresses itself in yeasty forms of protest.

On the other hand, a mechanistic model of humans, which views us as part of the natural world and therefore just as comprehensible as any other part of it, has some attractions. It relieves human suffering when we discover that a virus behaves biochemically in certain human tissues, instead of assuming that a diseased person is suffering as retribution for sinful thoughts. I doubt that Weizenbaum would deny this statement, but there's bias in his next step, the choice of mechanistic models to explain human mental functioning. His choice seems based less on science than personal preference. One could argue that the information-processing model of human mental functioning, which Weizenbaum attacks, is in fact more general and accounts for much more cultural and gender variation—and is a lot less bizarre—than the Freudian model, which Weizenbaum proffers without question, judging from his analysis of the compulsive programmer. In places in his book, and certainly in conversation, he concedes that the information-processing model has its uses. His objection, then, is that AI workers are single-minded about the model as the sole means of encompassing the human experience. That isn't my impression. Their writings are enthusiastic but tentative: they think they're onto a good thing, maybe one of the best so far, as an approach to explaining human cognition. But they read novels and poetry, compose and play music, see movies and write stories, and make the same noises about the value of doing those things as the rest of us.

Perhaps Weizenbaum means that by regarding humans mechanistically, we think we can escape our moral responsibility to one another and to the society in which we all live. That's an amazing leap, and when it gets made, suggests at the very least an impoverished moral education. Then again, human cussedness will always find an excuse for itself, call it original sin, bad company, disadvantaged background, or what we will.

The Eden Weizenbaum harks back to doesn't exist in human history, whatever its psychological implications for the individual. When—at what moment in the past—were we more careful about individual human dignity and autonomy than we are now? This is not to say that at this moment we deserve any prizes; it is to raise the question about what we're supposed to be in danger of losing if we take artificial intelligence, and by extension, science and technology, seriously. But we cannot lose what we never really had, or rather, what was the closely held private property of a small elite. As I see it, most of us are still struggling to attain autonomy, not lamenting its loss. The role technology plays in this struggle is complex. But if I'm asked to guess whether technology or good will ended the exploitation of six-year-olds in the mills and coal mines, I'd vote for technology.

Finally, Weizenbaum worries that to regard humans as another instance of information-processing machines seems to imply that "there's nothing unique about the human species." Strictly speaking, our definition as a species makes us unique. He doesn't say why regarding ourselves as another instance of information-processing machines automatically precludes regarding ourselves in other ways too. But the real question is whether humans as a species deserve more special treatment and consideration than other species. The humanist has traditionally answered a resounding yes to that question, but that yes is fraught with problems. Faith in our specialness has often been used as license to plunder and exploit. It's a good corrective, and in the long run self-preserving, to meditate on our connection with other things instead of our disjunctions. We are part, not monarchs, of the universe.

And if we introduce into our universe machines who think, will we then owe them our trust and love and respect? Will we, in my earlier phrase, be forced to extend the boundaries of The Other? AI researchers say of course not, and people outside the field are even horrified by the question. But I suspect we need all the practice in humility, in loving and respecting others that we can get. We'll find out soon enough if it's misplaced.

I opened this chapter with some questions. Should a machine think? Will an intelligent artifact grasp power over us? Will we cede it willingly? Is benevolence at issue? The Weizenbaum controversy addresses some of these questions. Weizenbaum wants to limit severely the range of intelligent machines if such machines can be made, but though I can imagine such limits in principle, I find

Weizenbaum's examples unpersuasive. McCarthy's declaration that what shouldn't be done, shouldn't be done by anyone or anything at all, seems to me the best we can say at present.

Will an intelligent artifact grasp for power over us? We're into the Coyote problem here, ready to assume that the worst about ourselves is also true of others. AI researchers dismiss as perfervid science fiction the notion that a will to power is necessarily concomitant with intelligence; this just happens to be so in the human species. AI can conceive of intelligences where power simply isn't at issue. As to the power we cede, whether to intelligent artifacts or other sorts of machines—automobiles, say, or county bureaucracies—the problems are delicate, complex, and hardly to be solved by exhortations or fiat. Our culture claims to put a high value on human freedom, independence, and self-sufficiency, but other human cultures do not, and the results of our own commitment to such values have been mixed. Again, McCarthy very reasonably points out that genuine artificial intelligences hold in themselves a possible solution to some of these problems—we can ask them about their best use in what we see as our interests.

But perhaps we ought to contemplate the possibility that an artifact far smarter than its human creators may turn out to look like still another splendid gesture of human adaptivity, as upright posture, opposable thumbs, and language were. We would have to worry no more about our vanity in surrendering to—or at least consulting—its intelligence to solve problems beyond our own solving though of our own making than we fret about using scuba gear to go where our human bodies otherwise can't.

In any case, while I honor the best qualities of my species, I should like to protest against becoming sentimental about ourselves, and against that vein of smarmy self-flattery that runs through much debate on Us versus Them—be it Us against the machines, Us against the extraterrestrials, or Us against the other humans, for example, the scientists, the military-industrial complex, the villain of your choice. Sentimentality is the least attractive aspect of the whole business. I understand its genesis in our self-doubt. But wouldn't we be better off admitting the doubt and looking around for help?

Regardless, the appearance of Weizenbaum's book, the issues it raises, the deeply felt motives out of which it was written, and the reactions it has evoked should remind us, I hope, that we're all in this together. Personally, I find that comforting, and, having squirted seltzer water in a few people's faces here, I wouldn't want anyone to miss the point. It's this: that we might—perhaps should—all invite ourselves into this debate. If the effort to make artificial intelligence has taught us one thing, it's that natural intelligence is a formidable and woefully underutilized resource.

So the question of whether an artificial intelligence can be made may be put aside with a tentative, qualified yes. Whether it should be done is still open for examination, and to ask how such machines will make us appear in our own eyes gets closer to the nub of what troubles us about AI. This question refers not to the technical feasibility of artificial intelligence, but to what we will have to face about ourselves, the possibility that we may have to undergo still another redefinition of ourselves as a species, another Copernican revolution that will move us further yet from the center of the universe.

With respect to the practical decision about whether to go ahead or abandon such research, John McCarthy's opinions are cautionary: "I think the only rational policy now is to expect the people confronted by the problem to understand their best interests better than we now can. Even if full AI were to arrive next year, this would be right. Correct decisions will require an intense effort that cannot be mobilized to consider an eventuality that is still remote" (McCarthy, 1976).

Shucks, John. That takes all the fun out of it.

Chapter Fourteen

⌐

Forging the Gods[1]

We're near the end of one of a handful of journeys that have persistently engaged the human intellect, a journey that has taken us from intuition to knowledge, from the particular to the general, from art to science. Nobody should read any judgments in such transformations, or the suggestion that science is somehow preferable to art. The meaning of the journey all depends on our purposes. This journey can profitably be a round trip.

Artificial intelligence has had a remarkable career as art. We've seen the impulse in Homer, the counterimpulse in Genesis; we've watched imitations of human bodies jerk awkwardly across the human landscape, imitations of human minds cross a chessboard or a theorem. There's something deep and central to the entire effort, which for all its carnival atmosphere is no sideshow of the human spirit but is in the center ring. Face to face with mind as artifact, we're face to face with almost more themes in the human experience than we can count or comprehend. And there's the added zest that this idea might turn out to transcend the human experience altogether and lead us to the metahuman.

Can a machine think? The answer, as we've seen, depends very much on what we're willing to admit as machine, and what we're willing to admit as thinking. The definitions of either aren't quite so simple as they looked when we embarked.

---

[1] This fine phrase was suggested to me by Jacob Schwartz of New York University.

For example, humans think—it's central to our notion of ourselves that we do, and correctly so. But are humans machines? Even machines that clank softly, meat machines? Is the mind a mechanism? We've said yes, then no. We've wanted to put ourselves inside nature that way, yet stumbled over the means, as the histories of both philosophy and psychology have shown us. And a big part of the problem has been that we always knew thinking when we saw it—call it consciousness, understanding, learning, reasoning—but we had a hard time with further identification.

Artificial intelligence sides with that waggish old Frenchman Julien Offray de la Mettrie, author of—and martyr to—*L'Homme Machine*, who was at pains to point out that one useful way of regarding human beings is as mechanisms—that is, all their functions are capable of being described in a logical, analytical way. Allen Newell, two centuries later, is more precise: "A mechanism is any determinant physical process. An abstract process constitutes a mechanism if, in principle, there are ways to realize it by a physical process" (Newell, 1973b).

Well, then, if the properties of mind can be realized by a physical process, they are mechanisms. Does that ease the resistance we feel to the idea of mind as mechanism? (For by now we're pretty comfortable with the idea of body as mechanism.) Again, only if we can see some mindlike behavior realized in a physical system. Apparently it isn't enough that those mechanisms all around us, our fellow humans, do thinking: historically, at least, we've preferred to postulate the mind-body split.

But what if some other physical entity, something that's indisputably a machine—for argument, let's say a computer—displays some intelligent behavior? Then are we prepared to agree that mind is mechanism, and that two instances of intelligent agents, namely humans and computers, realize those processes in a physical system?

And maybe only *some* mindlike behavior realized in a physical system isn't sufficient to allow us to regard humans as mechanisms. For if artificial-intelligence workers (and I) side with La Mettrie, our belief is, as La Mettrie's was, an act of faith—though these days we're at considerably less risk in holding it. For that's not all there is to us, he went on; there are other equally valid ways of regarding human beings. Moreover, to regard ourselves as mechanisms in no way demeans, or for that matter captures us completely. Perhaps no single approach ever will. And this view has not occurred only to me or La Mettrie. Newell, again, addressing himself to theories of artificial intelligence, writes, "The domain of such theories does not cover all of the phenomena surrounding man. As we might expect, it focuses most strongly on tasks that are highly symbolic and do not require continuous motor skills or an intimate dependence on sensory systems. It deals with phenomena of substantial complexity, but not with the full complexity of human affairs" (Newell, 1973b).

The domain of mechanistic theories, then, embraces neither the full complexity of human affairs, nor even the full complexity of the human mind, so far. All the evidence isn't in: there are curiosities about the human mind we cannot yet account for. I say yet, my bias showing that I believe we someday will. But plenty of serious scholars disagree. Perhaps, like Hubert Dreyfus, they hold that intelligent action is so intrinsic to functioning in a human body and living in a human world that the hope of modeling such behavior in any complex way is feeble: we cannot get outside ourselves in this way, we never will; and, furthermore, no means exist by which we could make the effort. That any intelligence worth talking about might exist elsewhere is simply preposterous, in this view, and stretches the word intelligence far beyond its proper meaning. The human being is the measure of all things.

I think Dreyfus's view is the majority one, and has been throughout history, or at least as long as we've believed human beings to be the measure of all things. But the dissenting view has also existed for a long time, and has recently picked up momentum. Workers in artificial intelligence have addressed some questions to which they believe that they, or their scientific descendants, will someday find full answers. Some such questions are these: What is intelligence? What is it about a system, man-made or begotten (in Warren McCulloch's phrase), that allows it to behave intelligently? What is consciousness? learning? understanding? What kinds of tasks require intelligent behavior?

Of course, we see at once that artificial intelligence wasn't the first to raise these questions at all. They're central to the Western intellectual tradition, and innumerable philosophers have dreamed up schemes to answer them, their reputations resting upon plausibility, personal persuasiveness, and sometimes sheer turgidity. But not upon empirics. And here's a major difference between philosophers and AI workers. The latter are dreaming up their schemes as scientists, securely within the empirical tradition.

When Galileo showed that physical phenomena need not be found by chance, but could be teased out of nature by theoretically well-founded experiment, he was not only on his way to revolutionizing physics and natural science in general, but he was also providing criteria by which one explanatory hypothesis or scheme might be accepted and another rejected. Of course, by taking this empirical approach to mental phenomena, psychology means to distinguish itself from philosophy, but until the advent of the computer, psychology's instruments—boxes, mazes, stopwatches, statistics—were too simple to capture any but a tiny fraction of such behavior, and work on the really tough questions got no further along than ever.

Artificial intelligence has changed that. The information-processing model includes a few potent ideas such as the belief that mind, however wonderful, is finally understandable by mind, and that mental phenomena can probably best be understood by decomposing them into the much smaller, simpler components that, acting together, produce complex, intelligent behavior. The proof of these hypotheses is the existence of intelligent computer programs—playing games, solving puzzles, having conversations—which are put together in just this way, from many well specified but relatively simple instructions, and which then produce complex behavior. These programs are simulations, not precise imitations. We do well to remember Vaucanson's wonderful eighteenth-century duck, which contained no detailed imitation of the digestive processes, but instead had a mechanism suggestive of the larger aspects, including intake, maceration, and an obvious chemical change in the product prior to excretion. From this mechanism we can derive an approximate answer to the question, What is digestion? We can, as we come to understand digestion better (partly by playing around with our approximation of it), make the model come closer and closer to matching the actual process. In a similar way, intelligent programs simulate intelligent behavior in its larger aspects, allowing us to answer approximately questions about such behavior, and giving us the opportunity to play around with our approximations so that we can more closely simulate the behavior as we see it demonstrated by humans.

For example, What is mind?

> An essential condition for intelligent action of any generality, [writes Allen Newell is the capacity for the creation and manipulation of symbol structures. To be a symbol structure requires both being an instance of a discrete combinational system (lexical and syntactic aspects) and permitting access to associated arbitrary data and processes (designation, reference and meaning aspects).
>
> (Newell, 1973b)

To the poet's ear, this sounds a bit flat. Whatever happened to mind as kingdom, as the uncopyable ("They copied all they could follow, but they couldn't copy my mind," wrote Kipling, champion of lost causes such as The Empire), as an inwardly growing root on the one hand and as a voyager on the other? Whatever happened to mind as a mountain range to be scaled; as a pantry to be filled; as great lever and great labyrinth; as marcher and breather of music; narrow as the neck of a vinegar cruet and wide as the troubled sea; forger of manacles and home of hobgoblins; marble for men and waxen for women (sentiments meant to flatter men, by the way)?

Well, such definitions live side by side now with Newell's, and we're free to select as inclination moves us. There's a difference, however. The fancies of poets are metaphors. They illuminate by comparing differences as well as similarities—nobody ever pulled a mackerel out of a mind and had it for supper, but we get what the poet means by calling a mind that wide and troubled sea. Newell's definition, however, aims at analysis rather than comparison. If we can put our hands on its components we can build a mind, whereas if we have our hands on the components of metaphors, we'll build a sea or a vinegar cruet, but not a mind.

Now, Newell's specifications—the capacity to create and manipulate symbol structures—admit more than just human minds. The digital computer has such capacities, and with it, scientists discovered they had an instrument for building and examining models of mind. Minsky writes,

> Furthermore, the general purpose computer had become necessary because theories of mental processes had become too complex and were evolving too rapidly to be built into ordinary machinery. Some of the processes we want to study make substantial changes in their own organization. The flexibility of computer programs makes experiments feasible that would be next to impossible in an "analogue" mechanical set-up.
>
> (Minsky, 1968)

But if the digital computer was the obvious instrument of choice for later researchers, we've also seen that two earlier men had other views. John von Neumann, fascinated as he was by the notion, never believed that a computer could be made to think, and though Alan Turing wanted passionately to believe it could be done, he had no concrete ideas about how it might happen.

How indeed was it done? The mere existence of the computer wasn't sufficient. No, it took this major shift in point of view missed completely by von Neumann and approached but never quite explicitly grasped by Turing. This move was the shift from the worm's-eye view, seeing the computer's on-off logic as somehow analogous to the on-off behavior of brain cells, upward to a grander, higher-level view, seeing the computer's ability to process information as analogous with the brain's ability to process information—never mind the hardware or the wetware.

Inspired by this higher level of comparison, this information-processing view, researchers could get to work and build systems that, at least by behavioral definitions, acted intelligently. And this has been the main research method of AI, an empirical one: build a system that can play chess, prove mathematical theorems, tell a bedtime story, and you'll have some sort of insight into intelligence. Put this way, it sounds a bit loony. In fact, we've seen that thinking machines aren't thrown together willy-nilly. They're constructed not only upon the best guesses of how human intelligence might work, but are open to exhaustive examination and rational analysis, thereby suggesting refinements for those best guesses about how humans think. As in all sciences, hypotheses are made, models are constructed and then fiddled with, in order that the original hypotheses might be improved upon. AI workers call that process generate-and-test. That they've sometimes acted more on faith than facts, and haven't necessarily approached their tasks with celestial disinterest, is no more than the truth about the way science works.

So goodbye to the original approach, derived from cybernetics and exemplified by all those charming mechanical mice, rats, turtles, and other simple beasts, which relied on physiological similarities between neurons and whatever hardware was then in vogue, whether tubes or transistors. Under this approach, they would be arranged in what Minsky calls "a very weakly specified structure," placed in an appropriate environment, and would gradually come to behave in an adaptive fashion. Though this work made considerable contributions to complex feedback-regulatory systems, nothing in the way of really intelligent behavior emerged, least of all a theory.

Instead, from a different set of circumstances, the information-processing model emerged and was richly productive. There were two ways of pursuing it. One was to imitate humans as closely as possible; the other aimed at producing intelligent behavior in a computer by whatever method—or as Minsky put it, "without prejudice toward making the system simple, biological or humanoid" (Minsky, 1968). This latter view was the one nominally pursued at MIT, while at Carnegie human psychology inspired research. Yet the division was never very rigid. Minsky and his colleague Seymour Papert, who had himself spent much time studying the development of children's intelligence, were always attentive to human problem solving as their most important model; Newell, especially later, would pour much energy into developing intelligent systems that might or might not have to do with human methods. As Edward Feigenbaum of Stanford discovered in the course of developing his intelligent chemist's assistant, such large, complicated programs almost *have* to mimic human processes, or humans won't understand them.

Somewhat lonelier, John McCarthy has taken a more theoretical view of what intelligence is all about, human or computer, and has spent most of his time looking for a mathematical way of expressing it. "Maybe mathematical rigor isn't quite what I'm looking for," McCarthy once said. "Maybe scientific rigor would be better, because it isn't a question of having some theorem that is proved mathematically—it's a question of having demonstrated the success of an approach empirically as well as by argument, saying that it ought to be the right thing to do."

Minsky, who glances longingly over to mathematics from time to time, shakes his head at McCarthy's stubborn characterization of this as the great unsolved problem of artificial intelligence:

It's only unsolved if you insist on solving it in a certain paradigm that John's got in his head about how it must be solved. I think it's been solved and now the problem is to figure out what the solution means. I think the structures we've developed solve the problem in principle, but people are quite confused about exactly what to do with the apparent solution. Students sit around paralyzed to some extent because there are so many alternative pathways to implement the idea of packages of knowledge. The difference really is that McCarthy has tried to isolate individual fragments of knowledge and see if he could avoid the somewhat ad hoc packaging of them into larger chunks. If he wins it will be nice, but he can't. I'm pretty sure of that.

Another time, Minsky paraphrased General Motors's Engine Charlie Wilson, who'd said it originally about scientists: "Mathematics should be on tap but not on top."

And so this twentieth-century version of our species dream, to make intelligent entities that aren't human, is summarized half facetiously by scientists at MIT as falling into three periods: the classical, the romantic, and the modern.

In the classical period, lasting from the early 1950s to the late 1960s, the search was for general principles of intelligence, regardless of the task at hand. Here was the first confirmation that the information-processing model was a rich one, that, although we had no physiological theory of mind (and still don't) intelligence could nevertheless be understood and expressed precisely enough that a computer could be programmed to behave intelligently. Here, in short, was the news that some other entity could be made to exhibit what heretofore we'd considered our exclusive, identifying property.

A romantic era replaced the classical one when it came to be seen that intelligent action required not only general principles but also a very large amount of highly diverse knowledge, custom-tailored to the task at hand. Humans, it seems, are veritable encyclopedias of specialized information; they pick up such information in many different ways (those ways are of intense interest just now to researchers) and store it with breathtaking economy and accessibility (also of intense interest to researchers just now). In other words, no matter what your native intelligence, you can't just pick up and do, say, brain surgery. You really must learn some things about anatomy, physiology, pathology, pharmacology, and surgical techniques. Then and only then will your acquisition of motor skills be worth anything. Early computers and computing techniques had been so primitive (relatively speaking, of course) that the hope of ever approaching that human virtuosity of storage and access seemed insurmountable. However do we do it? And what do we do it with? What does the knowledge in our minds look like?

For the crux of the matter was not in bigger and better hardware alone, though that too must come. How, researchers had to ask themselves, is all this special knowledge to be represented—not only how should it be encoded, but how is new knowledge acquired or assimilated, how is it accessed, and how is it derived from knowledge already acquired? It sounds as if we're talking about human learning, and so we are, for if we can answer these questions, we will know what learning is all about—the great nonproblem in artificial intelligence, researchers discovered. To ask what learning is asks the wrong question. Learning, it seems, is just a shorthand phrase for knowledge representation and manipulation—how knowledge is acquired, assimilated, restructured, and so forth. In the romantic era, the distinction between procedural and declarative knowledge— itself a key part of learning—came clearer and could be expressed in new computing languages, such as PLANNER and its offspring CONNIVER (may scientists never lose their sense of whimsy), that replaced the old step-by-step programming languages. These new languages had opportunities for contingent instructions, that is, do such-and-such whenever thus-and-so happens, which can be any time, any place; the sequential one-two-three commands much beloved by early computer users had been replaced by rules as precepts, advice, embodiments of knowledge. Programs could now rewrite themselves when something didn't work, which means of course that they had a sense of what not working means. (So it began to look as if Hubert Dreyfus was right about situations, contingencies, and goals being essential to intelligence, but wrong about the impossibility of modeling such processes on a computer. And the kinds of grand laws AI researchers had initially proposed were necessary but not sufficient to regulate intelligence, though the choice of computer as an instrument to model upon was quite right. A generous historian would award prizes all around.)

As tons of specialized knowledge poured into programs, the romantic age went into full swing. Programs were heterarchical instead of hierarchical—a bunch of loosely connected systems that nearly ran amuck, though in fact some impressive accomplishments appeared. But there seemed to be a barrier beyond which they couldn't go: SHRDLU, Terry Winograd's babe in toyland, didn't get any smarter.

AI has moved into the modern age, say the MIT observers. The near anarchy of the romantic period has been superseded by an emphasis on control. Control here refers to the division of programs into three parts—the data, the processes for transforming the data, and the processes for controlling the sequences of other processes—and the way these distinct elements interact. To change the metaphor from art to politics, it's as if a Mandarin aristocracy had been overthrown by a people's democracy, which has since been replaced by a people's republic. It seems that somebody, after all, has to make sure the garbage gets collected. Not everybody sees the history of AI as falling into these three eras, and the notion that research has reached some ultimate spot is the most unfortunate of all the implications of such a scheme. Future historians will probably consider the whole shebang to be prehistory, in the same way we regard the agricultural revolution.

This is not to say the last quarter century of AI research hasn't told us some fascinating and unexpected things. For one thing, disembodied rationality, which has headed the patriarchal aesthetic canon since the days of Aristotle as not only the most essential of human properties, but also the most valuable, turns out to be the easiest to simulate on a computer. If AI workers first aimed at mimicking what they considered cleverest in themselves, they managed to show the world that these activities—puzzle solving, game playing, theorem proving—are really pretty simple (as we nonpuzzle-solvers and nongame-players always secretly suspected). What are really hard are the things humans do easily—I nearly wrote "without thinking"—such as carrying on a conversation, telling stories, and getting about in the real world. This distinction was by no means apparent at the beginning. AI researchers genuinely believed they would penetrate the core of human intellectual endeavor if they could, say, create a chess machine that played splendid chess. And that belief rubbed some people the wrong way.

The poet Adrienne Rich once attended a lecture given by Herb Simon, where she remembers him saying, "The secret of problem solving is that there is no secret." This sentiment seemed outrageously complacent to her, and in her anger she wrote this poem:

Artificial Intelligence

To G. P. S.

Over the chessboard now,
Your Artificiality concludes
a final check; rests; broods—
no—sorts and stacks a file of memories,
while I
concede the victory, bow,
and slouch among my free associations.

You never had a mother,
let's say? no digital Gertrude
whom you'd as lief have seen
Kingless? So your White Queen
was just an "operator."
(My Red had incandescence,
ire, aura, flare,
and trapped me several moments in her stare.)

I'm sulking, clearly, in the great tradition
of human waste. Why not
dump the whole reeking snarl
and let you solve me once for all?
(Parameter: a black-faced Luddite
itching for ecstasies of sabotage.)

Still, when
they make you write your poems, later on,
who'd envy you, force-fed
on all those variorum
editions of our primitive endeavors,
those frozen pemmican language-rations
they'll cram you with? denied
our luxury of nausea, you
forget nothing, have no dreams.

(Adrienne Rich, 1961)

Now, we encounter a certain confusion of views here that ought to be disentangled. Poets articulate the individual view, reaching toward the universal. An individual Red Queen might indeed have incandescence, ire, aura, and flare. In fact, let's hope she does. But science is the universal view, reaching toward the particular. In that universal view, White Queens are justifiably regarded as operators, whatever their personal attributes. *Neither view obviates the other, which is something scientists and poets need to remember more often.*

Yet in another sense, Rich was right. What's really hard for computers is the ordinary business of being human—learning language, telling stories, storing memories, experiencing dreams. Intelligent programs don't brood (though it can be argued that they slouch among their free associations), have no mother with whom to have an ambivalent relationship, don't sulk, grow nauseated, or dream. Still, that's not to say they never will. For here's where artificial intelligence might fascinate us the most: that it serves as a bridge between art and science, that it has the capacity to be the grand synthesis we long for (and fear), that it opens up possibilities of line-crossing which have been hermetically sealed for centuries.

Reconsider, for example, the mind-body problem. We've been perplexed for centuries, unable to deny that mind and body are intimately connected, yet unwilling to believe they're somehow the same. Explanations both simple and tortured have sprung up to account for the problem, and they fill the textbooks of theologians and philosophers to the bursting point. Here's Dr. Samuel Johnson, that eighteenth-century pontificator: "Although God has made Nature to operate by certain fixed laws, yet it is not unreasonable to think that He may suspend those laws, in order to establish a system highly advantageous to mankind" (Boswell, 1968). Dr. Johnson talks about miracles, but the honest investigator would have to include the mind-body split in the same supernatural category.

However, artificial intelligence, like other sciences, believes the laws of nature are not suspended; they're simply unknown until they're discovered. It's an operating principle, but not dogma.

Both Newell and Minsky have addressed the mind-body split. Newell says that, first, mind is an information-processing system. And then minds, or information-processing systems, are realized in physical systems. "The brevity and simplicity of these answers reveals again the magician's trick—it was all over before it started. That is, once the information processing view is adopted, these answers (more or less) follow" (Newell, 1973b). He might make the same answer to those other pesky questions raised by the Western intellectual tradition. If we say that mind is an information-processing system, we can examine that system and ask what consciousness is. Consciousness, it turns out, is the ability of a system to hold a model of itself and its behavior—indeed, several models, if appropriate. And these models affect and change the shape of one another. This notion is pretty simple, and is demonstrated in intelligent systems that do indeed have models of parts and sometimes the whole of their systems, and that comment upon and revise such models (and thus the originals) as the circumstances require. A demonstration also exists, conversely, in systems that don't observe and revise, and thus break down, an event we observe sadly in human mental (and for that matter physical) systems.

Minsky writes about the mind-body problem: "When a question leads to confused, inconsistent answers, this may be because the question is ultimately meaningless or at least unanswerable, but it may also be because an adequate answer requires a powerful analytical apparatus." The mind-body problem, he says, is the latter kind of question, and some of the necessary conceptual tools are becoming available as a result of work on the problem of making computers behave intelligently (Minsky, 1968). He talks about consciousness too, models and models of models, which we carry around in our heads about the world and about ourselves. The bipartite model of mind and body has some useful aspects, but those useful aspects ought not to fool us into believing that such a model has anything to do with how things really are.[2]

---

[2] This notion offers Minsky a detour into free will. "If one thoroughly understands a machine or a program, he finds no urge to attribute 'volition' to it. If one does not understand it so well, he must supply an incomplete model for an explanation. Our everyday intuitive models of higher human activity are quite incomplete, and many notions in our informal explanations do not tolerate close examination. Free will or volition is one such notion: people are incapable of explaining how it differs from stochastic caprice, but feel strongly that it does" (Minsky, 1968). When we can discern no rules, he goes on, we would rather posit free will than chance.

Minsky concludes:

> When intelligent machines are constructed, we should not be surprised to find them as confused and stubborn as men in their convictions about mind-matter, consciousness, free will and the like. For all such questions are pointed at explaining the complicated interactions between parts of the self-model. A man's or a machine's strength of conviction about such things tells us nothing about the man or about the machine except what it tells us about his model of himself.

One by one they fall—the nature of mind, of consciousness, of intelligence, learning, understanding. Understanding, Simon says, is a relation among three elements: a system, one or more bodies of knowledge, and a set of tasks the system is expected to perform. "The development of understanding systems over the past twenty years has been a kind of two-part fugue, in which the proposal of a class of tasks to be performed generates a set of knowledge requirements for the system; while the knowledge a system acquires, in turn, enlarges its capabilities for understanding new tasks" (Simon, 1977).

As Newell says, it all becomes something of a magician's trick, and wouldn't persuade at all except that based on these principles, intelligent machines demonstrate what we have traditionally identified as consciousness, understanding, and learning.

The information-processing theory, Newell feels, will be as key to understanding the mechanism of mind as the theory of natural selection was key to understanding the mechanisms of biology. There are other comparisons to that splendid Victorian event. Much of the resistance to artificial intelligence has been on mystical grounds, as was the resistance to natural selection. "I need hardly go any further with these objections," wrote Adam Sedgwick, professor at Cambridge. "But I cannot conclude without expressing my detestation of the theory because of its unflinching materialism..." (Appleman, 1970). Professor Sedgwick was writing of natural selection and evolution, but his words are echoed a century later by critics of the information-processing theory of mind.

To cite another kind of comparison, nineteenth-century cell biologists, riding high on all their new discoveries about the cell, were deeply skeptical of a theory that had no basis in what was just becoming known about cellular structures. Natural selection and evolution seemed stratospherically unrelated to the earthly facts of nucleus and cell division. Where was the correspondence? As it turned out, much work was necessary to discover that correspondence, beginning with Mendel and culminating in Crick and Watson's discovery of the structure of DNA, whose various properties would show that natural selection and evolution were inevitable (Appleman, 1970). Neural physiologists are likewise skeptical of information-processing theories of mind. Certainly there are gaps to be filled.

Curiously, Newell muses, when philosophers and artificial-intelligence researchers turn their attention to the minds of computers, they part ways. Artificial intelligence sees the nature of the game as exploration, things to be done: let's make a computer analyze a scene, play a master-level game of chess. Philosophers, on the other hand, leap neatly over all that's relevant to AI to ask what *could* be known if AI were fully realized, and whether what could be known is a good thing to know. For instance, with respect to induction, AI wants to know how inductions are actually made, while philosophy is concerned with whether induction is justified and proper. It's an interesting distinction.

So the questions that have fueled the entire Western intellectual tradition are asked from within the information-processing view, and are asked in a scientific way, which is to say that many of them cannot yet be fully answered and will have to so remain until the time is ripe to answer them—for that too is how science works. Minsky adds that the distinctions we make between mind and body (or any other of the dimorphisms we find so irresistible) are useful everyday heuristics that we'll surely go on using.

Now, that conclusion seems to contradict a central thesis of this book, which is that the arrival of artificial intelligence has and will continue to have a deep effect on our view of ourselves. If we go on finding it useful to talk about spirit's being willing while flesh is weak, what difference will artificial intelligence make? Shouldn't it join the philosophers on the floor of the Chicago

Board of Trade, one more voice in the cacophony? Perhaps so. Certainly for everyday purposes we still carry around our sense impressions of astronomy, though that field has undergone some radical transformations. We still think of the moon as rising, though we know the earth is turning, and we gaze upon that dead planetoid with wonder as we sit beside a lover in its magic light.

Yet something has changed. Chaste Diana is chaste no more. Side by side with our former notions, we carry a new picture in our minds, not of the green hills of earth, but of the blue and white sphere that glows in the dark, a picture our privileged generation was the first to see. Our planet, awesome in its lonely beauty but comforting too, reduced to the appearance of a child's prize marble, our own in a way it never was before. When we picture it side by side with its ruddy neighbor Mars, our sense of what colors symbolize is suddenly realigned: blue and white are hospitable and life-giving; red is neither warm nor cheering.

I suspect that in a similar way artificial intelligence will change our view of ourselves. The change will be gradual: we'll continue to think of ourselves as the intellectual center of the universe at the same time we acknowledge a new species is on the horizon. The effort to discover and recognize intelligence wherever it might be—in machines we make ourselves, in our fellow creatures on this planet, or even in extraterrestrial phenomena—might indeed be humbling to us. It came as a blow to discover we weren't the physical center of the universe, too. But both experiences will have shown us a new truth about ourselves—which, to the credit of our species, we've always valued most in the end.

Margaret Boden makes another interesting argument, and makes it eloquently. She suggests that artificial intelligence will counteract what she sees as the subtly dehumanizing effects of natural science:

> It does this by showing, in a scientifically acceptable manner, how it is possible for psychological beings to be grounded in a material world and yet be properly distinguished from "mere matter." Far from showing that human beings are "nothing but machines," it confirms our insistence that we are essentially subjective creatures

living through our own mental constructions of reality (among which science itself is one).[3]

(Boden, 1977)

We are Red Queens, with incandescence, ire, aura, and flare; we are White Queens acting as operators. We contain a changing number of highly contingent symbol systems, the symbols sounding off one another with denotation and connotation, moving fluidly, hierarchically through time, alert to an inner and outer reality, connected with inner biological systems, with the outer symbols of our culture, but each of these connections idiosyncratic and often surprising. The message from artificial intelligence is that every individual consciousness is a poem, with all the elegance, economy, richness, and originality we expect from such a thing. Poets always knew it in their way. Now scientists know it in their way too.[4]

***

I've summed up where we've been and where we are. What next? The soberest minds in the field see AI as a dazzling helpmate to human intelligence. This is on the near horizon, with two likely forms. One is the intelligent assistant, already at work in chemis-

---

[3] Margaret Boden's *Artificial Intelligence and Natural Man* (New York: Basic Books, 1977) is a delightful and critical look at a handful of artificial-intelligence programs in much more detail than *Machines Who Think* attempts. It's written for the nonspecialist who would like a closer look at contemporary programs, emphasizing work of particular interest to readers with philosophical or psychological concerns. Though Professor Boden is often astringent, she is always lucid, and basically optimistic about the future of the field, especially as a counteraction to the dehumanizing influence of modes of natural science.

[4] And have known it for a while. Here, for instance, is Papert, in an essay entitled "Some Poetic and Social Criteria for Education Design" (1976);

"I am trying to say something like: the total experience of the child in learning must have something which I want to call poetic cohesion. I want to suggest that the total lack of the 'poetic' is a major (not the only) reason for the intransigent rejection by so many kids of the painfully prosaic stuff of the math class (new math and old math are scarcely different in this respect!). Now I have slipped over, you might observe, into talking about the Poetry Principle from the child's point of view. I find that it is easier to persuade people that the child needs poetry in his vision of mathematics than that the teacher, the educa-

try, mathematical research, and medical diagnosis, and sure to be put to work wherever intelligent assistants can be useful—in architectural design, economic planning, education. The second form is connected with the first, and is AI as a model of how humans think, and therefore how they learn, which will change our methods of education, including psychotherapy. All these developments are very near, and very welcome.

On a more distant horizon is another set of prospects. We can go to science fiction for a look at them, though most such tales disappoint, for they are undernourished in imagination and steeped in gloom. Nearly every one of them has the machines taking over by force or stealth, conflicts of plot having to do with how humans resist or succumb.[5] John McCarthy also points out that science fiction artificial intelligences, usually robots, tend to be of one kind, whereas it's quite possible that artificial intelligences will be of many kinds. The real developments of AI will probably differ from science fiction versions in at least three ways. First, he says, it's unlikely that there will be a prolonged period during which it will be possible to build machines as intelligent as humans but impossible to build them much smarter. If we can put a machine capable of human behavior in a metal skull, we can put a machine capable of acting like ten thousand coordi-

---

tional psychologist and the curriculum designer all need it. I believe they do. And so does our society at large! And all this is a plea for not being trapped into thinking that being 'scientific' means rejecting the Poetry Principle on any of these levels.

"I have only apparently strayed from computers to poetry. Or rather: the opposition many people see between computers and poetry is quite profoundly false. In fact my present thesis would not suffer much from being re-stated as: the embodiment of mathematics in properly designed computers is the most powerful means we have for giving it poetical, cultural and personal-human dimensions which are a necessary condition for it to be accepted and absorbed in a natural and easy way by billions of children."

[5] Marvin Minsky was a consultant for Stanley Kubrick's film *2001*, and managed to persuade the filmmakers at least to tone down the flashing lights and spinning wheels of the original HAL's appearance, even if he couldn't make HAL's behavior more plausible. I saw the film *Colossus: The Forbin Project* with a computer scientist who laughed himself silly at the premise of locking up a computer system in a way that it could never be improved, let alone be tampered with. The widely successful *Star Wars*, where humans have the last word, seems to say more about the vanities of those lining up at the box office than about a likely future.

nated people in a building. Second, although the present stock of ideas is inadequate to make programs as intelligent as people, there's nothing to prevent the required new ideas from coming very soon, in five years or five hundred. Finally, present ideas are probably good enough to extend our ability to have very large amounts of information at our fingertips, which will create its own changes (McCarthy, 1972).

Now, we probably shouldn't be surprised that scientists at work in the field already see things differently from outsiders—predicting the social impact of technology is one of the skills humans are feeblest at, and within those feeble limits, the best prognosticators are usually people who are steeped in the technology itself. When the telephone was invented, everyone save its inventor had difficulty imagining what to use such a device for. It took Alexander Graham Bell years of exhausting promotion to convince the world his vision of a telephone in every home, someday linking all the civilized world, wasn't sheer lunacy. Immersed in it, he alone saw its possibilities (Pool, 1977).

Thus, the best predictions of the distant future probably come from people who've been deeply immersed in the creation of such a technology, who are aware of its profoundest consequences, and who are then maybe slightly removed. Consider Edward Fredkin, a professor of electrical engineering at MIT, who came to know Minsky and McCarthy in the late 1950s and became their good friend. Though his own research efforts haven't gone directly into artificial intelligence, he's been a long-time and enthusiastic supporter of such research, especially when he directed Project MAC during much of the time it housed MIT's artificial-intelligence efforts. Fredkin is close enough to understand the vast potential of artificial intelligence, and intimate enough with McCarthy and Minsky to have the sort of late-night talks in which the most exotic ideas can flourish, safe from cool daylight reason and the withering skepticism of funding agencies. At the same time, Fredkin is removed enough from the field to be unencumbered by the day-to-day frustrations of getting robot arms to move blocks, or intelligent programs to make conversation. All this has had its effect on his thinking.

I've said again and again here that the impulse to make artificial intelligence has always been central to our project as a species. But Fredkin takes a step I'm not altogether prepared to take myself, and thrusts artificial intelligence firmly into biology. In a big way.

Simply put, Fredkin believes that artificial intelligence is the next stage in evolution. His argument runs like this. Until now, biological evolution has taken place in a single mode, a sort of genetic-message relay race, where each species carries forward the message with information about itself and its predecessors. Individual species fade away, but life itself carries on. Humans stand at the end of this biological relay race, but they aren't the end of evolution altogether.

Artificial intelligence is the next step in evolution, but it's a different step [Fredkin says]. And it's very different, because the genetic-message concept has disappeared with it. One artificially intelligent device can tell another not only everything it knows in the sense that a human teacher can tell a student some of what he knows, but it can tell another device everything about its own design, its makeup—its genetic characteristics, as it were—and about the characteristics of every other such creature that ever was.

Part of Fredkin's view of artificial intelligence is colored by his view of human intelligence:

There's a popular view that the human mind is this fantastic thing that most of us are just barely using—5 or 10 percent of its capacity. If we could only unleash the whole human mind and all of its powers, we'd be supermen. Now my notion is that for an ordinary person to get along in society in a conventional way requires about a 110 percent of the capacity of the human mind, causing breakdowns and trouble of various sorts. Basically the human mind is not most like a god or most like a computer. It's most like the mind of a chimpanzee, and most of what's there isn't designed for living in high society but for getting along in the jungle or out in the fields. Our response to aggression and everything else like that is really not keyed for dealing with thermonuclear war but for dealing with life in the jungle. We're tuned to dealing with local, not global situations, and our biggest problems turn up when global problems emerge. We try intellectually to think our way through global problems, but we don't do very well, so we run into disasters. World War II was such a disaster. There's no rational excuse for having World War II. We tell ourselves we had to fight off Hitler. Well, why did it have to be Hitler? And on and on. The fact that we're able to deal at all with the complexity of society is really amazing, but it taxes us a great deal.

To me, the human mind suffers in many ways. I mean, we forget things, learning is incredibly slow and difficult. We try to teach something to somebody twenty times, and they still don't get it. If you did that to a computer, you'd have to complain and throw it out. On the other hand, we're versatile and can deal with diverse things.

And one of the diverse things Fredkin himself has spent a long time dealing with is exploring with his friends what might happen if and when fully realized artificial intelligences are upon us. The consequences could be good or bad, he concedes. His vision of the good is that when artificial intelligence comes, it will assume the heavy thinking for us, and solve the problems that are beyond our own capacity for solutions.

He's calculated how fast advanced artificial intelligences might think:

Say there are two artificial intelligences, each about the size of a small table. When these machines want to talk to each other, my guess is they'll get right next to each other so they can have very wide-band communication. You might recognize them as Sam and George, and you'll walk up and knock on Sam and say, "Hi, Sam. What are you talking about?" What Sam will undoubtedly answer is, "Things in general," because there'll be no way for him to tell you. From the first knock until you finish the "t" in about, Sam probably will have said to George more utterances than have been uttered by all the people who have ever lived in all of their lives. I suspect there will be very little communication between machines and humans, because unless the machines condescend to talk to us about something that interests us, we'll have no communication.

For example, when we train the chimpanzee to use sign language so that he can speak, we discover that he's interested in talking about bananas and food and being tickled and so on. But if you want to talk to him about global disarmament, the chimp isn't interested and there's no way to get him interested. Well, we'll stand in the same relationship to a super artificial intelligence. They won't have much effect on us because we won't be able to talk to each other. If they like the planet and don't want to leave, and they don't want it blown up, they might find it necessary to take our toys away from us, such as our weapons.

And Fredkin has other scenarios he's dreamed up, all of which illustrate that the future has possibilities that are beyond our experience. The possibilities are not merely expansions or obvious developments of the past, but are likely to be quite beyond what we can imagine, though Fredkin and his late-night story swappers have a high time trying.

If we have a small sense of *déjà vu*, it probably comes from the recollections of brazen heads, those remarkable artifacts said to be owned by medieval sages, which were fabricated as both evidence *and* source of sagacity. You had to be very bright indeed to fashion one, but once you had it, it bootstrapped itself into superhuman intelligence. It's a beguiling idea.

Where Fredkin's optimism fails him, however, is at what he calls the intermediate stage of AI development:

That problem occurs when we have semiintelligent programs. That is, we have experts who are very advanced in given fields, like DENDRAL in chemistry or MYCIN in internal medical diagnosis, only these are, let's say, foreign policy experts. But such a machine might be full of bugs. Such a limited expert could bring about catastrophes we aren't even able to cause ourselves—ones we aren't smart enough to concoct. This danger is a temporary one, existing only when the machines are somewhat, but not very, intelligent. For the really intelligent ones will be able to understand our motives, what we want; will be able to look at themselves and understand their own operations, and know that they're working towards solving the problems we want solved. Whereas the semiintelligent ones will have no such introspection, and will operate in ways that have nothing to do with our highest-level goals.

Can we assume that artificial intelligences will share human goals?

There are ways to arrange that, at least in the interim stage [Fredkin says]. Eventually, no matter what we do there'll be artificial intelligences with independent goals. I'm pretty much convinced of that. There may be a way to postpone it. There may even be a way to avoid it, I don't know. But it's very hard to have a machine that's a million times smarter than you as your slave.

Fredkin's ideas for getting through that interim stage of semismart artificial intelligences with our human hides intact bear examination:

We have to work on AI like crazy [he says]. Now, why? Why not take our time? These days, AI can only be done at big laboratories like Stanford, MIT, and Carnegie. Today it costs a quarter of a million dollars to get a machine good enough to do AI research. But two years ago it was five million dollars. Today I can buy a machine for five dollars that's better than one costing five million dollars twenty years ago. The trends are very clear—they'll continue to the end of the century, close to a factor of two every year. Which means every ten years a factor of a thousand, every twenty years a factor of a million. Today's quarter-million-dollar system will cost $250 in ten years, and will be better besides. A paper boy with his route money will be able to save up in a month and buy such a machine. Thus anybody will have the necessary hardware to do AI pretty soon; it will be like a free commodity.

Now, under those circumstances, it's possible that some mad genius, some Newton-like person, even a kid working by himself, could make tremendous progress. He could develop AI all by himself, relying on what others do, but building it in private rather than at a big institution like MIT. And the application of such a machine would be irresistible. How could you avoid this? You can't license computers; that never was practical. Electricity is a key, of course. If you made the use of electricity in any way a capital offense, worldwide and suddenly, and you did it immediately—otherwise there are ways around such a prohibition—then perhaps you could prevent this from happening. But anything short of that isn't going to do it, because you won't need a laboratory with big government funding very soon—

that's only a temporary phase we're passing through. So what Joe Weizenbaum would like to do is impossible—it's bringing time to a halt, and it can't be done. What we can do is make the future more secure for human beings by being reasonable about how you bring AI about, and the only reasonable course is to work on this problem in a way that promises to be best for all of society, and not just for some singular mad genius.

Fredkin summarizes his argument, his dark brown eyes moving with excitement:

Look, we see our knowledge of artificial intelligence is slowly growing. Meanwhile, the cost is a handicap that's about to disappear, first for the laboratories and then for everyone. What's equally frightening is that the world has developed means for destroying itself in a lot of different ways, global ways. There could be a thermonuclear war or a new kind of biological hazard or what-have-you. That we'll come through all this is possible but not probable unless a lot of people are consciously trying to avoid the disaster. McCarthy's solution of asking an artificial intelligence what we should do presumes the good guys have it first. But the good guys might not. And pulling the plug is no way out. A machine that smart could act in ways that would guarantee that the plug doesn't get pulled under any circumstances, regardless of its real motives—if it has any. I mean, it could toss us a few tidbits, like the cure for this and that.

I think there are ways to minimize all this, but the one thing we can't do is to say well, let's not work on it. Because someone, somewhere, will. The Russians certainly will—they're working on it like crazy, and it's not that they're evil, it's just that they also see that the guy who first develops a machine that can influence the world in a big way may be some mad scientist living in the mountains of Ecuador. And the only way we'd find out about some mad scientist doing artificial intelligence in the mountains of Ecuador is through another artificial intelligence doing the detection. Society as a whole must have the means to protect itself against such problems, and the means are the very same things we're protecting ourselves against.

Say there's artificial intelligence being developed in a laboratory with a benevolent environment, where the greater good of society is the enterprise's main concern. Can they protect themselves? The answer, I think, is yes. It's complicated, but the only way it can be accomplished is with another machine, secured from the influence of the first one. If the machines can communicate, they can conspire. Some of the concepts developed in computer science for privacy and security turn out to be exactly relevant to issues of trying to make a secure artificial intelligence—where one machine is being monitored by another, but can't influence the one that's monitoring it. It's kind of a risky business.

Fredkin proposes that an international laboratory for artificial-intelligence research be established, designed to be cooperative instead of competitive. "You don't want competition in this field. That's almost deadly. There isn't much now because of the hardware advantage the United States has enjoyed for so long, but that advantage is going to disappear shortly." Fredkin himself tried to set up such a laboratory in 1961, but for several reasons, it didn't happen. In a way, he sympathizes with Weizenbaum, but he sees himself as an activist, trying to bring about the Golden Age and getting the human race through the Dark Age that must inevitably precede it, instead of merely grumbling how terrible the Dark Age must be. Fredkin is not always greeted with delight by his colleagues.

I can't persuade anyone else in the field to worry this way [he tells me]. They get annoyed when I mention these things. They have lots of attitudes, of course, but one of them is, "Well yes, you're right, but it would be a great disservice to the world to mention all this." Oh, computer scientists are always afraid of being taken as fools. Scientists in general are very fearful of this. [He grins.] I'm unusual in that regard. The ideas sound crackpot, maybe. But my colleagues only tell me to wait, not to make my pitch until it's more obvious that we'll have artificial intelligences. I think by then it'll be too late. Once artificial intelligences start getting smart, they're going to be very smart very fast. What's taken humans and their society tens of thousands of years is going to be a matter of hours with artificial intelligences. If that happens at Stanford, say, the Stanford AI lab may have immense power all of a sudden. It's not that the United States might take over the world, it's that Stanford AI Lab might.

I respond that the last is a sobering thought.

Right. And maybe they have in mind taking over benevolently. They Just want to change the curriculum so that everyone learns computer science in kindergarten worldwide. Maybe they only want to influence the stock market to make Stanford's endowment bigger than Harvard's. Who knows? Whatever it is, the consequences may be bad, and it can't be allowed.

And so what I'm trying to do is take steps to see that such an international laboratory gets formed, and that these ideas get into the minds of enough people. McCarthy, for lots of reasons, resists this idea, because he thinks the Russians would be untrustworthy in such an enterprise, that they'd swallow as much of the technology as they could, contribute nothing, and meanwhile set up a shadow place of their own running at the exact limit of technology that they could get from the joint effort. And as soon as that made some progress, keep it secret from the rest of us so they could pull ahead. McCarthy may be right. He's often right, more than anyone I know. He has his biases. Yes, he might be right, but it doesn't matter. The international laboratory is by far the best plan; I've heard of no better plan. I still would like to see it happen: let's be active instead of passive. I'll say one thing. The highest level of Soviet scientific policy making is very much aware of and appreciative of the concepts of AI. The same is certainly not true in this country. In this country it's almost an embarrassment.

We come to speak of the psychological consequences of AI and Fredkin says,

People are in anguish today, even in primitive societies; they're in anguish because of the emotional burdens they have to bear. Things can be helped by creating a society that poses fewer such problems, and by helping people to be able to deal with the problems. The point is that artificial intelligence can help to run society in ways that are more beneficial for everyone.

Fredkin leans back, laughing.

Humans are okay. I'm glad to be one. I like them in general, but they're only human. It's nothing to complain about. Humans aren't the best ditch diggers in the world, machines are. And humans can't lift as much as a crane. They can't fly at all without an airplane. And

they can't carry as much as a truck. It doesn't make me feel bad. There were people whose thing in life was completely physical— John Henry and the steam hammer. Now we're up against the intellectual steam hammer. The intellectual doesn't like the idea of this machine doing it better than he does, but it's no different from the guy who was surpassed physically. So the intellectuals are threatened, but they needn't be—we should only worry about what we can do ourselves. The mere idea that we have to be the best in the universe is kind of far-fetched. We certainly aren't physically.

The fact is, I think we'll be enormously happier once our niche has limits to it. We won't have to worry about carrying the burden of the universe on our shoulders as we do today. We can enjoy life as human beings without worrying about it. And I think that will be a great thing.

There are three events of equal importance, if you like. Event one is the creation of the universe. It's a fairly important event. Event two is the appearance of life. Life is a kind of organizing principle which one might argue against if one didn't understand enough—shouldn't or couldn't happen on thermodynamic grounds, or some such. And, third, there's the appearance of artificial intelligence. It's the question which deals with all questions. In the abstract, nothing can be compared to it. One wonders why God didn't do it. Or, it's a very godlike thing to create a superintelligence, much smarter than we are. It's the abstraction of the physical universe, and this is the ultimate in that direction. If there are any questions to be answered, this is how they'll be answered. There can't be anything of more consequence to happen on this planet.

And we get back, slowly, to evolution.

Most people think that evolution means a better human will evolve. If what appeared was only a slightly better human, then we'd disappear, because it would want the same niche we have. If that were the way evolution always proceeded, there'd be exactly one kind of creature on the planet. But there's always lots of kinds. The turtle has been around for a long time because the turtle is better than anything else is at being a turtle. Now with AI coming along, if it were a slightly better human, then we'd disappear. But it's not going to be a slightly better human. It's going to be a completely, a totally different thing, which leaves us our niche. We'll be the best creatures at being humans on the whole planet. And you know what? We might enjoy it if we only had to be humans. Today we have to be humans and gods because we're worrying about everything in the universe. Whenever there's a war and people die or are maimed, that's a sign of our great incompetence at dealing with the situations.

I get uneasy. Maybe it's no more than reading Aristotle's *Poetics*, which taught me that a *deus ex machina* (literal then—and suddenly again) is a second-rate way of saving the situation. I'd hate to think of human history as a second-rate drama.

I don't expect Utopia [Fredkin goes on]. If we got rid of wars, and weapons of mass destruction, and famine, it wouldn't change our lives at all, hardly, except to take away certain long-range worries. That's not Utopia, but it's a change. All humans are equally precious, and somehow the place can be better managed for them, but not by us. We might become capable of it, but it's a fantastic burden and a lot of the circumstances, like natural disasters, famines, and so forth, are out of human hands. But what could happen is that the vast majority of people can live happier, better lives, provided we manage our planet well. And AI can help manage the planet well. I'm convinced of it.

A glorious promise, fraught with such danger.

"And wherefore was it glorious?" Victor Frankenstein once cried out to the frightened. "Not because the way was smooth and placid as a southern sea, but because at every new incident your fortitude was to be called forth and your courage exhibited, because danger and death surrounded it, and these you were to brave and overcome."

Ah. We nearly forgot.

To back off is to turn away from an essential human project, whether we see it as fanciful impulse or biological destiny. Like any human project worth doing, it will call forth our fortitude and make us exhibit our courage. Thus it must be to set freedom free, Simone de Beauvoir's definition of the good.

If we're somewhat breathless at the turn of the discussion, Fredkin's ideas sound positively Main Street compared to the fancies of some of the younger people in the field.

Here's Hans Moravec, presently a graduate student in computer science at Stanford. After a fairly plausible discussion of how an individual human might be amplified by the transfer of consciousnesses, a transfer which will make the concept of personal death meaningless—techniques for that process to come from the study and building of artificial intelligences—he goes on to speculate about other possibilities:

> The amount of memory storage an individual will typically carry will certainly be greater than humans make do with today, but the growth of knowledge will insure the impracticability of everybody lugging around all the world's knowledge. This implies that individuals will have to pick and choose what their minds contain at any one time. There will often be knowledge and skills available from others superior to one's own. The incentive to substitute those talents for native ones will be overwhelming most of the time. This will result in a gradual erosion of individuality, and formation of an incredibly potent community mind.

The notion is of a community mind that will embrace all animals, higher and lower, with the simpler organisms contributing only the information in their DNA, if that's all they have, until we have a synthesis of terrestrial life, and perhaps Martian and Jovian life as well. The synthesis would be, in Moravec's words,

> constantly improving and extending itself, spreading outwards from the solar system, converting non-life into mind. There may be other such bubbles expanding from elsewhere. What happens when we meet another? Well, it's presumptuous of me to say at this tender stage of the evolution, but fusion of us with them is certainly a possibility, requiring only a translation scheme between data representations. This process, possibly occurring now elsewhere, might convert the entire universe into an extended thinking entity.
>
> (Moravec, 1977)

Well.

I suppose I'd be more skeptical if the same message weren't dinning in my ears from a number of places and in a number of ways. There's something of the same point in Lewis Thomas's

Lives of a Cell (1974), and Thomas is no wide-eyed graduate student, but runs one of the world's leading centers of cancer research, the Sloan-Kettering Institute in New York. And, though he sets the teeth of his fellow astronomers on edge with such speculations, Carl Sagan makes a similar point in his *Cosmic Connection* (1973). But it's no wonder that traditional humanities departments in universities feel beleaguered: their man-is-the-measure-of-all-things assumption sounds suddenly quaint and parochial in the extreme.

I've said we're near the end of one of a handful of journeys that have engaged the human intellect. And yet we're also back where we began, so to speak, in the forge of Hephaestus. However we choose to read that word forge—as a place of manufacture or of counterfeit—we're unquestionably in the business of forging the gods. The field belongs, as it always has, to the visionaries. Oh, the visions change character. Restless now with the mere replication of human intelligence, the new visionaries look out toward other, better intelligences. Anyone who considers that impulse ridiculous had better recall how silly the all-but-realized visions of earlier times once seemed.

And we're also bound to confess once more that these visions are after all our own, born of our human yearning for the transcendent. For that's the important thing. Whether artificial intelligence is indeed the next great step in evolution, or I've just finished chronicling the history of one of the most wrong-headed human follies in existence, in some sense doesn't matter. We only live—we only survive—as individuals and as a society and as a species by reaching out beyond ourselves.

Minsky and I once talked about such matters:

Often [he said], we have people who say we've got to solve problems of poverty and famine and so forth, and we shouldn't be working on things like artificial intelligence. That's an issue of expertise about which I feel very strongly. They have to be very sure that they're the right person to do that. There's no use in everybody worrying about poverty because there's no point to a culture that has only interior goals—it becomes more and more selfish and less and less justifiable. No, we should have a certain number of people worrying about what happens when the sun gets too hot or too cold, and whether artificial intelligence will be a huge disaster some day or be one of the best events in the universe. I don't know. I would feel silly about spending all my time worrying about short-range things because somehow they eat themselves up. Did I tell you that quote from Auden? "We're all on earth to help others—what I can't figure out is what the others are here for."

He played delightedly with that phrase, and then went on:

No, you may do more harm than good when you go into somebody else's garden with only good moral intentions and not good technical judgments. You might be the only one who can help with the disaster that's going to happen in twenty years, and if you don't prepare yourself, and instead just go off into some social welfare project right now, who will do it then? So taking responsibility, paradoxically, is abrogating it, and some people, particularly radicals, don't feel that stress, or don't understand the postponement of their gratification about trying to make things better. Yes, I feel that there's a great enterprise going on which is making the world of the future all right. It's sort of funny that it isn't being done by people who know what they're doing, but that's the way it always is.

Most of us respond only to our interior goals, but we recognize and honor those few who reach outside themselves, wittingly or unwittingly making the world of the future all right. The enterprise is a godlike one, rightfully evoking terror in the hearts of those who think the borders between humans and gods ought to be impermeable. But the suspicion has been growing for some time that gods are a human invention. Whether they take the form of randy old souls in tunics waiting in the shrubbery for unsuspecting female bathers or miserable, punitive patriarchs, demanding more of us than they can manage themselves, whether they're women whose pleasure fructifies the fields or all-forgiving matriarchs, they are surely one thing: powerful manifestations of our own fears and hopes and dreams. The invention—the finding within—of gods represents our reach for the transcendent at the same time we reach to understand ourselves as humans in a human world.

And so with the reflexive enterprise of artificial intelligence. We are as gods in the exercise, counterfeiting aspects of the human just as we always have, whether in theology or the arts, and for pretty much the same reasons of self-enchantment. That we might be forging the gods in the other sense—deities to rescue us from our own overreaching—is an idea allowing degrees of accord. Certainly artificial-intelligence builders fall all along the line from dubious no to enthusiastic yes. So might we all.

And yet. At the beginning of this history, I quoted Isaiah's complaint about the Judeans: "Their land also is full of idols; they worship the work of their own hands, that which their own fingers have made." And I asked then whether it was the creation or the worship which is objectionable. In the act of forging the gods, that question is more pertinent than ever.

Me, I breathe easy on the whole subject, being of a Hellenic, rather than Hebraic turn of mind. The accomplishments have been significant, and the promises are nearly beyond comprehension. I pause just now, before I have to call forth fortitude and exhibit courage. I pause to savor the thrill of sharing in something awesome.

Part Six

Afterword

The Following Quarter-Century of Artificial Intelligence

Afterword

—

The Following Quarter-Century of Artificial Intelligence

25 Years in a Nutshell

In the late 1970s and early 1980s, artificial intelligence moved from the fringes to become a celebrity science. Seen in the downtown clubs, boldface in the gossip columns, stalked by paparazzi, it was swept up in a notorious publicity and commercial frenzy. Stoking the craze, the Japanese government began its Fifth Generation project, to bring smart computing to the masses. Soon, the Defense Advanced Research Projects Agency (DARPA) of the United States government launched a grand plan known as Strategic Computing, which, within a decade, aimed to deliver intelligent solutions to very difficult problems. Those bedazzled by (and perhaps investing in) AI's premature industrialization and subsequent commercial collapse were led to confuse science and commerce. As the next big thing seized the public imagination, artificial intelligence was declared dead.

Within the field, the situation was more complex, but far from clear. The goals once articulated with debonair intellectual verve by AI pioneers appeared unreachable, and their methods seemed, if not exhausted, not quite scalable either. Subfields broke off—vision, robotics, natural language processing, machine learning, decision theory—to pursue singular goals in solitary splendor, without reference to other kinds of intelligent behavior. Incremental improvements on what already existed, using well-understood statistical methods and logic theorems—theorems that were proved, buttoned down, impregnable—seemed the most prudent strategy.

The major exception to this trend appeared to confirm just how dubious a proposition artificial intelligence was. A burgeoning family of robots with no claims to symbolic intelligence, but superb sensors and reactors, began to achieve what no robots with "brains" had yet been able to. At the decade's end, one of artificial intelligence's main US patrons, the Defense Advanced Research Projects Agency, drastically curtailed that research. The 1980s began to be known as the "AI Winter."

But all sciences move in rhythms, thriving, then lying fallow (or simply assimilating what has gone before), then growing once again. In the 1990s, shoots of green broke through the wintry AI soil. Researchers began to consider intelligent behavior in different ways. Drawing new lessons from their research, they boldly expanded the idea of intelligence, saw it now as a collective effort among multiple agents, with multiple systems and sources of knowledge (in the human world, we call them cultures), a far more nuanced view than before. The single-agent, single-task model of intelligent behavior reached its apotheosis (and perhaps the symbolic end of its life as a significant AI goal) when a computer beat the world's current chess champion, Garry Kasparov. AI had achieved its early goal; now it harbored bigger scientific ambitions.

Robotics boomed, and head-to-head competition showed that robots with symbolic capabilities, "minds," could work as well as the reactive kind: Minds and sensors fused beat either one on its own. Robotics exploded not because, as some claimed, real intelligence must above all be embodied in a human-like creature (in fact, the tribe of robots claimed some bizarre morphologies), but because humans wanted these agents to perform tasks in a human world.

Throughout these two and a half decades, artificial intelligence research became genuinely international, with first-rate groups in North and South America, Asia, Europe, the Middle East, and Australia and New Zealand. In the first years of the new millennium, AI's traditional patrons, national governments, began to return, announcing even more ambitious goals for the field, betting that here lay a significant part of the future of an information economy, and that the science was poised to soar again. At the same time, almost in proportion to new successes, public controversy grew.

The '80s Open with an Identity Crisis

In the early 1980s, researchers could take satisfaction from significant successes in artificial intelligence. For example, intelligent robotics systems were being installed in factories; natural language processing was moving forward, with the first real-world applications underway; computers were getting steadily better at playing chess, once thought to be the *sine qua non* of intelligent behavior; and expert systems were being introduced to dozens of real-life situations from medicine to large-scale construction projects.

That very success, however, seemed to beg not only for some unified theory of intelligence, artificial or otherwise, but also for some clear best way of achieving it. Such a unified theory would surely elucidate the next steps a young science ought to take. In part, researchers were hoping to cut down to manageable proportions the possibilities of action, or, in AI terms, limit the search space. In part, the field was having some difficulty defining just what artificial intelligence was or should be (though natural intelligence was also poorly defined and was undergoing serious reevaluation: psychologist Howard Gardner had started to question the common assumption that intelligence consisted of the small set of problem-solving skills measured by a century of IQ tests, arguing instead that intelligence came in many flavors (Gardner, 1983); and psychologists like John Anderson, who paid close attention to work in AI, were beginning to make rigorous what had been mainly hand-waving.[1])

The coming 25 years would not yield final answers to these questions (nor will the next 25 years, probably). But we can see

---

[1] The reciprocity between cognitive psychology and artificial intelligence is fascinating and deep, but beyond the scope of this essay.

how the simple question of "Can a machine think?" which opened the first edition of this book, morphs brilliantly into a rich set of questions about many complex patterns in the radiant texture of intelligence—no matter who, or what, thinks.

A sign of the disordered state of AI research at the beginning of the 1980s was a survey taken by AI researchers Ron Brachman and Brian Smith, then at Bolt Beranek and Newman, who asked the scientists actually working in the field what they thought AI was, or should be. Not surprisingly, the Brachman and Smith survey yielded no substantial consensus from the scientists who responded. In his 1981 address as the first president of the newly formed American Association for Artificial Intelligence, Allen Newell of Carnegie Mellon University mused on this survey: "What is so overwhelming about the diversity is that it defies characterization," he said. "There is no tidy space of underlying issues in which respondents, hence the field, can be plotted to reveal a pattern of concerns or issues."[2] But this hardly indicates a crisis, he added with sanguinity. "Science easily inhabits periods of diversity; it tolerates bad lessons from the past in concert with good ones. The chief signal ... is that we must redouble our efforts to bring some clarity to the area. Work on knowledge and representation should be a priority item on the agenda of our science." He added, "No one should have any illusions that clarity and progress will be easy to achieve"(Newell, 1981).

The research reports of that era bear out Newell's words, and leave a reader with two distinct impressions. The first is that the field, having just enjoyed some initial practical successes, felt poised for greater accomplishments. The biggest problem facing it seemed to be which of many promising paths to take. At the same time, those papers can be read to suggest that though the initial successes were welcome, suspicion lurked that they weren't necessarily extendible or more generally applicable, that something more com-

---

[2] Newell set the superb precedent of giving a presidential address of intellectual substance, which has been more or less followed by his successors. These presidential addresses are a helpful set of guideposts to the field's intellectual preoccupations over time, and the reader will see that, in this essay, I have taken advantage of that.

plex, perhaps deeper, lay beyond the horizon, just out of reach. The best path ahead wasn't at all apparent.

Danny Hillis recalled visiting the MIT AI Lab as a freshman in 1974, and finding AI in a state of "explosive growth." Could general-purpose intelligence be far off? By the time he joined the Lab as a graduate student a few years later, "the problems were looking more difficult. The simple demonstrations turned out to be just that. Although lots of new principles and powerful methods had been invented, applying them to larger, more complicated problems didn't seem to work." One drawback was computing speed, where data on the scale needed to simulate intelligent behavior slowed computing intolerably. Another part of the problem was computer architecture. Hillis would go on to design a massively parallel computer called The Connection Machine to test some of the theories of his mentor, MIT professor, and friend, Marvin Minsky[3] (Hillis, 1998).

Just what should AI be, anyway? Researchers had come to believe that the great lesson from the 1970s was that intelligent behavior depended very much on dealing with knowledge, sometimes quite detailed knowledge, of a domain where a given task lay. And if, as it now seemed clear, large amounts of knowledge were essential to intelligent behavior, how should that knowledge be acquired in the first place, and then represented so that the machine could act appropriately on it? Lively debates arose in those old journal pages.

For example, Nils Nilsson, who succeeded Newell as president of the AAAI in 1983, (and who changed his affiliation from the Stanford Research Institute to Stanford University in 1985), used the occasion of his presidential address to congratulate the field for its accomplishments. But he was concerned, he said, that major AI architectures consisted of a declarative knowledge base plus an inference engine. For the field to mature, "Much of the knowledge we want our programs to have can and should be represented declaratively in some sort of declarative, logic-like formalism. Ad

---

[3] Parallelism came to dominate high-end computing at the end of the 20th century. But the firm Hillis had started while he was still finishing his graduate work in 1983, Thinking Machines, was ahead of its time, and eventually failed.

Nils Nilsson (Courtesy of Chuck Painter/Stanford News Service)

hoc structures have their place, but most of these come from the domain itself." He cited the use of PROLOG, a logical formalism (which was to be used by the Japanese Fifth Generation Project) as a good example of such a powerful formalism. With such a formalism, the answers AI systems offered to complicated problems could be proven correct—long gone were the days when you could tell just by looking whether the solution to a problem was correct or not. He acknowledged the wide disagreement in the field about its core content. Was AI empirical (the Scruffies?) or was it a theory-based technical subject (the Neats)[4]? This, he said, was a nonissue; the field needed both. In the long run, Minsky might prove to be right, claiming that natural intelligence is a kludge and thus the search for unifying theories is futile, but it was far too early to concede the point. AI should continue to try to simplify, organize, and make elegant models (Nilsson, 1983).

In response, Alex P. Pentland and Martin Fischler of MIT argued that, if anything, it was far too early to constrain AI to any particular way of doing business. Much interesting and important AI research was being done outside of the logic-and-theo-

---

[4] We owe these pointed and colorful terms to Roger Schank.

rem-proving paradigm. The problem must dictate the tool (as in science) rather than the tool defining the problem (as in logic) (Pentland and Fischler, 1983).

Like Newell, Nilsson was concerned that the field was uncertain about its own definition, and with this Pentland and Fischler also took issue. "AI has already defined for itself a set of 'core topics': the study of the computational problems posed by the interrelated natural phenomena of reasoning, perception, language, and learning. These phenomena may, of course, be viewed from many other vantage points including those of physics, physiology, psychology, mathematics and computer science. AI has continued as separate from these other sciences because *none of these other disciplines focus on developing computational theories for accomplishing intelligent behavior* [my emphasis]. It should not bother us, therefore, if our study of intelligence borrows results, observations, or techniques from these other disciplines; or if these other disciplines occasionally address some of the same problems, and use some of the same techniques. Their central interests remain quite different."

Well, yes; but this couldn't resolve another odd paradox. Practical AI successes, computational programs that actually achieved intelligent behavior, were soon assimilated into whatever application domain they were found to be useful, and became silent partners alongside other problem-solving approaches, which left AI researchers to deal only with the "failures," the tough nuts that couldn't yet be cracked. Once in use, successful AI systems were simply considered valuable automatic helpers. MACSYMA, for example, created by MIT's Joel Moses, building on work by James Slagle, had originally been an AI program designed to solve problems in symbolic algebra. It became an indispensable workhorse for scientists in many disciplines, but few people credited artificial intelligence for having borne and nurtured it. If you could see how it was done, people seemed to think, then it couldn't be intelligence—a fancy that many people entertain to this day.

Pentland and Fischler concluded that intelligent behavior requires rational, and not merely deductive, reasoning. "There is no question that deduction and logic-like formalisms will play an important role in AI research; however, it does not seem that they are up to the Royal role that Nils suggests. This pretender King, while not naked, appears to have a limited wardrobe" (Pentland and Fischler, 1983).

To an outsider, these two positions didn't seem to be very far apart. Each agreed that the other was legitimate and deserved attention. It was just that Nilsson was hoping for a more formalistic emphasis, and soon, while Pentland and Fischler were saying, not so fast. If their definition of AI's core interests didn't please everybody, it seemed commodious enough. What that definition failed to foresee was the rough shattering of AI into subfields—vision, natural language processing, decision theory, genetic algorithms, and robotics, to name only a few, and these with their own sub-subfields—that would hardly have anything to say to each other for years to come. (Worse, for a variety of reasons, not all of them scientific, each subfield soon began settling for smaller, more modest, and measurable advances, while the grand vision held by AI's founding fathers, a general machine intelligence, seemed to contract into a negligible, probably impossible dream.)

This division between formal and informal structures could be seen not only in how programs were implemented, but how knowledge was to be represented. Assuming knowledge could somehow be acquired (not a trivial assumption), how could it then be represented in all its buzzing blooming diversity so that a machine could act upon it appropriately? A number of schemes were proffered and implemented, and their degree of formalism varied from simple if-then rules, what were called production systems, to rigorous, logical, mathematical structures. Each had its partisans. But which was the most effective? Allen Newell argued that different schemes were appropriate for different tasks: For instance, if the task at hand was analyzing a problem, formalisms were useful, or if the task was solving a problem by machine, then less formal methods were preferred (Newell, 1981).

In 1984, as if to settle the argument, three scientists at Hitachi Corporation, Kyoshi Niwa, Koji Sasaki, and Hirokazu Ihara, set four different knowledge representation schemes to work on the same problem, namely risk management of large construction projects, and compared their effectiveness (Niwa et al., 1984). The four schemes were a logic system, a simple prod   uction system, a structured production system, and a frame system. Niwa and his colleagues showed that logic systems might give you precision, but at a very high cost. You had to trade off among cost, speed, and accuracy.

They summarized their findings by observing that in a poorly understood domain, whose knowledge structure could not be well-described, modular knowledge representations, for example, simple production and logic systems, should be used. But this caused low runtime efficiency. Structured knowledge representations, however, increased runtime efficiency, and reduced the effect of knowledge volume on running time—you got your answer sooner—but system implementation was more difficult. Mathematical completeness made logic systems more difficult to implement and less efficient in runtime. But, they reported with some patriotic delicacy, "Our problem was too simple to adequately demonstrate the advantages of logic representation." The Japanese Fifth Generation Project, a major multicorporate and government investment in AI, launched three years earlier, rested on PROLOG, a logic programming language.

Other Important Research in the '80s

Machine learning also received continuing attention in the 1980s. Its goal, to push programming techniques so that computers would approach human performance in certain tasks, was driven by both practical and scientific considerations. Constructing an expert system might be difficult, but programming a robot to perform a task was particularly tedious: if only a machine could simulate human cognitive processes, and do theoretical analyses independent of the task domain. If only a machine could learn to perform a task by example, or by analogy to a previously solved task. If only a machine could learn from past mistakes, or learn by watching and imitating an expert. If only.

Advances in machine learning, researchers hoped, would illuminate scientific issues as well. The nature of learning itself was—

and continues to be—a central question in the disciplines of philosophy, psychology, and education. In the late 1970s and 1980s, a series of schemes were proposed and doggedly tested to address those possibilities. For example, at the University of Michigan, John Holland, working quietly and almost alone, save for a few graduate students, sought to implement machine learning in ways that mimicked a vastly simplified version of evolutionary processes. Faced with an environment to search and survive in, a population of simple algorithms would evolve according to rules of selection and other operators, referred to as "search operators," such as recombination and mutation. Each member of the population received a fitness measure, and reproduction focused on high-fitness individuals. Holland wasn't necessarily claiming that humans learned in this Darwinian way; he was just hoping that this was a way for machines to learn independently.

The US Strategic Computing Project

Whether the field was foundering or whether it was a cornucopia of opportunities, AI was certainly in transition in the early 1980s, and seemed ripe for some kind of big push. In 1983, the Defense Advanced Research Projects Agency, which had sponsored so much of artificial intelligence research in the past,[5] announced its Strategic Computing (SC) Initiative, focusing on supercomputers, microelectronics, and artificial intelligence. Robert Kahn, the architect of this initiative, and head of ARPA's Information Pro-

---

[5] I regret that the first edition of this book hardly mentions the (Defense) Advanced Research Projects Agency of the US Defense Department, (D)ARPA, founded during the Eisenhower administration as a response to Sputnik. This is partly because I was focused on the history of the science and not necessarily its sponsors, but it is also partly because J. C. R. Licklider, the psychologist who had so much—everything—to do with ARPA's early nurturing of AI, refused to be interviewed. Even a direct plea from Allen Newell did not move him to see me. Luckily, several books have remedied that omission, notably *The Dream Machine: J.C.R. Licklider and the Revolution That Made Computing Personal*, by M. Mitchell Waldrop (New York: Viking, 2001), and *Strategic Computing: DARPA and the Quest for Machine Intelligence, 1983-1993*, by Alex Roland with Philip Shiman (Cambridge, Mass.: MIT Press, 2002). I have drawn on the latter for much of my discussion of that initiative here.

Robert Kahn
(Courtesy of Robert Kahn
Photograph by Bachrach)

cessing Technology Office (IPTO), believed that fast machines did not necessarily equal smart machines, and smart machines were needed (Moore's Law, which observes that computing power will double, and costs will drop by half, approximately every 18 months, was marching inexorably and quite splendidly on).

In Kahn's original vision, the AI part of the program would proceed in three steps. The first group of technologies was ready and ripest, including distributed data bases; multimedia message handling; natural language front ends; display management systems; and common information services. A second group of technologies would follow, including language and development tools for expert systems; speech understanding; text comprehension; knowledge representation systems; and natural language generation systems. The third and last group would include image understanding; interpretation analysis; planning aids; knowledge acquisition systems; reasoning and explanation capabilities; and information presentation systems. Then would come an ultimate integration, melding all these capabilities into a single whole, a machine capable of human-level intelligence.

Kahn's boss at DARPA, Robert Cooper, and *his* boss, Richard DeLauer, who headed the Defense Science Board, liked the pro-

posal, but they needed deliverables, in the jargon; clear military relevance, sure payoffs on investment, metrics, target dates, even a crisper
definition of success. Neither Cooper nor DeLauer was rigid—they
accepted that the path toward this goal must be flexible, but they
knew that such flexible paths would fail to convince on Capitol Hill,
where Congress was accustomed to a well-defined and concrete goal
(put a man on the moon), with explicit intermediate steps, before it
would be willing to appropriate the extra $600 to $700 million the
project would cost above DARPA's usual budget. It was the age-old
tension between research and development, between the practical
goals of a patron and the scientific goals of the researcher.

Kahn revised his program into 12 areas (down from 16). Seven
areas would be developed in generic software packages, independent of application (vision, natural language, navigation, speech,
graphics, display, and a combination of data management, information management and knowledge-base technology); five areas
(planning, reasoning, signal interpretation, distributed communication, and system control) were more application-dependent.
Metrics were introduced, goals and subgoals defined, and timelines
proposed. The six-year plan became a ten-year plan, and was revised yet again, to become more applications-driven, with work to
be carried out on an autonomous land vehicle, a pilot's associate (to
help an aircraft pilot in the cockpit), and battle management. The
cost rose from $700 million to one billion dollars, and to repeat,
this was over and above DARPA's normal budget.

But as Alex Roland with Philip Shiman writes, the emphasis
on practicality in the new scheme "failed to address the two most
important issues of process in the whole undertaking ... First, how
could managers of application programs begin designing their systems when the new technology on which they were to be based had
not yet been developed? ... Second, if applications were going to
rely on developments in the technology base, how would the two
realms be coordinated, connected? By what mechanism would
demand-pull reach down through the layers of the pyramid to shape
microelectronics research or computer architecture? By what mechanism would new research results flow up the pyramid into waiting

applications? By concentrating on what was to be done, the SC plan had neglected how to do it" (Roland, 2002). It also had the unintended consequence of setting up the field of artificial intelligence for a possible catastrophic crash.

By 1983, Congress had been persuaded to appropriate the money, and contracts had been let, but dissension between Bob Kahn, who had envisioned Strategic Computing, and Lynn Conway, who had been brought to DARPA to make the program happen, along with some inherent difficulties in the program's original design, caused SC to drift leaderless inside DARPA for more than a year. Two years later, in 1985, Conway, Cooper and Kahn each left DARPA for a variety of reasons, and some program managers left, too. Though the front-line researchers understood what was expected of them, neither a plan nor management structure was left in place to coordinate the flow of research results. Furthermore, new laws intended to keep government procurement clean also kept DARPA's program managers from seeking advice from researchers on both the possibilities of their research, and the merits of others who might be potential contractors. Nevertheless, by the end of 1985, SC had already spent more than $100 million, and 92 projects were underway at 60 different institutions, about half of them universities and government labs, and half in industry.

Saul Amarel, of Rutgers University, succeeded Bob Kahn as head of IPTO in September 1985. He was enthusiastic about AI, and about Strategic Computing in general, but concerned that AI stick to task-specific instead of generic systems, and that those tasks be military ones. He was above all concerned with the organizational problems he'd inherited at IPTO. He was forced to fight just to maintain the status quo, and the grand ambitions he harbored went unfunded, despite the promises made to him when he'd taken the job. Cuts came from Congress and were forced by other branches of DARPA with different priorities. Some of IPTO's program managers responded by cutting all contractors equally; some funded one part of their particular program at the expense of other parts.

⟞⟝

Two years later, in September 1987, Jack Schwartz, a mathematician on leave from New York University's Courant Institute, took over from Amarel. Schwartz could detect no unifying principles in AI, and he dismissed expert systems as no more than "clever programming," a bad direction for AI to take, he believed, since expert systems only worked when they narrowed the traditional goals of AI research. Citing unsatisfactory performance, Schwartz terminated SC contracts with two AI startups, IntelliCorp and Teknowledge, which were forced to survive as consulting firms, building custom expert systems, and he also terminated a number of university contracts. You had to feel for the researchers. They were accustomed to being berated on all sides for their "extravagant claims" on behalf of the field. Now came an IPTO manager who dismissed programs that actually worked as not extravagant enough, a betrayal of AI's traditional goals, and who pulled the plug.

Schwartz believed that DARPA was using a swimming model—setting a goal, and paddling toward it regardless of currents or storms. DARPA should instead be using a surfer model— waiting for the big wave, which would allow its relatively modest funds to surf gracefully and successfully toward that same goal. As a consequence, he eviscerated Strategic Computing, a swimmer model in his view (though Kahn's original vision certainly seemed to be premised on catching the wave that was beginning to swell). Schwartz thought that in the long run, AI was possible and promising, but its wave had yet to rise, so a number of sites working on AI and robotics found their funding cut suddenly and brutally. Schwartz's own interests lay in new architectures, which he favored as the swelling wave, and so he funded the revival of connectionism and machine intelligence, based on new findings in neural modeling.

Schwartz's tenure was essentially the demise of the Strategic Computing program, though funds were disbursed through 1993, bringing total costs to more than a billion dollars. Roland and Shiman do not call the program an outright failure, but they point out the lack of systems integration that plagued the project from the beginning. "Component development is crucial; connecting the components is more crucial. The system is only as strong as its weakest link. Machine intelligence could not be achieved unless all levels of the pyramid worked and connected." Management schemes came and went over those ten years, and most were hardly management at all: even the best managers did not leave behind an institutional structure that others could follow. Personnel turnover alone "was enough to disrupt the careful orchestration that its ambitious agenda required."[6] Furthermore, unlike "put a man on the moon," the goals of Strategic Computing depended on science and technology that was to be invented during the program, while the Apollo program, complex as it was, depended on technology that was on hand or nearly complete.

Some components in the SC program were developed, but were never connected into a system. The pilot's assistant and the autonomous land vehicle were not delivered, though research continued on aspects of both after the program ended. Battle management did succeed—the Navy got two expert systems for useful planning and analysis functions. SC also succeeded, as Bob Kahn had hoped it would, in filling out the national computer research infrastructure, in seeding new centers of excellence, both academic and industrial, and in educating a new generation of computer scientists. But networking, in both its technical and social senses, was absent. "Fine-grained" AI applications, such as speech understanding, found their way into dozens of commercial products ("Ironically, AI now performs miracles unimagined when SC began, though it can't do what SC promised," Roland and Shiman write.) Above all, a general machine intelligence did not emerge.

---

[6] DARPA is still committed to four-year tenures for its managers, for many good reasons. I'll say more about this later.

Artificial Intelligence's Deep Winter

It all came to be known as the "AI Winter." Raj Reddy's presidential address in 1988 to the AAAI was less the usual presidential state-of-the-science assessment and more frankly defensive, divided between what the field had accomplished, and what remained to be done (Reddy, 1988), a direct response to the new DARPA frostiness. Accomplishments, he protested, weren't negligible. Chess, AI's *e. coli*, (he meant that chess was to AI research as *e. coli* was to biological research, a testbed for hypotheses, a small-scale way to seek solutions to bigger problems) had not only begun to win games over 70% of expert human players, but had led to the development of techniques that AI programs could use elsewhere, such as the alpha-beta pruning algorithm, the B* search, and others. The more complicated problems of speech and vision had notched some significant, though not final, wins, giving AI researchers insights into how to use incomplete, inaccurate, and partial knowledge within a problem-solving framework. Finally, expert systems were at work in many practical domains.

Thirty years of "stable, sustained and systematic explorations" had not only yielded performance systems, but had offered fundamental insights into the nature of intelligent action, principles (though hardly laws) that characterized intelligence. Among them was an important addition to bounded rationality, first enunciated by Herb Simon in his book, *Administrative Behavior* (Simon, 1976), which tells us that humans cannot know everything, and therefore we settle for a "good enough" solution to problems. Artificial intelligence had shown that bounded rationality implies opportunistic search, and as silicon intelligence was achieved, differences in memory access time and bandwidth might have different computational constraints than human intelligence (another way of saying that the machines might shine at different tasks than humans, but what they did well, they would soon do better than any human).

Another principle was the physical symbol system, necessary and sufficient for intelligent action (though Reddy mentioned competing hypotheses from brain science and other fields). This was the idea first enunciated by Newell and Simon, that an essential condition for intelligent action of any generality was the capacity for the creation and manipulation of symbol structures, which are always instantiated in a physical system.

Furthermore, it had been an AI discovery that across many disciplines, an expert knows 70,000 "chunks" of information, plus or minus 20,000, whose acquisition usually takes about ten years of intense full-time study and practice in the domain.[7] It had become a working principle in AI that search (such as trial-and-error) compensates for lack of knowledge, and knowledge compensates for lack of search: Within bounds, tradeoffs between them can take place.

These principles of intelligence had first been elucidated in chess-playing programs, and had been found to hold in other task domains. If they sounded obvious now, they had not been at all obvious 30 years earlier. "When the Fredkin Prize for the World Chess Championship is won, it will probably be by a system that has neither the abilities nor the constraints of a human expert; neither the knowledge nor the limitations of bounded rationality. There are many paths to Nirvana." It was an idea he felt impelled to repeat several times, that silicon intelligence would surely be different from human intelligence.[8]

Reddy made a special, and unusual, plea at the end of his talk, that sharing the wealth of knowledge would be far more important to the world's disadvantaged than any sharing of food, and he urged his colleagues to get involved in such international projects (he spoke from his own background as a boy in a village in India whose life had been transformed). When he gazed about,

---

[7] Reddy recalled George Miller's classic paper of cognitive psychology, "The Magical Number Seven, Plus or Minus Two," and noted the parallel between this property of human short-term memory, and the 70,000 "chunks," plus or minus 20,000, that constitute human expertise in most fields studied so far, and asked aloud whether it was coincidence, or a significant pattern.

[8] Here he was touching on a paradox that continues to plague AI. When, as Reddy predicted, a chess program did defeat the world's champion human chess player in 1997, reactions were mixed. Some people thought it was a negligible win, because after all, it had "only been achieved through brute force methods," not "real intelligence." I'll say more about this further on.

he concluded, he saw not the "mythical AI winter," but a spring with flowers blooming.

Artificial Intelligence as Celebrity Science in the '80s

As the Strategic Computing Initiative illustrates, the science of artificial intelligence exists inside a human culture, and to understand its scientific journey better, we need to examine some of those social circumstances, the field's public face.

Surely one of the most striking aspects of AI during the late 1970s and 1980s was its move from obscure, almost fringe, science to hot public issue, and its subsequent popular (and perhaps scientific) swoon.

I use the term hot public issue in several senses. First, the topic itself raises controversy, and more must be said about that further on. But second, even as the field was in scientific flux in the 1980s, its most public and aggressive face was on Wall Street. Every practitioner, it seemed, whether professor or graduate student, engineer or programmer, was being ardently wooed by venture capitalists who, in turn, helped their new celebrity clients to establish startups and find investors for these new enterprises— hot public issues—that promised a very great deal.

For example, expert systems, described in embryo in the 1979 edition of this book, had suddenly become a very hot issue in every sense. By the early 1980s, startups and old-line companies competed with each other to use and sell their knowledge about systems that assisted experts, or codified their expertise so that it could be diffused anywhere and everywhere it was needed. Initial public offerings of stock in such firms were as heady (and almost as well-attended) as rock concerts. Robotics and speech understanding firms sprang up, along with firms that would manufacture special-purpose symbolic processing computers. The industrialization of artificial intelligence seemed unstoppable. If expert systems were too cumbersome, too tedious to be programmed, or the problems a company faced were unsuitable for expert systems, then other researcher-entrepreneurs stepped in to offer robotics, or speech understanding, or neural nets (a set of statistical learning algorithms that could be trained, and would learn from experience how to solve problems automatically). It seemed too good to be true.

It was. A young and fragile science was being asked to perform miracles that were then beyond its powers. And so the industrialization of AI in the early 1980s came to prefigure, more modestly, the industrialization of the Internet a decade later (and railways and automobiles a century and more earlier) in feverish boom, followed inevitably by deflated bust, a typical pattern as new technologies are commercialized.

Which is not to say it was all smoke and mirrors. The science was young and fragile, not fraudulent. Expert systems were based originally on Ted Shortliffe's 1974 PhD dissertation on MYCIN, which demonstrated the power of rule-based systems for knowledge representation and inference in the domain of medical diagnosis and therapy. They indeed saved corporations substantial sums of money (though they seldom created new revenue streams, the business executive's Holy Grail). The international credit card system, for example, would simply be impossible without a combination of neural nets and expert systems to approve credit and detect fraud (Feigenbaum et al., 1988). Natural language generating programs offer us directions as we drive, and relief from pounding a keyboard at our computers. Fuzzy logic can be found everywhere from washing machines to rice cookers. These early AI applications have become internal or background technology nearly everywhere. On the other hand, as a business opportunity, specialized AI machines were nullified by Moore's Law: instead of buying a special-purpose AI machine, it was easier and cheaper to wait for mass-produced general-purpose chips, and then program AI applications into the software.

The earliest market for AI products had been found in industrial research and development groups, whose members understood the novelty (and the difficulty) of the new systems. But elegant, handcrafted solutions to industrial problems were difficult to build and expensive to maintain. They came programmed in languages nobody in the mainstream part of the firm had ever heard of, and required tireless champions to usher them across internal borders. Less elegantly crafted systems simply could not meet the demands made on them, or the high expectations for them in either performance or return on investment.

So it was that nearly all the early AI firms, formed with the intention of selling products and tools, became mainly consulting companies, helping people use these new tools, and if eventually they didn't fail, larger firms swallowed them. The initial fever had broken.

The Fifth Generation

I confess my own contribution to all this. In October 1981, Japan's Ministry of International Trade and Industry announced an audacious national plan to develop what it called a fifth generation of computers, computers for the 1990s and beyond, computers that would be intelligent. Fifth-generation computers would be able to converse with humans in natural language and understand speech and pictures. They would be able to learn, associate, make inferences, make decisions, and otherwise behave in ways that we have always considered the exclusive province of human reason. Another part of the plan was to push Japan's supercomputer effort.

It was a plan for a decade, two decades out, a sortie into a part of computing that had been largely ignored, even scorned, by the dominant American and European computer firms. The plan seemed to offer a plausible path for a country whose natural resources were nearly nil, and whose industrial might of the 1960s and 1970s was rapidly being copied, then done better and cheaper, by other countries rushing to industrialization. Japan's great resource was—and for that matter remains—a well-educated, diligent citizenry. Why not exploit that resource through the development of intelligent computing that would increase productivity, not only in industry, but also where productivity increases had proved difficult to achieve, such as fishing and agriculture, or services, design and management (Feigenbaum and McCorduck, 1983).

Japan faced other serious problems: Its society was aging rapidly (with a concomitant drop in births), yet it did not welcome immigrants. It was not energy-independent, nor could it hope to be. Fifth-generation machines would address these problems, too, by offering care and companionship to the aged and by amplifying the work of the young.

My friend Ed Feigenbaum, who traveled to Japan regularly and watched the beginnings of this plan form, suggested that we write a book about it, and so we did, called *The Fifth Generation*. It got a lot of attention and inspired a lot of argument. The Japanese, it was said, couldn't possibly do this. In fact, nobody could, since artificial intelligence was impossible, but certainly the Japanese couldn't. Not for that miniscule an investment, anyway. Feigenbaum and I were accused of being alarmist and self-serving. After all, if Feigenbaum got the government to believe the Japanese were trying to leap over us at our own game, it would increase his grants (which made me a double dupe, I suppose, first of the Japanese and then the sock puppet of my friend Ed Feigenbaum).

Rereading the book after two decades, I can say that compared to the publisher's hyperbole, the book is quite restrained. We celebrate the Japanese vision, but we admit that we don't know if they can do it in the 10 years they hope, or even in 20. We have technical reservations about logic programming, and wonder if the Japanese firms involved will give up their autonomy and competitive spirit quite so easily. We offer some possible scenarios for an AI future that now make me blush for their poverty of imagination. But we believed then, and still believe, that the Japanese were right in their aims, if not in the details of their plan.

Important people took the message seriously, including Congress. Bob Kahn and Richard DeLauer of DARPA claimed later to be skeptical of Japanese abilities to pull this off, but they weren't above exploiting congressional and military alarm to help acquire funds for the Strategic Computing Initiative the same year the book was published. The professional skepticism wasn't universal—many people remembered that the early history of AI in the US had some significant similarities to the Japanese plan. A small group of smart young people would be steadily nurtured (and funded) over a sustained period of time. Their ideas would be systematically transferred to firms that could make products out of those ideas. It had happened once. It could happen again.

But the best story of this episode, in my view, revolves around the geriatric robot. Permit me, as participant-observer, to tell this delicious little tale in some detail.

I'd thought the manuscript was getting a bit heavy, so I inserted a jokey little turn I often used during my talks to college students. I wrote:

> For years McCorduck has been nagging for, promoting, advocating the geriatric robot. She'd all but lost hope, watching her friends in AI create physician-machines, intelligent geologist-machines, even intelligent military spy-machines, but never anything down-home useful. Time is getting on. The geriatric robot might soon be a matter of immediate personal concern.
>
> The geriatric robot is wonderful. It isn't hanging about in the hopes of inheriting your money—nor of course will it slip you a little something to speed the inevitable. It isn't hanging about because it can't find work elsewhere. It's there because it's yours. It doesn't just bathe you and feed you and wheel you out into the sun when you crave fresh air and a change of scene, though of course it does all those things. The very best thing about the geriatric robot is that it *listens*. "Tell me again," it says, "about how wonderful/dreadful your children are to you. Tell me again about that fascinating tale of the coup of '63. Tell me again …" And it means it. It never gets tired of hearing those stories, just as you never get tired of telling them. It knows your favorites, and those are its favorites, too. Never mind that this all ought to be done by human caretakers; humans grow bored, get greedy, want variety. It's part of our charm.
>
> McCorduck felt a slight jolt a few years ago when she heard Yale's Roger Schank muse in a lecture that he didn't believe a machine could be considered intelligent *until* it got bored, but he reassured her later that the art of programming was already refined to the point that a never-bored robot could be fashioned.
>
> Now here were the Japanese, those clever people, claiming their Fifth Generation would alleviate the problems of an aging society. She read

the reports eagerly: lifetime education system; medical care information; other rubbishy pieties. She flung down the proceedings in disgust. She is reconciled that she may have to turn AI from spectator to participant sport and whip one up herself before it's too late."
(Feigenbaum and McCorduck, 1983)

Ed Feigenbaum called me and expressed some doubts. Leave it in, I said. The editor deleted the vignette without asking, and I insisted it go back. No reason why a book about technology had to be altogether tedious.

I thought I'd hoisted all the rhetorical flags—a joke, folks, maybe not a thigh-slapper, but a little levity to keep us all going. Imagine our astonishment a few weeks after the book came out to find ourselves worked over, flayed, skin peeled off like victims of an Aztec sacrifice, for something going on five pages in the sober *New York Review of Books*. In the review's opening paragraph, Joseph Weizenbaum, the Savonarola of artificial intelligence whom readers met in *Machines Who Think*, compared us to Hitler, Mussolini, and Pinochet, and the only comfort to be taken was that at the time, at least Pinochet was still alive. Of course, I was the culprit. Poor Ed had wanted to take the piece out. So I took it all rather personally.

Why the outrage? Why the comparison with monstrous evil when all we'd done was write a book? To someone looking to find fault (or utterly innocent of a sense of humor), I was apparently making a serious proposal. It was a sign, Weizenbaum fumed, of my appalling lack of humanity. Looking after the elderly was a human's job, and must properly be done only by humans. (It wasn't that he himself expected to take care of any elderly persons, but *someone* ought to. Twenty odd years ago, that meant a woman at minimum wage. These days, it still does.)

Weizenbaum went on to say that the geriatric robot was not only a sign of the authors' lack of humanity; it symbolized the utter lack of humanity in anybody associated with the field of artificial intelligence. This, of course, was his main agenda; he'd been literally rubbing his hands together (I heard from Michael Dertouzous, one of his colleagues) as he roamed the halls of MIT, where he was then on the faculty. "This'll get them," he said to whoever would listen. (Okay, I'd suggested in *Machines Who Think* that much of his very loud self-righteousness about artificial intelligence was driven by his faltering scientific manhood, and he was mad. But that's a different story.) In any case, the review continued with the same kind of stuff. Readers will recall that in the first edition of this book, I had drawn an early distinction between the Hellenic and Hebraic points of view with regard to the making of intelligent machines. The Hellenic view welcomed the machines as a marvel; the Hebraic view considered the very idea sinful. Weizenbaum's entire review was canonical Hebraic jeremiad.

My coauthor called me in some pain. He wished he'd prevailed, and I hadn't insisted on leaving the passage in. We might've gotten a good review instead of five pages of excoriation. But given what was then the *Review*'s consistent antipathy toward any technology more recent than the telephone, I doubted it. Our editor was unhappy. I tried to soothe them both, and composed a cheerful and brief reply, which my coauthor and I signed. It was duly printed in the *Review*, and there, I thought, the matter ended. Luckily, the book was selling well anyway, but it was a bit odd to go to a New York City book party for someone else's book and hear a stranger, who didn't even have the gist of the review right, tell me what monsters those people must be. I volunteered that I was one of the monsters in question, and instead of smiting me on the spot, he flushed and slunk away, mortified.

Issues the book raised could be argued. We were writing about an ambitious national Japanese effort in artificial intelligence, with stated intentions to do many things, though ameliorating some of the problems of old age—loneliness, boredom, homecare—was pretty far down on the national wish list. Serious questions could be asked about whether the grander goals could be met. If so, could they be met using the means the Japanese proposed? If so, were the Japanese—in particular, their guiding agency called the Institute for New Generation Computer Technology, and their firms—the ones who could do it? And could they do it within 10 or 20 years?

Two and a half decades later, we can see that the Japanese didn't quite meet all of those ambitious goals, though they have had many admirable accomplishments. For one thing, those goals became international, not just Japanese. People who had scoffed got busy. Computer technology continued to improve, and small general-purpose machines remained the rule rather than special-purpose AI machines. The World Wide Web, first made accessible by the Mosaic browser, turned the Internet into a utility for anyone with a computer and a telephone line or even a wireless attachment (not to mention a social force of its own). The participating Japanese firms did prove highly competitive with each other, and there were complaints that some of them pulled their best people out of the project as soon as they could. Five years into the project, researchers began to see that however elegant their logic programming was, they still needed piles of dirty real-world knowledge to produce a system that would do something useful in the real world. They had to scramble to acquire and represent it, so that useful demonstration programs could be presented at the end of the project.

The parallel machines they'd hoped to build were commercially neutered by Moore's Law, which allowed serial processors to overtake parallel processors relentlessly. However, another part of the Fifth Generation effort did form the foundation of their supercomputer efforts, at which they are present world champions. For example, an NEC supercomputer at Yokohama called the Earth Simulator is the envy of earth scientists everywhere.

Most importantly, perhaps, they educated hundreds of young people in AI techniques and technology, who went back to their firms with great exploitable ideas. Unfortunately, Japan itself plunged into a period of economic stagnation that persisted for at least 15 years, with grave consequences for its industrial sector, a stagnation that was the fault in great part of an interlocking gerontocracy that grew more timid (but not less powerful) with each passing year. Still, many of the goals that the Japanese proposed with their Fifth Generation Project have indeed been met, not only in Japan, but elsewhere, or are well within reach.[9]

The AI Winter Summarized

As the 1980s drew to an end, yet another AI paradox arose. Business people, politicians, and the military were clamoring for AI; the media had made it into a nine-day wonder, but the field's new big ideas seemed fewer and further between. (Or alternatively, its founding fathers had laid down such a rich agenda that what seemed to remain was to push that agenda to its utmost.) AI's major funding agency in the US, DARPA, was cutting support and asking for precise, measurable, and therefore incremental results, and researchers elsewhere were just climbing the learning curve. AI research didn't stop, but it became more "normal," to use Thomas Kuhn's description of one kind of science, as distinct from "revolutionary," the other kind. I hesitate to mention the Kuhnian paradigm, because it doesn't quite capture another recurring theme in AI, that ideas are picked up, exploited to the maximum extent allowed by available hardware and software, abandoned, and then picked up again as major improvements in hardware and software (the latter often from AI research itself) allow a new round of exploitation.

For example, neural nets, which found many applications in the 1980s, was a resumption of early work on "the perceptron" in the 1950s and 1960s, an automatic pattern-perceiving machine, which itself was based on work by Pitts and McCulloch in the 1940s. In the mid-1980s, Rodney Brooks's seemingly radical turn on the robotics path was really a return to the principles of cybernetics, a model proposed in 1948 by his distinguished predeces-

---

[9] With regard to the geriatric robot itself, 20 years ago, the Japanese saw the demographics clearly: a rising proportion of the aged, with the numbers of those above 80 years of age soon to be equal to the numbers of those below 20. This is now even apparent to the rest of the developed world. For Japan's planners to think of using computers to help the problems of an aging population was not only compassionate, but prescient. And now, in many places, versions of the geriatric robot are here, too, as we'll see in the discussion of socially interactive robots later on.

sor at MIT, Norbert Wiener, who himself had built on the work
of the 19th-century physicist James Clerk Maxwell. A new effort
that began in the 1990s called "narrative intelligence" would build
on work done in the 1970s and 1980s by Roger Schank and his
colleagues, but that was abandoned for various of reasons. Danny
Hillis's pioneering 1980s work on parallel computers is yet another example; it was really only developed a decade or more later.
As one of Allen Newell's maxims observed: "Everything must wait
until its time. Science is the art of the possible."

You Can't Keep a Philosopher Quiet

In the period originally covered by *Machines Who Think*, artificial
intelligence as a scientific field had two main public antagonists.
One was philosopher Hubert Dreyfus, at the University of California at Berkeley, and one was the computer scientist and former
AI practitioner, Joseph Weizenbaum, at MIT, who had been so
exercised by *The Fifth Generation*. I don't think it distorts their
respective stands to say that Dreyfus argued strongly that AI could
not be done, but even if it could be, the field was going about it
in entirely the wrong way; whereas Weizenbaum argued that AI
probably could be done, but should not be, owing to the ways a
totalitarian government, such as Nazi Germany's, might abuse it.
But in the 1970s, these debates were more academic (in the literal
sense) than popular. They were staged at universities and scholarly meetings; their audiences were academic, and they generated
more heat than light.[10]

By the start of the 1980s, other philosophers came to address
the question of whether or not AI is possible. John Searle, also of
Berkeley's philosophy department, had no objection to "weak AI,"

---

[10] I let Lotfi Zadeh persuade me into participating in one of these at Berkeley in the late
1970s, though I began my presentation by stating that scientific issues were not settled
by rhetoric, but by doing science. However, when I pointed out that many of the
reasons my adversary offered against the possibility of artificial intelligence were eerily
parallel to the arguments made in the 1800s against the possibility of women thinking
(their bodies, their emotions, their "different" brains), I do believe I won the day. Entertainment, but not science.

a phrase he apparently coined to describe computer programs that could behave intelligently in limited ways. But in his famous Chinese room, where a man is confined, transferring one set of symbols (English words) into another set of symbols (Chinese) using only a set of rules, or more elaborately, reading stories in Chinese and answering questions about them in Chinese sufficiently human-seeming to pass the Turing Test, he argued that the man cannot be said to actually *understand* Chinese, nor be *conscious* of using Chinese, no matter how well he follows the translation rules. Without real knowledge, real understanding, or real consciousness, intelligence does not exist (Searle, 1980).

Searle provoked many responses, including one from another philosopher, Daniel Dennett, of Tufts University. He and his computer scientist coauthor, Douglas Hofstadter, of the University of Indiana, replied to Searle, that first, he had committed "a serious and fundamental misrepresentation by giving the impression that it makes any sense to think that a human being could do this. By buying this image, the reader is unwittingly sucked into an impossibly unrealistic concept of the relation between intelligence and symbol manipulation[11]" (Hofstadter and Dennett, 1981).

But an important part of their reply was an idea shared by other scientists, cognitive psychologists, anthropologists, and computer scientists alike, that intelligence—understanding and knowledge—is enacted only within a larger system. None of these qualities is born, or resides, solely in a single human's head, but rather, these concepts describe reciprocal relationships between the individual and the surrounding culture, a culture built over many generations, to which the individual may contribute innovations, but whose totality is collective, distributed, or even emergent.

---

[11] The authors note that Searle had finessed the entire issue of level of implementation, whether simulation (approximate modeling) or emulation ("in a deep sense exact"). Minds worth calling minds exist only where sophisticated representational systems exist; they exist in brains and may come to exist in programmed machines. If this happens, their causal powers will derive from their design and the programs that run on them. "And the way we will know they have those causal powers is by talking to them and listening carefully to what they have to say."

Intelligence as the property of a large system, whether embedded or emergent, wasn't something that had occurred to most of us, so fixed were we on the idea of intelligence residing only in a single (human) head, but this point of view was consonant with new research in dynamical systems, such as complex adaptive systems—which could mean anything from a brain to the global economy. This notion would later come to be known as distributed intelligence (Bruner, 2002). Searle, however, was unconvinced, and continued to make arguments to appreciative audiences for whom the idea of intelligence as a collective effort was psychologically disconcerting, and for whom thinking machines were only ludicrous.

The Society of Mind

In 1985, Marvin Minsky published his *The Society of Mind*, a summary of what he'd learned, been thinking of, and speculating about for many years[12] (Minsky, 1985). I mention the book here because it found an audience far outside AI, and became a brisk and popular seller. Its ideas are deep and complex, but shrewdly camouflaged below surfaces of engaging simplicity and blithe style. Sections of no more than a page, sometimes only a half page, cluster into chapters that tackle some of the most enduring puzzles of philosophy and psychology: the self, individuality, insight and introspection, and so forth. The book not only presents the idea that intelligence is the result, the emergent property, of many unintelligent processes known as "agents" conjoined in agencies that form a "society," but it embodies that idea, too. Simple and comprehensive segments in the book connect with other segments in surprising ways; together, they mean to present a coherent model of mind that can slowly be built and tested. Along the way, concepts such as multiple hierarchies and scales, learning, remembering, sensing similarities, emotion, emergence, and frames (which are mental images of things, or mental narratives of procedures in situations, that can be combined, contrasted, and built upon) are made explicit.

---

[12] "Who's publishing it?" Minsky's elderly mother asked him. Simon and Schuster, he told her. "Liked Simon," she said thoughtfully. "Never liked Schuster." She'd known them both.

"Good theories of the mind must span at least three different scales of time: slow, for the billion years in which our brains have evolved; fast, for the fleeting weeks and months of infancy and childhood [where experience is being built upon daily]; and in between, the centuries of growth of ideas through history," Minsky writes. This explicitly recognizes the contributions that the physical, the experiential, and the cultural make to individual intelligence. *The Society of Mind* addresses each contribution, connecting them back and forth in a combination of Socratic query, storytelling, conversational dialogues, and occasional simple diagrams.

For example, geniuses? Not so different from the rest of us, except for possessing some kind of "higher-order" expertise that allows them to organize and apply the things they learn, hidden tricks of mental management that produce those works of genius. How do geniuses acquire that expertise? It's a mystery at the moment, but one possibility may be a childhood accident: While one child earns praise for rearranging bricks, another child has been rearranging his thinking.[13] (A similar, though not identical, idea would be proposed by Simon, and by Schank and his colleagues.)

Knowledge-lines, K-lines for short, constitute Minsky's theory of memory, which rests on the idea that "we keep each thing we learn close to the agents that learn it in the first place. That way, our knowledge becomes easy to reach and easy to use." A K-line is a type of agent itself, a wire-like structure that attaches itself to whichever mental agents are active when you solve a problem or have a good idea. Activating that K-line later arouses the agents attached to it, putting you in a mental state much like the one you were in when you solved that problem or got that idea before, making it relatively easy to solve new, similar problems. Memory can only recall our minds to prior states, by putting back what was in the mind before.

---

[13] Would that explain Mozart, you ask? Mozart-like AI's have been proposed and even implemented, but nobody seems eager to listen to their music. Perhaps the product isn't good enough; perhaps in the 21st century, new versions of 18th-century forms are more a curiosity than a satisfying aesthetic experience. If that, it's more evidence of the larger milieu a successful intelligence must operate within.

"We're told that by their nature, all machines must work according to rules. We're also told that they can only do exactly what they're told to do. Besides that, we also hear that machines can only handle quantities and therefore cannot deal with qualities or anything like analogies. Most such arguments are based upon a mistake that is like confusing an agent with an agency"—what the machine does inside itself with how it appears to the outside world. But how does that admit the illogic of some thinking? Nothing prevents us from using logical language to describe illogical reasoning. "Logic no more explains how we think than grammar explains how we speak; both can tell us whether our sentences are properly formed, but they cannot tell us which sentences to make."

The Society of Mind is not a prescription for how to build an artificial intelligence. "Since most of the statements in this book are speculations, it would have been too tedious to mention this on every page. Instead I did the opposite—by taking out all words like 'possibly' and deleting every reference to the scientific evidence. Accordingly, this book should be read less as a text of scientific scholarship and more as an adventure for story for the imagination. Each idea should be seen not as a firm hypothesis about the mind, but as another implement to keep inside one's toolbox for making theories of the mind."

It is a theory of how mind *might* work, sufficiently vague to entice graduate students to attack; sufficiently detailed and grounded in experimental experience so it could not be dismissed as armchair philosophizing. That was a lesson Minsky had learned long ago, and had never forgotten. But in addition to seducing graduate students to try to turn some of its speculations into reality, it also struck the fancy of a larger audience, specialists and nonspecialists alike, became part of the public face of artificial intelligence, and remains in print more than fifteen years later.[14]

---

[14] Danny Hillis built The Connection Machine to test some of Minsky's theories. In 1983, he founded a company, Thinking Machines, Inc., to market massively parallel computers. Though he had customers, the idea was ahead of its time—programming the machine was very, very difficult. Twenty years later, massive parallelism is a vigorous part of computer science research though programming a parallel machine is still stupefyingly difficult.

A Physicist Explains It All to You

By 1989, another scoffer had emerged, this time a distinguished physicist, Roger Penrose, who argued in his best seller, *The Emperor's New Mind*, that real intelligence grows out of some undiscovered behavior at the quantum level of the human brain, and AI will never be able to achieve that with computers constructed using current silicon technology (Penrose, 1989). He surely didn't mean that computers have no quantum level; in fact, it still isn't clear to me (or, I bet, to most of his readers) exactly what he did mean, though I thought I recognized an old formalist argument once more "proving" that since machines didn't employ certain formalisms to think, they couldn't be said to think. (Of course, neither do humans, though formalisms sometimes model those human processes.) Penrose was adamant that consciousness, and its hallmark, judgment-forming, was something AI researchers would have no concept of how to program on a computer, and lacking this, no machine could ever tell us how it felt about being intelligent.

John McCarthy replied: "In fact, most of the AI literature discusses the representation of facts and judgments from them in the memory of the machine. To use AI jargon, the epistemological part of AI is as prominent as the heuristic part." McCarthy concluded that some future programs would be able to answer what it feels like to be a computer "based on their ability to observe the reasoning process that their programmers had to give them in order that they could do their jobs. The answers are unlikely to resemble those given by people, because it won't be advantageous to give programs the kind of motivational and emotional structure we have inherited from our ancestors" (McCarthy, 1990).

In 1999, I happened to sit in on a meeting of computational mathematicians at Oxford University, who'd invited Penrose to speak to them about artificial intelligence. The audience had spent days discussing their own intellectual and professional concerns, such as strong tractability, the average cost of the simplex algorithm, and other arcana. Artificial intelligence was very distant, in fact and in spirit. "Amazing," I murmured to one of the program organizers. "I don't believe I've ever been to an AI meeting with a plenary speaker who asks whether path integrals are real, or only a figment of someone's imagination."

But the Oxford University meeting was yet another manifestation of the fierce proprietary interest nearly everyone lays claim to in artificial intelligence. We think, therefore we are—authorities. We know in our hearts that with enough work and sufficient breakthroughs, it can be done. No, we know it can't be done, because our human intelligence is unique. Oh, it can be done all right, but it mustn't be, because something catastrophic will happen to us if we permit this alien intelligence to penetrate our world. Too late, we know it's already here, so let's at least try and learn from it, even try to make it useful to us.

It's not only fear of the unknown that makes people uneasy. Much human self-esteem is wrapped up here. In this book's first edition, I wrote that humans were like the Wicked Queen in the tale of *Snow White*, peering anxiously into their magic mirror, seeking reassurance that they were still the smartest of them all. It's still the case, at least with those of a certain age. Younger people, whose familiarity with the computer is lifelong, and whose vanities perhaps lie elsewhere, don't seem to worry too much about whether computers can *really* be said to intelligent. If anything, they long for computers to be smarter faster.[15] But not all of them, as we'll see.

AI Keeps at It Anyway

Throughout this downright theatrical burst of public activity, the science of artificial intelligence was proceeding quietly in its laboratories—which themselves were growing in numbers and spreading out across the world: North and South America, Europe, Australia, and Asia.

Allen Newell at Carnegie Mellon, for example, was pushing on themes he'd been preoccupied with for many years, the nature and theory of knowledge and symbols. In his 1981 presidential

---

[15] Though they also love the staple Hollywood product of Us versus Them, whether it's the wildly popular *Matrix* series, the nonterminating *Terminator* series, or even Stephen Spielberg's sad and somewhat incoherent film called *AI:Artificial Intelligence*. These are films where Us always wins. I like them, too, but films nearly never touch on issues that haven't already been explored in detail in written science fiction or by philosophers, which is why I have nothing more to say about them here. Still, these films bring the old issues to a mass audience for the first time.

address to the AAAI, he proposed the idea of another computer system level immediately above the symbol (or program) level, which he called the knowledge level. This was "an abstract level of analysis of a computational system (human or artificial) where predictions of behavior could be made by knowing just the system's knowledge and goals (where prediction includes explaining behavior, controlling behavior, or constructing something that behaves to specification)" (Laird and Rosenbloom, 1992). The knowledge level's medium was knowledge, and its laws of behavior were about reaching goals (Newell, 1990).

In 1981, Newell and his graduate student, Paul Rosenbloom, set out to investigate certain regularities in human cognitive performance at many tasks that had appeared in the psychological literature, regularities they were able to confirm, and eventually attribute to the architecture of the human brain. Meanwhile, with another graduate student, John Laird, Newell began to construct a large production system called SOAR (derived from state, operator, and result) organized around problem spaces. After some difficulties, Newell, Rosenbloom and Laird "decided to place their intellectual bets with SOAR as an architecture for general intelligence and make it the center of their research for the immediate future" (Laird and Rosenbloom, 1992). From 1983 until his death in 1992, Newell focused on SOAR with these colleagues and others.

First, SOAR established a feasible architecture for complex, real-world tasks, beginning with expert systems. Newell and his colleagues then moved on to implement integrated problem solving, learning, and even algorithm discovery within SOAR. They thought that SOAR might effectively be used to model human cognitive behavior, even be the basis of a unified theory of cognition. Newell didn't consider SOAR to be the only unified model, but instead a vehicle that demonstrated what a unified theory would look like. He hoped it would encourage the field to embrace unified theories as an appropriate goal for psychological research, and encourage others to join in the search for unified theories. After Newell gave the William James Lectures at Harvard University in the spring of 1987, research on SOAR as a theory of cognition became international. Just before his death in 1992 (the year he also won the National Medal of Science), he was engaged in developing a theory of social agents. SOAR became used widely enough to merit its own annual international conferences, which are still being held.

Newell's maxims were widely quoted among his graduate students. On at least one occasion, he even gave a lecture where he numbered them (17) and commented upon them (Newell, 1991). "Science is in the details," he would say (and anyone who knew Newell knew he meant *details*), or, "There is no substitute for working hard—very hard." (His working hours were legendary.) "Diversions occur, make them count. Salvage what is possible for the main goal." "Embrace failure as part of success. But use it for the main goal." And more. He lived his work; the maxims only described it. They feel so true because they grow not out of distant philosophizing, but arise from the experience of a gifted and hardworking scientist who happened to be graced with extraordinary insight into the nature of that work.

Herb Simon, Newell's Carnegie Mellon colleague, and research partner in earlier days, had worked all his scientific life on the nature of the cognition problem, too, and in the 1980s, he continued to elaborate on some of his old favorites: EPAM (the Elementary Perceiver and Memorizer), the General Problem Solver, and familiar problems like the Tower of Hanoi. "Using these well-tried tools is fitting," he remarked in his autobiography, *Models of My Life* (Simon, 1991). "Old dogs should not be learning new tricks after their sixty-second birthdays." But he immediately confessed that his research had in fact taken new directions. In a flurry of international collaborations, he and his colleagues elucidated short-term memory and learning processes, especially how students can learn from worked-out examples and how this process can be modeled by adaptive production systems. He was experimenting with visual imagery in thinking, and its underlying mechanisms. Finally, he was beginning to ask questions about scientific discovery, as represented by collaborative work on a program that designs sequences of experiments, adapting each new experiment to the findings of the previous one. Though he watched with deep interest the efforts of his colleagues to build grand theories of cognition (Newell and SOAR, John Anderson's Act*, and Jay McClelland's connectionist nets), he was, he said, more interested in "theories of the middle range," a phrase he'd borrowed from the sociologist of science, Robert Merton. "Theories of the middle range" were programs that simulated human behavior over a significant range of tasks, but did not pretend to model the whole mind and its control structure. As for whether cognition was serial or parallel, he wrote: "These 'connectionist' architectures have a role to play (for instance in simulating visual and auditory sensory processes) but ... they will not replace physical symbol systems as models of higher mental processes." (He added: "I am now for the first time learning how it feels to be a target of the attacks of Young Turks, to have one's cherished beliefs challenged and the permanence of one's life work threatened. So far, I have not felt any painful anxiety, perhaps because I am not convinced that the ramparts will crumble.")

In attempting to simulate scientific discovery, he noted that it was important to use tasks that could not be dismissed as trivial or humdrum. The experimental tasks were taken from great moments in the history of science: Kepler's discovery of his Third Law of planetary motion; Ohm's law of electrical conduction, Dalton's theories of chemical reactions; the discovery of atomic and molecular weights; the conflict between the phlogiston and oxygen theories of combustion; Krebs's explication of the synthesis of urea in living organisms. Simon and his colleagues had chosen these moments because they were genuinely creative moments of great significance. What they accomplished "by no means completes the job of explaining the processes of science, but it takes several long steps toward that goal and sharpens the questions that remain. It supports strongly the proposition that scientific discovery is achieved by the normal problem-solving processes that have been observed in less formidable problem domains." (Recall Minsky's assertion that genius is us, only more so.)

Simon also took up the problem that had divided the AI (and cognitive science) community from the beginning: Is thinking best viewed as a process of reasoning from premises (the Neats?) or as a process of selective search through a maze (the Scruffies)? He had no final answer to that when he died in 2001, though we can assume he leaned toward the Scruffy point of view, given his book, *Human Problem Solving*, that he'd published with Allen Newell in 1972.

From the beginning, Simon found big lessons in small places: In 1935, studying the decision processes of the Milwaukee municipal recreation system, he had asked how people make decisions, or more formally, how do people reason when the conditions postulated by neoclassical economics are not met? They bring decisions within reasonable bounds by identifying the partial goals for which their own organizational units are responsible, his justly famous "bounded rationality," which became a label for the computational constraints on human thinking.

The broad problem of accounting for human rationality kept him busy for a lifetime. In the course of trying to find the answer to that question, he discovered that scientific discovery is incremental; surprises only strike the well-prepared mind. To find laws that fit empirical data requires good empirical data to begin with (to make rabbit stew, first catch the rabbit); one works backward, one uses probabilities. Nothing magic, nothing mysterious.

Empiricism was the key. It might seem as if Simon weren't discovering anything "new" (after all, Kepler and Ohm had got there long ago), but he was. He was discovering and making explicit the empirical substance of the process of scientific discovery.

Though Minsky, Newell, and Simon were deeply attentive to the details, their work could fairly be described as top-down, that is, generalized theories about intelligence both artificial and human compared to other work that was also taking place at this time. These three pioneers of AI held strong beliefs about mental models as fundamental to higher intelligence. But younger researchers had different ideas.

Rodney Brooks
(Courtesy of Donna Coveney, MIT)

The Tribe of Robots

In the first edition of *Machines Who Think*, I conflated robotics with the problem of general intelligence because our ideas about intelligence then were very much in the form of a single, mobile, perhaps humanoid agent that can roam about the world acting and reacting as it encounters obstacles, works at tasks, and tries to achieve goals. The robots I described had inside their "heads" a fairly detailed representation of the very simple and custom-built outside world where they moved about, a representation that they modified and acted upon dynamically (or else they were tethered to an off-board computer which contained such a representation). This class of robot, descended from Shakey, and described in Chapter 10, "Robotics and General Intelligence," included robots at JPL (Jet Propulsion Laboratory) in Pasadena, others at the Laboratoire d'Analyse et d'Architecture des Systèmes in Toulouse, France, and even a cart-like vehicle originally intended for moon exploration, but prematurely retired to the Stanford Artificial Intelligence Laboratory before it had a chance to travel.

Rodney Brooks, now the director of MIT's Laboratory for Computer Science, describes how "the Cart" became a testbed for mobile intelligence. It traded Shakey's full internal world model for a limited model plus sensors that allowed it to "see" where it

Hans Moravec
(Courtesy of Ken Andreyo, Carnegie Mellon University)

had to go (Brooks, 2002). Eventually, the Cart was inherited by Hans Moravec, whom *Machines Who Think* readers met as a graduate student at the Stanford AI Lab (he is now a research professor at Carnegie Mellon's Robotics Institute). Moravec believed that for the Cart to act intelligently, it needed an accurate three-dimensional model of the world, which he proposed to endow it with. That 3D model resided in a computer to which the Cart transmitted camera images, which in turn gave it commands to steer and drive, all of this resulting in a forward lurch of a few feet about once every 15 minutes (though during peak computer usage, the lapse between lurches might be 3 hours). For computational economy, Moravec had programmed in the assumption that the Cart's world was static, but in truth, people, furniture, and doors were always moving. Shadows lengthened. Batteries of the era soon ran down.

As Moravec's fellow graduate student in robotics, Rodney Brooks couldn't help but be disappointed when he compared the Cart's behavior to Grey Walter's turtles of 1947, which had moved about autonomously for hours on end, interacting with a dynamically changing world and with each other. Watching the Cart struggle along, thinking through every move, Brooks asked himself whether that multimillion-dollar contraption was worth it.

"Were the internal models truly useless, or were they a down payment on better performance in future generations of the Cart?"

Five years later, in September 1984, and now at MIT's AI Lab, Brooks set about building a cheap robot. His team's first, Allen (after Allen Newell), was a three-wheeled cylinder about 25 centimeters high. It had been bought ready-built from a new and short-lived robotics company run by enthusiasts, one being Grinnell More, son of Trenchard More, an attendee of the original Dartmouth Conference. Grinnell was a high school dropout who hung around Brooks's lab, worked on Allen, and eventually was to become a senior vice president of Brooks's iRobot Company. Allen had cheap off-the-shelf sensors (sonar finders adopted from Polaroid cameras, for example), but its brain resided in an attached Lisp machine (by cable; reliable wireless connections were years away).

As Brooks thought about how to coordinate the sensory data with the motor processes, he imagined what he called a "cognition box," the heart of thinking and intelligence. And then, in a bold insight, "The best way to build this box, I decided, was to eliminate it. No cognition. Just sensing and action. That is all I would build, and completely leave out what traditionally was thought of as the *intelligence* of an artificial intelligence" (Brooks, 2002).

It was a bet that went in the opposite direction from most of what AI had done so far. Intelligence, Brooks writes, was for his colleagues "best characterized as the things that highly educated male scientists found challenging"—chess, symbolic integration, proving mathematical theorems, and solving complicated word algebra problems. "The things that children of four or five years could do effortlessly, such as visually distinguishing between a coffee cup and a chair, or walking around on two legs, or finding their way from their bedroom to the living room were not thought of as activities requiring intelligence. Nor were any aesthetic judgments included in the repertoire of intelligence-based skills."[16]

---

[16] I'd remarked on this myself, in a draft of *Machines Who Think*, but I phrased it so tendentiously that the editor strongly suggested I take it out. I did. *Eppur si muove*, as Galileo did or did not say.

Herbert A. Simon
(Herbert Simon collection)

This contrarian bet derived from Brooks's own research heuristic. "I would look at how everyone else was tackling a certain problem and find the core central thing that they all agreed on so much that they never even talked about it. Then I would negate the central implicit belief and see where it led." He knew that insects could do much more than any robot that had yet been built, though they had nearly no cognitive processing. Brooks eventually saw that by getting the robot to react to its sensors quickly, it didn't need to construct and maintain a detailed internal model of the world. In the natural world, complexity built on top of simplicity, and that would be his way, too, the layers to be added one after the other, emulating the historical process of evolution.

Allen's first public performance (by video in 1985 at a robotics conference in France) was met with disbelief. Okay, it was doing things other robots hadn't been able to do, moving briskly, centering itself in the middle of the hallway, evading people in its way, discovering the exit when it was surrounded on three sides by people, but it was doing it in such a simple-minded way! Clever engineering, but surely not intelligence.

Brooks was nevertheless convinced he was on the right track. Next came the robot Herbert (for Herb Simon), equipped with an arm, and programmed to find and collect the empty soda cans around the lab. Herbert was followed by Genghis (able to walk over anything in its path), again inspired by insect behavior, with six legs, and the ability to get up when it fell down, as its sensors drew it to the invisible infrared glow of every warm mammal, its "prey."

From time to time, Brooks had been sitting in on the artificial life symposia at the Santa Fe Institute, whose dictum was that complexity arose out of simplicity, and this suited him perfectly: Each robot he built had very simple rules that led to complex results. "The software itself was certainly not profound. It was rather straightforward, in fact. The software's behavior, however, was profound." Genghis appeared to have intentions, but no intentions were internally represented. It all raised some wonderful philosophical questions.

Brooks describes what happened next as "a Cambrian explosion" of robots—the group built robots with different bodies and different capabilities, some offering tours of the lab to strangers (anybody who stopped in the middle of the hall), and some that cooperated without explicit communication, much as social insects and even birds in flight do (the Nerd Herd). In 1989, Brooks and his colleague Anita Flynn published a paper, "Fast, Cheap, and out of Control," which became an Internet slogan (Brooks and Flynn, 1989), and another part of the public face of artificial intelligence. On the strength of their experience with Genghis, Brooks and Flynn were arguing that instead of sending one thousand-kilogram robot to Mars, we'd be better off sending a hundred one-kilogram rovers. A small rover could accomplish much of what a large rover could; it would be cheaper and faster to develop, cheaper to transport to Mars; and redundancy would allow ground controllers to send a single rover off on a path that seemed attractive, but possibly dangerous.[17]

---

[17] Doug Lenat once made a similar discovery playing many hours of a war game. Hundreds of agile two-man vessels were more likely to win a war than a score of ponderous, huge vessels. I mentioned this finding to an admiral who was Chairman of the US Joint Chiefs of Staff at the time, and he roared with laughter. Maybe so, he conceded, but who'd want to be captain of a two-man vessel instead of a major battleship? You couldn't keep professionals in the Navy.

Hans Moravec

Though simulations suggested that a fast, cheap, and out of control minirobot would work beautifully for extraterrestrial exploration, it took some convincing (and waiting) to get such a robot out of earth orbit; eventually, on July 4, 1997, *Sojourner* rolled out of its bay and onto the surface of planet Mars, an autonomous robot explorer.

Brooks went on to found a robotics firm whose most public successes at the moment are a toy in collaboration with the Sony Corporation, the AIBO dog, an autonomous robot creature that appeals to very rich uncles who wish to indulge very lucky nephews and nieces[18], and the "Roomba," a relatively cheap, but effective, home vacuum cleaner that roams a room sucking up dust, and evading table legs, pets, and other obstacles. Less publicly, iRobot's vehicles seek landmines, negotiating through real mine fields, without human risk. Brooks's "simple" robots also helped search the World Trade Center for victims of the 9/11 attack. Other robot companies have followed this lead, making robot lawnmowers, for example.

---

[18] As I see this for the first time, I'm standing in F.A.O. Schwartz on New York's Fifth Avenue, listening to a toyshop clerk tell me that AIBO uses "a combination of neural nets, machine learning, and some heuristics." A *toyshop* clerk. I'm speechless.

The Mind of a Robot

Was Hans Moravec, Brooks's former fellow-graduate student, standing by idly while Brooks and his team were filling the world with sassy robots? No, he was not. After receiving his PhD, he'd moved from Stanford to Carnegie Mellon in 1980, to become a researcher at the Mobile Robot Laboratory of Carnegie's Robotics Institute. Moravec and his team adopted what they thought was useful from Brooks's robot Allen (for example, the sonar range sensors Allen was equipped with to avoid obstacles), but Moravec wanted his robot to do more than just avoid collisions: He wanted those sonar range sensors to build a map of the surroundings that could direct the robot's navigation. Which, after some fancy programming footwork, is what they did.

Robots were designed that employed the best of both approaches: they had control systems that maintained two-dimensional representations of their environment based on the data continuously provided by their sensors, representations they could use to plan future moves. These robots could answer questions like, "Where are you?" or "Why did you do that?" with answers like, "I'm in an area of about 20 square meters, bounded on 3 sides, and there are 3 small objects in front of me," or "I turned right because I didn't think I could fit through the opening on the left." Replies were both verbal and pictorial; Moravec would think of them as a direct window into the robot's mind. "In these internal models of the world I see the beginnings of awareness in the minds of our machines—an awareness I believe will evolve into consciousness comparable with that of human beings" (Moravec, 1988).

Moravec's 1988 book, *Mind Children*, also became part of the public face of AI in the 1980s. He made some audacious projections, for instance, that by 2030, computers would have the capacity for human-level intelligence (he has since allowed it might be as late as 2050 in his 1998 update on the topic, *Robot: Mere Machine to Transcendent Mind*) (Moravec, 1988). Based on projections from past progress, he argued that evolution, which had allowed millions of years between significant changes, was happening in machines in only decades. He also elaborated on the ideas he'd expressed as a graduate student, that we would find a kind of individual immortality by transferring our minds into much hardier machines, although now the plans for that transfer were a little more concrete. An historian might have some difficulty reconciling Moravec's intriguing speculations with the simultaneous fact of DARPA's guillotining of AI at the end of the 1980s. To be sure, some of his more conservative colleagues just rolled their eyes and groaned, but it all seemed to me part of the texture and appealing verve of AI as we've come to know it.

Robots Everywhere

The robot tribes would grow wondrously in the next years. Small, insect-like robots were developed alongside big guys designed to withstand hazardous territory (underseas, deep space). They came singly and they came in swarms, they slithered, rocked, rolled, galloped, lurched, and floated (in the form of "smart dust") out of the US, Japan, Germany, France, Iran (Sharif University's robots won a number of international competitions), Brazil, Israel, and other places besides. Robot vehicles drove themselves across the US and made road trips in other countries. At Xerox PARC, small morphing robots changed their shapes depending on the task they had to tackle; other, more stable, robots donned black tie and jacket to serve canapés at cocktail parties, or a nurse's cap to assist the elderly; some limbered up and played in soccer tournaments.

Moravec himself took advantage of vastly increasing computational power (for a given price, he calculates, power roughly doubled annually in the 1990s)[19]. He began to develop three-dimensional mapping capability inside his robots that yielded "dense, almost photorealistic 3D maps of their surroundings. Navigation techniques built around this core spatial awareness

---

[19] AI research has benefited greatly by increasing computer power, but those computers now used in such research are still in the midrange of power. Very high-end computing remains the territory of military applications, finance, and computational scientists (astrophysicists, physicists, and earth scientists, for example) who use them to model large-scale natural phenomena.

will suffice, I believe, to guide mobile utility robots reliably through unfamiliar surroundings, suiting them for jobs in hundreds of thousands of industrial locations and eventually hundreds of millions of homes. Such abilities have so long eluded [us] that only a few dozen small research groups pursue them" (Moravec, 2000).

True enough. In earthquake- and hurricane-prone Acapulco, I watched the 2003 robot urban search-and-rescue competition. I'd visited Carnegie's Robotics Institute and read widely in the robotics literature: I was hoping to see the robots race into the urban catastrophe course undaunted, administer CPR (and maybe a tot of brandy), then drag the crash-dummy victims out by the scruff of their necks. This was not how it was. Mirrors and transparent walls often befuddled the competitors (not to mention dangling pipes, wires, and overturned chairs); sometimes they just quit for no apparent reason at all. Robotics is difficult business. If I'm unlucky enough to get caught in an urban disaster and they can't send in the elite robots from some of those few dozen small research groups Moravec mentions, I hope they send in old-fashioned furry canines instead.

In fairness, I must add that the urban catastrophe search-and-rescue competition is a relatively new one, and the challenges are formidable. The goal is to locate victims, allowing humans to rescue them, and this is how mobile robots were used in the World Trade Center disaster, where they could penetrate to places that humans could not. Both tethered and wireless robots routinely penetrated 20 to 45 feet into the rubble pile versus the 8-foot limits of traditional search cameras. They were used to determine whether voids were safe for human penetration, or worth rubble removal for further investigation, and were responsible for finding at least five victims.

Robots have taken on some bizarre shapes and sometimes no shape at all. Self-reconfiguring robots, at an early stage of research in several laboratories, are small modular structures that autonomously organize and reorganize their shape depending on the task they're deployed at and the terrain they must travel. "Smart dust" is under development (and has even been deployed) in several kinds of tasks—academic, military, and industrial. Inside a cubic millimeter of skin, about the size of a grain of sand, a complete sensor and communications system is able to report on its environment—useful for applications as mundane as inventory control and meteorology to systems as important and immediate as keeping the Super Bowl safe from terrorists (Garreau).

Robots 'R Us

The increasing power of computation in the 1990s, plus an accumulating body of knowledge, permitted some robotics researchers to turn their attention to socially interactive robots, robots that helped and were at home with people and each other.

Cynthia Breazeal at MIT, one of the leading researchers in social robots, and formerly a student of Brooks, has suggested that how effective such robots are depends on how well they support the social model ascribed to them (a good museum guide, but not necessarily a good taxi driver in addition), and on how complex the interaction might be. This suggests a continuum of robot social behavior. Simplest are robots that are socially evocative, meaning that they rely on our tendency to anthropomorphize.[20] More complicated robots have a natural-seeming presence that employs human-like social cues and communication; and yet more complicated ones must be socially receptive, that is, able to imitate human behavior appropriately. A further complexity is sociability, the ability to initiate engagements with humans to satisfy internal social aims (drives, emotions, etc.). Other researchers add that successful social robots must also be socially situated (surrounded by a social environment that they perceive and react to); socially embedded (they can interact with their en-

---

[20] Breazeal's robots fall into the first few categories, but Manuela Veloso's soccer robots, which have nearly no social presence, evoke passionate cheers from human spectators with their wired-in "victory dance" when they win at robotic soccer.

vironment, with other agents and with humans, as in turn-taking); and socially intelligent, based on deep models of human cognition and social competence (Fong et al., 2003).

Certain socially interactive robots (mainly at the simpler end of the socially interactive spectrum) have been designed and implemented, notably Breazeal's own work with Kismet, a bug-eyed expressive creature with surgical tubing lips and big pink ears. Breazeal intentionally designed Kismet to imitate the kinds of interactions a human and a baby might have. "My insight for Kismet," she told Claudia Dreifus of the *New York Times*, "was that human babies learn because adults treat them as social creatures who can learn; also babies are raised in a friendly environment with people. I hoped that if I built an expressive robot that responded to people, they might treat it in a similar way to babies, and the robot would learn from that. So if you spoke to Kismet in a praising tone, it would smile and perk up. If you spoke to it in a scolding tone, it was designed to frown" (Dreifus, 2003). As her advisor, Rodney Brooks, wrote: "Kismet is alive. Or may as well be. People treat it that way." Kismet was retired to the MIT Museum and, incorporating what she'd learned from human-robot social interaction, Breazeal has moved on to her next project, Leonardo, which has arms, a torso, legs, and skin, with even more humanoid facial expressions and gestures.

Are these illusions at all plausible? Ed Feigenbaum reports that on one of his recent periodic visits to Japan, he was dismayed by a schedule at the NEC Laboratories that gave him an entire hour with its humanoid robot, PaPeRo, which stands for Partner-Pet-Robot. He emailed me: "When I saw they booked me for an hour with PaPeRo, I thought I would be bored to tears after twenty minutes. But that was not the case. The experience was riveting, fascinating, attention-grabbing. How could that be? PaPeRo was just a bunch of plastic and electronics running around the floor, and I knew how it worked, so there was no magic. So here's the philosophical and technical slant, my take-away lesson. Maybe 'being human' is simply knowing how to do the thousands of 'little things' that constitute the behavior of human interaction. Maybe the achievement of the NEC engineers was brilliant knowl-

Nursebot Pearl

edge engineering to capture the first thousand of those things and put them into an expert system that is expert at 'being a playful loveable little child'."[21]

The Geriatric Robot Revisited

Earlier, I reported on my public scourging, punishment for having playfully suggested the geriatric robot. "The geriatric robot sounds like *it's* the one that's old," objected Sarah Kiesler, a social psychologist who works on a Carnegie Mellon robot known as Nursebot Pearl, designed to aid the elderly at home or in nursing homes. She's right, but then she couldn't defend Nursebot Pearl's name either. Pearl had replaced an earlier robot, Flo, for Florence Nightingale, and Kiesler was hoping the newest version would be named Clara, for another famous pioneering nurse, but her students and research team overruled her, and Nursebot Pearl it is.

Nursebot Pearl illustrates some of the surprising difficulties that social robots carry along with them. Kiesler and her team (of roboticists, nursing faculty, gerontologists, and others from

---

[21] PaPeRo did not become a consumer product and has been discontinued, a victim of the long Japanese economic crisis, but that doesn't diminish its achievement.

Carnegie, the University of Pittsburgh Nursing School, and the University of Michigan) had originally been inspired by roboticist Sebastian Thrun's disappointment that his beloved grandmother was sent to a nursing home from her familiar surroundings because she needed more help than she could have at home. With a little unintrusive help, Thrun reasoned, his grandmother might have remained at home another two or three years. Thus, the Nursebot project was born, raising technological challenges, but offering social ones, too.

Pearl is a fusion of many AI technologies—speech understanding, computer vision, dialogue management, reasoning under uncertainty, embedded sensors, mobility, flexibility, planning, scheduling, and so forth. Her[22] whole purpose is to help older people stay in their own homes as much as several years longer than they might otherwise be able to. She's an intelligent reminder, navigation aid, and general assistant. She has to be intelligent enough to work in a physically uncertain environment, (noise and surprise obstacles), but she must operate in a psychologically uncertain environment, too. She must understand an older person's speech, and she must allow for the fact that many older people cannot express their wishes clearly or unambiguously. She has to know when she doesn't understand, and figure out what's really needed. Many humans can't do that.

Nursebot Pearl reminds her patients to go to the bathroom or take medicine, or that a favorite TV show is about to begin. She's her patient's arms and legs, when they can't move because of illness—her intelligence is designed to integrate with that of her patients, so she can turn appliances, including the TV, off and on. She can lead her patients, acting as a kind of smart walker, or she can follow them as they move around the house. Patients communicate with her by voice or touchscreen. Nursebot Pearl is even tentatively being programmed to "take over certain social functions," in other words, to be company for the lonely elderly who miss social contacts.

---

[22] Nursebot Pearl is feminine, for many reasons.

Meanwhile, Nursebot Pearl is also a mediator with the world outside, a way of allowing elderly patients to interact from home with their physicians and other specialists. Just like a human nurse, she collects data and monitors her patient, heading off emergencies by assessing the information she collects, and informing others, like a physician or the family. She takes responsibility and acts.

To accomplish all this isn't just a technological task, though it's certainly that as well. Nursebot Pearl must also inspire confidence, even comfort, in her users. The Nursebot's creators are wary that she not be so human-like that her clients form inappropriate attachments to her, a particularly Western fear, and maybe one of the reasons why Pearl doesn't look very humanoid. Western culture is full of lurid stories about robots seizing power and destroying their human creators. The Japanese, on the contrary, feel warm and affectionate toward robots, because their cultural stories are all about friendly, helpful ones.[23] Their real-life robots are not only often humanoid looking, but are built explicitly to help, entertain, and keep humans company, and no one thinks that's at all inappropriate. The Japanese don't welcome immigrants to refresh their less-than-zero population growth, so they face practical problems of looking after a large elderly population, which makes a robot an ideal solution.

Having pointed out this difference between Western and Japanese attitudes, I must report that much research is going into Nursebot Pearl's face—she will eventually have a repertoire of facial expressions and head gestures that signify happiness, unhappiness, approval, disapproval, puzzlement, and so forth. Carnegie Mellon's research team has discovered that if the Nursebot is too machine-like, her human clients ignore her, and won't exercise or take their pills. Like Kismet's, these facial and head gestures are tapping into millions of years of human evolution, the hardwiring that compels us to pay attention to other humans, the fact of sociability.

---

[23] True enough, but a vast oversimplification. The *ningyo* tradition of robot-like puppets in Japan is, in fact, far more psychologically shaded. For the complex cultural role these puppets have played over history, see *Puppets of Nostalgia* by Jane Marie Law. Princeton, NJ: (Princeton University Press, 1997.)

Nursebot Pearl is under development to answer a real need with real expertise. For example, expert caregivers know the difference between pushing an individual patient to do too much, raising constant frustration, and doing too much on the patient's behalf, thereby infantilizing him—a delicate, ongoing assessment that escapes unskilled or mediocre human caregivers. Technologically speaking, the Nursebot's immediate forebear is a robot that guides people through a museum, so much of the technology is already well-developed. A cousin robot was, at the time I visited Carnegie, mapping abandoned coalmines in the state of Pennsylvania.

The elders who've alpha- and beta-tested the Nursebot like her just fine. Why not? Enormous care is going into her design and development. Is eldercare a job that ought to be done by humans? Maybe, if they're well trained and experienced, but the realities of world demographics say it isn't going to happen. In this case, better to have skilled and reliable robotic care than nothing, or dismally unskilled human care. I've heard informally that the upkeep of the early Nursebot has made at least one site abandon the use of robots and go back to minimum wage human workers. It happens that I'm old enough to remember when everybody said machine translation would never happen because, first, it couldn't be done, and second, human translators were cheaper. I argued then that this only said human translators were underpaid; I didn't even reckon on the immense amounts of work that would soon face translators of any kind. Even at minimum wage, we couldn't afford to pay humans to translate as needed these days. This doesn't mean just the European Union, generating hundreds of documents a week which must, by statute, be available in 11 languages, but it's also for ordinary people: log on to a foreign-language website via Google and let Google's automatic translator do its job. It's far from perfect, but it's very useful. I cannot say whether the linguist Yehoshua Bar-Hillel, foremost among those who, 40 years ago, said it could never be done, would be delighted or chagrined (McCorduck, 1979). In any case, sheer demographics suggest a similar necessity for automated eldercare eventually.

Robot Competitions

In 1992, the AAAI was persuaded to host an "AI-centric" mobile robotics competition, which grew to be the most enduring robotics competition yet (Balch and Yanco, 2002). The contest organizers faced the interesting problem of devising challenges that were worthy of world-class research, and yet not so difficult that no robot could complete the task. The competition grew to include multiple events, some robotics research better addressed to certain tasks, some better at others, including navigation and manipulation, and finally, social interaction.

A number of robotics achievements were first unveiled at the annual competition, and competitors everywhere were inspired to enter. In the earliest days, most competitors had come from among the usual suspects—Carnegie Mellon, Stanford, Georgia Tech, and MIT. In later years, undergraduates competed with graduate researchers, and their home institutions ranged from major research universities to teaching colleges to high schools. As top prizes went alternately to "reactive" sensor-laden robots, and then to robots that relied on internal representations of the world, and planning, the competitions defused some of the religious fervor that had threatened to divide these two schools of thought, suggesting that each approach will play a role in the successful robots of the future.

In 1995, all robots that competed in the events accomplished the tasks, and so by 1996, competition tasks could be less structured, more dynamic. By 1997, the Hors d'Oeuvres, Anyone? competition drew robots that served food during the cocktail hour of the conference banquet. When competition organizers worried that challenges were being met by integrating proven technologies rather than inventing new technologies, a harder goal was set: a robot that can attend the national meeting and present a talk about itself. It hasn't quite happened yet, though in 2002, a robot called CEREBUS from Northwestern University would, like certain boors at cocktail parties, give an interactive talk about itself to passersby.[24] Yet another competition, which I'll say more about later, the RoboCup, engages scores of competitors and thousands of spectators. Competitions get the juice going, inspire younger researchers to take a crack, and showcase research. They can be enormous fun or tedious beyond the telling, but they are here to stay.

A Reconciliation with Fuzzy Logic

Another technique emerging at the end of the 1980s was fuzzy logic. It had actually been invented in the mid-1960s by Lotfi Zadeh at UC Berkeley to be a representation and calculus for dealing with partial, vague, or uncertain concepts. Fuzzy logic is a kind of Boolean logic, dealing with that murky—well, fuzzy— area between the clearly true and the clearly false.

Zadeh had originally intended it to solve problems in the soft sciences, especially those that involved interactions between humans, or between humans and machines. For many years, fuzzy logic seemed somewhat distant from classical AI. It may not be fair to say that the AI community in the US snubbed Zadeh, but he was not routinely invited to give talks at AI meetings, and AI practitioners in the US expressed puzzlement as to how his oddly named fuzzy logic had anything to do with them. Even those AI researchers predisposed to formal logical systems were lukewarm to the idea. It must surely have been painful for him to be treated so negligently, quite without honor in his own country.

In contrast, in Europe and Japan especially, fuzzy systems had become very prominent by the late 1980s and early 1990s, found

---

[24] CEREBUS has clever semantic mechanisms to bring the conversational topic back to itself. When its human interlocutor changes the subject, it replies with transitional cues such as "but first ..." or "getting back to ..." which brings the discussion back to itself. No! you cry, we've certainly heard *those* before. Do we really need artificial examples? (I will have my little joke. CEREBUS is, in fact, an admirable accomplishment and won the Nils Nilsson Prize in 2002 for the integration of its technologies.)

in intelligent robotics, speech and image understanding, control systems, and in expert systems. One Japanese researcher suggests that fuzzy logic is sweetly consonant with important precepts of Japanese culture (human nature is vague, and thus all human concepts belong partially to contradictory sets). Perhaps so; certainly the Japanese took up fuzzy systems with great enthusiasm and success. Applications in Japan alone run into the hundreds, including robot control, camera aiming for televised sports events, optimized planning of bus timetables, a prediction system for early recognition of earthquakes, medical technology, backlight control for camcorders, and so on. Zadeh received the Honda Prize in 1989, and has collected many other honors in both Japan and Europe. He was known to treat his marginalization by the AI community in the US as a bittersweet good-news, bad-news joke: The good news is that AI works; the bad news is that it's fuzzy logic.

But this changed. Zadeh was formally welcomed into the artificial intelligence community in 1998 when he received first, the Edward Feigenbaum Medal from the International Society for Intelligent Systems, and the same year, the Information Science Award from the Association for Intelligent Machinery. In the late 1990s, his research continued apace on "computing with words, and manipulating perceptions" as he put it in an award-winning paper. In 2000, he won the Allen Newell Award from the Association for Computing Machinery for seminal contributions to AI.

Collaborative Intelligence

Earlier, we saw that the practice of artificial intelligence had begun to change in the 1980s as researchers came to believe that intelligent behavior was not just the property of a lone agent, but instead a reciprocal relationship among agents (some equal, some not, all requiring training); knowledge from many sources; and a way of transforming that relationship into a process of symbolic reasoning toward a goal. (In 1977, Simon had proposed a characterization of "understanding" as a relation among three elements: a system, one or more bodies of knowledge, and a set of tasks the system is ex-

Daniel Bobrow
(Courtesy of the Palo Research Center,
Brian Tramontana, photographer)

pected to perform (Simon, 1977), but this was somewhat oblique compared to the notion that was gaining a wider view in AI in the 1980s.)

Two AAAI presidents made this general topic the center of their presidential address in the early 1990s, though each from a slightly different point of view. In his 1990 presidential address, Daniel Bobrow directly addressed the idea of intelligence as a collective effort. Ten years earlier, he recalled, Allen Newell had presented what was then the dominant AI paradigm, symbolic reasoning programs executed by an agent that would act rationally to achieve a goal. Knowledge was fixed; the goal was specified. "The agent was disconnected from the world with neither sensors nor effectors, and more importantly with no connection to other intelligent goal-driven agents. Research results in AI consisted primarily in the determination of the principles of construction of such intelligent, but deaf, blind, and paraplegic agents." Humans would formulate the problem in a previously defined language that the artificial agent understood, presenting a problem statement to it that included background knowledge, a description of the state of some world, operators to use in that world, and a description of a desired state (a goal). The agent would reason logically to determine a sequence of operations to achieve the desired goal, and a more complete description of the resulting desired state (Bobrow, 1991).

A useful and still relevant model, Bobrow said, but now comes increasing recognition of how important it is to design and analyze systems that do not act in isolation. He proposed a next step, active agents (human and machine) with multiple goals, communicating among themselves, and interacting directly with the world, all of which would require communication and coordination among them, and integration with conventional systems. These capabilities were at the moment largely absent from AI programs, but necessary if intelligence among agents was to be achieved.

In traditional AI programming, computational agents used a common built-in language and vocabulary, and humans in the system were assumed to understand that. But what if a knowledge-based system attempted to use facts from two or more different knowledge bases? "To use a set of facts from another KB system, it is necessary to understand how those facts were intended to be used in a reasoning process." (This is what other researchers call its ontology, a term borrowed from philosophy.) Once those facts from different, even competing, knowledge bases were automatically transformed into arguments, reasoning at the meta-level could determine which arguments were more compelling. Then, "The full reasoning engine can be used to reason about the reasoning process itself."

Automated mediators might be added to systems to help bridge the differences between two or more agents—and also to allow more effective partitioning of tasks in a human/machine system. They could locate and bring together data from multiple sources, while they bridged semantic mismatches. Or, if unable to make that bridge, they could warn the end user explicitly. Bobrow cited as an example an expert system in use at Consolidated Edison, the electric utility that serves metropolitan New York, a densely populated 593-square-mile area. During an emergency at the operations control center, the system filters out nuisance alarms, analyzes the system state to determine which components are failing, and recommends action to the human agents, all information coming from disparate sources.

Integration of AI with standard programming is overdue, already having shown its usefulness in tasks as different from each other as the daily bake orders for the hundreds of Mrs. Fields Cookies outlets, or the warranty administration by Ford Motor Company for its network of dealerships. IBM had integrated AI technology with more conventional technology to schedule, decide, and report on a vast complex of tools and processes in a semiconductor factory, saving significant sums of money, and improving human performance as well by empowering people close to the work to make good real-time decisions. Bobrow concluded that one major challenge for AI in the decade of the '90s would be to deal with the issues of interaction between agents, human and artificial. This would call on fields such as cognitive psychology, anthropology, economics, neural modeling, and the rest of computer science. AI must cross the boundaries, he declared—to be intelligent is to be intelligent about something: success would come from tasks outside the field.

Picking up explicitly the theme of multiple agents, Barbara Grosz of Harvard also made collaborative systems the subject of her 1994 AAAI presidential address (Grosz, 1996). Though Grosz drew much of what she had to say from her own research in computational discourse, natural language front ends to data bases, and models of collaboration between humans and computers, she was also referring to the research of a larger group of AI scientists who had been exploring interactions, teamwork, and collaboration among artificial agents.

Intelligent, collaborative, problem-solving partners, she said, were an important goal for both the science of AI and its applications. To develop the underlying theories and formalizations needed to build collaborative systems would raise interesting questions and intellectual challenges across AI subfields. Perhaps more importantly, results of such research promised to have significant impact not only on computer science, but also on the general computer-using public.

Softbots, software robots, which were just beginning to make searches on the Internet more focused and less time-consuming for human queries, were one important step, but human users also needed other kinds of intelligent collaborators. Deeper problem-solving and more difficult tasks could not yet be approached automatically: systems were dumb servants rather than helpful assistants or partners. Using Newell's terminology, Grosz said, "It wouldn't be too far-fetched to say that the mouse frees users from having to tell the system what to do at, or perhaps below, the symbol level, but it fails to provide any assistance at the knowledge level. Mice and menus may be a start, but they're far from the best we can do to make computers more accessible, helpful, or user friendly."

For systems to collaborate with humans, the kinds of techniques AI has developed are necessary. But that collaboration also requires "that we look beyond individual intelligent systems to groups of intelligent systems that work together." Moreover, collaborative behavior is interesting in its own right, an important part of intelligent behavior. Even if individual intelligent behavior isn't yet completely understood, we must still try to understand collaborative behavior, for such capabilities cannot merely be patched onto a system, but must be designed in from the start.

For example, a patient arrives at a hospital with problems affecting his heart and lungs. Three specialists, a cardiologist, a pulmonary specialist and an infectious disease specialist, each provide different expertise, but they are a team of equals—no single doctor is the manager; they must come to a consensus about what to do, who will do it, and in what order. They must plan jointly. Grosz compared this with the dialogue between a user and a system in network maintenance, where the system doesn't simply follow orders, but reports subproblems which it offers to fix, or has already fixed as the problem is being solved. Shared information and trust are the hallmarks of both these collaborations.

Nature itself is full of collaborations—social insects cooperate to obtain food, build nests, and raise the next generation. They do this—and AI systems might be designed likewise—using simple individual systems that still accomplish complex goals. But collaboration is not the same as interaction. Interaction only means acting on someone or something else; collaboration is inherently working (the "labor") with ("co") others. AI needed to characterize that "jointly with" more precisely.

Why take on this challenge now? Grosz offered three reasons. First, modeling collaboration presents a variety of challenging research problems over AI subfields. "But this would be irrelevant were it not for the second reason: progress in AI over the last decade." A substantial scientific base had been established by research across most of the areas involved in meeting this challenge. Specifically, "Work in both the natural language and the DAI [Distributed Artificial Intelligence] communities has provided a range of models of collaborative, cooperative, multi-agent plans." Researchers in planning who addressed uncertainty and the pressures of resource constraints had rich ideas for the modeling problems that arise in the design of collaborative systems. The third and final reason for taking this up now was compelling applications needs. Again, collaborative plans are not simply the sum of individual plans: They must be built into systems from the beginning of their design.

At last, Grosz posed different kinds of questions: "What capabilities could we provide if systems were to collaborate rather than merely interact? What difference would they make to system performance?" (Grosz, 1996). What if AI systems were built collaboratively with other computer science systems? Would they benefit from what AI researchers knew, by exchanging (moderately) user-friendly languages for an intelligent agent that would consider what the user was trying to do and needed, rather than demanding to be told how to do what was needed? If AI could be

Barbara Grosz

helpful to the rest of computer science, even to other fields of science, so the social sciences might help AI.

AI and Molecular Biology Collaborate: A Success Story

A rich example of what Grosz proposed in 1992 would be the focus of a lecture a decade or so later by Larry Hunter, of the University of Colorado School of Health Sciences, who in August 2003 gave the first Robert Engelmore Memorial Lecture at the International Joint Conference on Artificial Intelligence in Acapulco, Mexico. A thriving collaboration between artificial intelligence and the molecular biology community had been so successful, he said, that many molecular biologists, who grasped how vital AI was to accomplishing their tasks, had taught themselves to program using AI languages and techniques. Biologists built their own gene ontologies, curated knowledge bases, and were eager consumers (and creators) of software.

For example, hidden Markov models (not an AI invention, but certainly a mathematical workhorse exploited to the utmost by AI in the 1980s and 1990s) were the main mechanism used to represent patterns in DNA and protein sequences. Information extraction from enormous data streams, another AI technique, was central to molecular biology, since interpretation of results could overwhelm human powers of inference. Understanding gene expression, the multiple genes that contribute to certain traits and even diseases, and that can change in context, required novel statistical and data management techniques, novel ways of testing and winnowing some 10,000 hypotheses at a time. In short, methods pioneered in AI could be found across the spectrum of molecular biology, from drug design to the design of novel organisms to the evolution of life. Molecular biologists used AI techniques daily for literature searches to retrieve, extract, and analyze many documents in a timely way.

If the AI researchers in his audience didn't know this, it was because the results of these projects went not into the AI literature, but directly into mainstream biology and medical literature. AI and molecular biology had enjoyed a long partnership, beginning with the earliest expert systems like MYCIN and MOLGEN, and that collaboration continues to this day. It was an example, Hunter said, "of what no human could do, nor even a group of humans, in any timely way." To really understand life, he said, we need the help of machines (Hunter, 2003).

Teamwork

It was all surely part of the Zeitgeist as the 1990s began. In the late 1980s, Allen Newell had declared that it was time for the subfields of AI to merge and create "complete intelligence agents" capable of perception, action, and cognition.

In 1992, Alan Mackworth at the University of British Columbia suggested that soccer-playing robots might be an interesting testbed for collaborative intelligence. A little later that year, a group of Japanese researchers independently suggested the robot soccer problem as a fruitful way to push AI technology, and in June 1993, researchers who included Minoru Asada, Yasuo Kuniyoshi, and Hiroaki Kitano, decided to launch a robotic competition called the Robot J-League (J-League was the name of the newly established Japanese professional soccer league). The first public announcement of the initiative came in September 1993, and regulations were soon drafted. International response was so overwhelming that they renamed the project the Robot World Cup Initiative, "RoboCup" for short, and invited teams from all over.

Manuela Veloso

Other research groups had also been working on robot soccer players independently in the early 1990s, including one headed by Manuela M. Veloso at Carnegie Mellon. Veloso points out that in chess, the physicality of the game is irrelevant—tables do not shake, pieces do not fall. But in soccer, the environment is only partially observed, the effects of the players' actions in the presence of opponents are uncertain and difficult to model, and the cycle of perception, cognition, and action must run in real time. All this produces much more complexity with which to cope than a chess match. The robot players of Veloso's 2002 team, called CMPack, were four Sony AIBO dogs, but with considerably enhanced intelligence (especially mental models) and communication among the "players." They were continually in motion and moved much more quickly than they had just a few years ago. They were capable of maneuvering around obstacles, scoring goals, and localizing themselves, all autonomously, and became RoboCup World Champions in their class in 2002 (Veloso, 2003). The University of New South Wales's rUNSWift team took it away in 2003.

In 1997, a long-range goal was publicly set: *By the mid-21ˢᵗ century, a team of fully autonomous humanoid robot soccer players shall play against the winner of the most recent World Cup, comply with the official rules of the FIFA, and win.*

To the organizers of RoboCup, a 50-year goal seemed right. It had taken 50 years from the Wright Brothers' first flight to the successful Apollo landing on the moon; it was 50 years from the invention of the digital computer to Deep Blue, which beat the human world champion in chess (I'll talk more about that later). A robot team to challenge—and beat—the best human soccer players could be accomplished in 50 years, too. Meanwhile, more modest subgoals were set. Among the first was to build both real and simulated robot soccer teams that played reasonably well with modified rules, just the way chess machine designers had begun. A later goal would be to develop a robot soccer team that played like human players. As with chess, success was clearly defined, the competition had great public appeal (far more than chess, I'd bet) and now that the chess challenge had been met, it was time for a new grand challenge.

Unlike chess, robot soccer would need to handle real-world complexities (if only the limited world of a soccer pitch). To succeed, the task would have to employ real-time sensor fusion, reactive behavior, strategizing, learning, real-time planning, multiagent systems, context recognition, vision, strategic decision making, motor control, and intelligent robot control.

It Isn't Just Man versus Machine; It's the End of a Paradigm

The first 50-year challenge, beating the world's top-ranked human chess champion, had indeed been met, though not as unambiguously as might be thought. "Yes, it won," a scientist in another field grumbled to me, echoing a perpetual theme, "but it didn't win by thinking the way humans think. You can't call *that* intelligence." No? Then what should it be called?

On the afternoon of May 11, 1997, before hundreds of spectators in New York City, an IBM machine called Deep Blue beat Garry Kasparov, the world's chess champion in a six-game match. Though my husband, Joseph Traub, had given up playing chess in his adolescence when he "discovered girls didn't play chess," he was deeply interested in this match as a computer scientist and a former chess enthusiast. And so he sat among the spectators in an auditorium where they watched by remote hookup the game being played elsewhere in the building.

As Raj Reddy had predicted, the program that won the $100,000 Fredkin Prize for being the first to beat a human world champion did not quite play like a human being.[25] Deep Blue,

---

[25] "When the Fredkin Prize for the World Chess Championship is won, it will probably be by a system that has neither the abilities nor the constraints of a human expert; neither the knowledge nor the limitations of bounded rationality. There are many paths to Nirvana." –Raj Reddy

the brainchild of IBM scientists, had specialized chess processors and extremely high processing speed. But it did not, as *The Economist* would later sniff, "simply resort to mindless number crunching" (*Economist*, 2003). In fact, Kasparov himself repudiated that view. "Yes, Deep Blue was 100 times faster [than other machines] but so what? Sheer power means little in chess because it is a mathematically near-infinite game. The only way to measure the strength of a chess-playing computer is to analyze its moves. While putting Deep Blue's six games to the test with current [2003] top programs—Deep Junior and Deep Fritz—we discovered that they consistently play better than Deep Blue. The only exception is when Deep Blue showed a stroke of genius in one game (when I suspected certain interference)" (Kasparov, 2003). IBM, of course, denied any interference in Deep Blue's playing, and as Joseph Traub remarked, serious human chess players change their strategies from game to game, so how was this different? Or, if it was human interference, it had to be somebody pretty good to outplay Garry Kasparov, said to be one of the best world champions ever. Nevertheless, Traub thought Kasparov behaved badly about his loss. He refused to shake hands with the IBM programmers, and although he lost the match, Traub thought that with his slashing style, "he won the press conference." The machine, exhibiting better sportsmanship, kept silent.[26] In any case, IBM immediately retired Deep Blue, would not agree to a rematch, and according to Kasparov, did not permit him to study the machine's games either before the match or afterward.

It was a significant milestone, the dream of Shannon, Turing, von Neumann, and countless others. By hook or by crook, machine could beat man, and Deep Blue won the $100,000 Fredkin Prize.

In February 2003, Kasparov accepted the challenge of another computer program, this time called Deep Junior, the work of two Is-

---

[26] Return if you will, to the last chapter of *Machines Who Think*, called "Forging the Gods," and rereading Adrienne Rich's poem, "Artificial Intelligence: to GPS."

raeli programmers and amateur chess buffs, Amir Ban and Shay Bushinski. In six games, Deep Junior played Kasparov to a draw. During the sixth game, Kasparov didn't seem to the spectators (in an auditorium elsewhere in the building) to be in very big trouble, but he offered a draw, which Deep Junior refused. Five moves later, Deep Junior in turn offered Kasparov a draw, and to the dismay of the spectators, who were unaware of the tremendous stress Kasparov felt, he accepted it. *Chessbase News* said, "Kasparov was feeling the intense pressure of the match, the fear of repeating his experience in game six against Deep Blue in 1997, and the unpleasant results of the previous four games. To add to his apprehension, Kasparov had completely missed Junior's clever 25.Bc1-a3 maneuver and was no longer enthusiastic about his position." "Of course I wanted to win," Kasparov told a press conference afterward, "but the top priority on my agenda today was not to lose" (*Chessbase News*, 2003). The next day, he wrote in the *Wall Street Journal*: "It was a tough battle and even though I was much better prepared than I was for Deep Blue in 1997, the match finished in a 3-3 tie."

What had changed? Kasparov credited new programming techniques combined with superior chess knowledge (though the Israeli programmers were amateur chess buffs, they had consulted a ranking chess player). And I think he was correct to say that being able to reconstruct the computer's decision-making process makes for better science. "Chess," he said, "offers the unique opportunity to match human brains and machines. We cannot do this with mathematics or literature; chess is a fascinating cognitive crossroads" (Kasparov, 2003).

Humans, staring intently into that looking glass to see who's the smartest of them all, still had the last word. "This match," said the human commentator, "is between man the chess player, and man the tool maker." Indeed. "We now know that chess-playing skill does not, in fact, equal intelligence," said *The Economist*. No, perhaps not. But then what does? Some propose the Japanese board game, *Go*, which combines pattern recognition plus logic, plus intuition acquired over many years of play. *Go*, it is claimed, shows much more clearly how *people* think. That may be so. As the case was in computer chess for so many years, nearly everyone working on the *Go* problem has day jobs, so we won't hold our breath. And we can almost bet money that if a machine *Go* champion emerges, the bars will be reset.

"You can't call *that* intelligence." No, you certainly can't, if you insist that thinking, by definition, can only take place inside the human cranium, using the methods humans have evolved over the millennia. Perhaps confusion arises because computers—symbolic processing machines—can be programmed to model more and more aspects of human thought. But they *need* not be programmed that way, and to attempt tasks that humans cannot do, they *must* be programmed differently.

I find the arguments a bit sterile, about at the level of "But is it *really* art?" Elsewhere I've said that the whole thing reminds me of Columbus bringing back the news of the Americas to old Europe (McCorduck, 1985). "It's the Indies," he insisted until the day he died. It wasn't the Indies (and they weren't Indians, even if the US government has never renamed its Bureau of Indian Affairs), but what Columbus found was something at least as interesting. Put more scientifically, the characterization Newell and Simon made long ago has yet to be improved upon: The essence of intelligent behavior is symbolic functioning, and so far as we know, two entities exhibit the capacity for it, humans and computers.

But Deep Blue's win over Kasparov in 1997, and the draw the grandmaster accepted gratefully in 2003, was something more. It was the apotheosis of the single agent, single task intelligence. It was an important accomplishment, and AI had shown it could be done, but the definition of intelligent behavior had already expanded marvelously to include collaboration with other agents, the dynamic environment, and multiple tasks in progress simultaneously.

Narrative Intelligence

In the early 1990s, another AI effort had got quietly underway that addressed some of the same issues raised by collaborative intelligence, but with a different point of view. It would come to be known as "narrative intelligence," and its beginnings remind me of the kinds of impatience and grand vision that characterized early artificial intelligence research. Although the study of narrative, especially story-telling, had been an active research area in the 1970s and 1980s, notably in the work of Roger Schank and his group at Yale University, the emphasis then had been on natural language processing. Schank soon saw that sentences could not be understood or generated in isolation, and so he and his students were also trying to elucidate the structures and processes humans needed to understand and generate stories.[27] The research, which relied on intense knowledge engineering, fell out of favor as funding for natural language processing became more focused on statistical methods, with problems that lent themselves to discrete, measurable objectives.

In the autumn of 1990, Marc Davis and Michael Travers, then graduate students at MIT's Media Lab, began meeting weekly with other students (eventually including some from other MIT departments, Harvard, and Brown University) to see what issues might be explored at the intersection between artificial intelligence and literary theory. Davis, a humanist and new to computing, "wanted to build programs that could automatically assemble short movies from archives of video data." Travers was a computer science student with an interest in literary theory, and wanted "to program software agents that could understand a simulated world, each other and themselves." Though they guessed that their respective projects had commonalities, they couldn't be sure, since the language each used was largely unintelligible to the other. If they read the core texts in each other's disciplines, then they might be able "to construct a common language and a useful discourse" (Davis and Travers, 2002).

Over several semesters, this is exactly what they did. They could identify stark differences in practice between literary theorists and computer scientists ("at the time we founded NI, mainstream artificial intelligence seemed bogged down in a view of mind based on

---

[27] I once took some stories generated by Schank's group to my beginning creative writing students at the university where I was then teaching. They scoffed: How simple these stories were! How easy to see they'd been generated by a computer! Their instructor kept a tactful silence about the similarities between the products of novice human and novice computer storytellers.

mathematical logic and objective representation"), but they also found important overlaps. Less formal forms of knowledge, such as those proposed by Minsky, Newell, Schank, and Brooks, among the earlier AI researchers, and, outside the field, the work of Daniel Dennett, George Lakoff, and others, were influential. From literary theory, they found useful ideas in Aristotle, Roland Barthes, Wolfgang Iser, and Frances Yates. They drew also on media studies, psychology, and user interface theory, especially the work of practitioners in the field such as Abbe Don, Brenda Laurel, Tim Oren, and others. Software and social computing came under study, as did constructionism in science and learning.[28]

The Narrative Intelligence Reading Group gradually refined a theory and practice of "analyzing, designing and building computational media" that paid attention to humanistic disciplines. A central thesis was that human intelligence, at least, is organized into a series of narratives, personal and social: I am a woman (with all that this implies in my culture); I've had special experiences over time that I remember happily or unhappily; I have unique interests; and, along with spiritual and emotional events, these stories form who I am, here and now. If AI were to move beyond its present view of mind as only logical (however elaborate that logic might be) and only objective, to the next level of accomplishment, the power of narrative must be incorporated into research.

It took perseverance and patience, but after four years of meeting underground (literally—they met in a Media Lab basement), the group began to affect the MIT curriculum: interdisciplinary approaches became less novel. After six years, young researchers were degreed and scattered into the field, bringing fresh perspectives to artificial intelligence research.

Phoebe Sengers at Cornell, from whom I first learned about the narrative intelligence effort, writes, "The divide-and-conquer methodologies currently used to design artificial agents result in a fragmented, depersonalized behavior, which mimics the fragmen-

---

[28] A full bibliography of the books that the narrative intelligence group used appears in (Davis and Travers, 2002).

tation and depersonalization of schizophrenia in institutional psychiatry." She is at work on an architecture for agents called the Expressivator that allows artificial agents to behave so that they show visible markers of narrative. That is, they are context-sensitive and able to negotiate a situation as new interpretations are called for; their motivations, thoughts, and feelings are clear; their behavior in one situation influences their behavior in another; and a coherent picture of all that emerges over time to the user. She is not eager to jettison the objective mechanistic point of view in AI (or in all of science); she argues only that it needs to be balanced by a commitment to narrative.

It's too early to say how successful this new point of view will be at reforming the practice of AI, never mind whether it can lift AI achievements to a higher level. Most striking to me, however, is how similar it is in spirit to the first generation of AI research, the refreshing disrespect for boundaries, the strong belief that formal logical proofs, however useful, are not the sole nor even the most important signature of human (or any other) intelligence.

Neats versus Scruffies Yet Again

Here is a dilemma. Some AI applications are so deeply complex, deployed in tasks that lie so far beyond the capacities of human intelligence in terms of speed and complexity, that only formal proofs can assure us that the answers or advice we receive are actually correct. For example, the Semantic Web, which I'll discuss later, will rely on logical proofs to verify information it offers to its users.

Victor Lesser at the University of Massachusetts, Amherst, argues that logic "provides a language for talking about and dealing with uncertainty in the environment and in our knowledge which is at the heart of AI. In some sense, AI has become one of the most formal disciplines of computer science. However, it will probably take one or two more generations of students to debug the myth that AI is ad hoc." He adds: "For me personally, the major change in AI over the last 25 years is my understanding that anything I do these days must, of necessity but also for reasons of intellectual honesty, be more formally based either in connecting my research to some formal system or in its analysis through formal techniques" (Lesser, 2003).

In addition to using formal logical systems to achieve AI goals, much of the progress of AI in the 1980s and 1990s was based on mathematical techniques drawn from fields such as probability, statistics, and decision theory. Using these techniques, AI successfully found its way into all kinds of nooks and crannies, from airline reservation systems, where pleasant humanoid voices offer information about arriving and departing flights, and even book your seat, to video games, where virtual opponents (or teammates) anticipate, understand, or cooperate with the human player, to remarkable summaries of news bulletins from disparate sources, to machine translations done on the fly by automatic translators attached to search engines like Google—a search engine which itself uses many AI techniques, and is begging AI researchers for more.

These approaches achieved results (some worked very well indeed) and their designers felt no particular obligation to be faithful to human models of intelligence. As I write, AI enjoys a Neat hegemony, people who believe that machine intelligence, at least, is best expressed in logical, even mathematical terms. Their good work lies all around them, and is in daily use. They are looking forward to the next generation of theorem provers, yet more powerful, with yet more interesting results; they borrow probabilistic techniques from statisticians to cut down search spaces, and call on automata theory and graph theory. Some AI researchers, like John McCarthy and Nils Nilsson, were always of that mind.

As his title, *What Are Intelligence? And Why?*, made clear, Randy Davis of MIT, making his 1996 AAAI presidential address, would argue that intelligence has many aspects, and four fundamental behaviors: prediction, response to change, intentional action, and reasoning. But reasoning itself had different flavors, logical, psychological (human behavior), and social (emergent behavior). AI had taken on the task of exploring the "design space" of intelligences.

Why does intelligence exist? he asked. To piece together the evolution of intelligence in our own species is extremely difficult, and we must "be wary of interpreting the results of evolution as nature's cleverness in solving a problem. It had no problem to solve; it was just trying out variations." Nature is not an efficient engineer, and often rediscovers its discoveries in different ways; nature is a satisficer, not an optimizer. It sometimes converts an organ serving one purpose into one serving a new purpose, and it is conservative: it adds new layers of solutions to old ones rather than redesigning. What's true of our anatomy may also be true of our cognitive architecture. Thus, seeking the minimalism beloved of engineers may be a futile diversion: It may not be there.

Because of intelligence's evolutionary origins, AI is more likely to resemble biology and anatomy than mathematics or physics. A task lay ahead of collecting those almost accidental mechanisms that humans use to think. If AI should begin to conceive of itself as the study of the whole design space of intelligences, it might discover a way of synthesizing the logical and the psychological perspectives, the elusive fusion of the Neats and the Scruffies (Davis, 1998).

Kathy McKeown heads a highly successful natural language processing project at Columbia University called "Newsblaster," which crawls web sites for news bulletins in multiple languages, and synthesizes and summarizes those news bulletins into a single presentation which can be customized according to a user's needs. Though it is not yet a commercial project, it has been carefully scrutinized and is widely admired by professional editors. It has even been adapted (not effortlessly, for some issues of implementation here are much more formidable) to summarize medical texts. As McKeown says, it uses off-the-shelf parsers and relatively shallow techniques and still achieves extremely useful results. But, she says, statistics alone will not work to achieve really human-level performance.

Or consider Takeo Kanade at Carnegie Mellon, one of the world's foremost experts on computer vision. In his keynote address to the International Joint Conference on Artificial Intelligence in 2003, he summarized the rocky relationship between AI and vision. We got married too early, he joked, so no wonder we got an early divorce. After the divorce, researchers decided computer vision was a geometry problem. And once the geometry problem was solved, computer vision became an optics problem. Then an algebra problem. Then a statistical problem. Then an engineering problem, with no need for heuristics. But, he asked, what does understanding and recognition entail? Vision is certainly grounded in all of the problems above, but it is finally a problem of what the viewer *knows*. "It's time for us to remarry," he concluded, laughing.

Something more is needed to achieve human-level performance (and beyond). "Computer scientists, with their typical backgrounds from logic and mathematics, enjoy creating elegant and powerful reasoning methods," Ed Feigenbaum wrote in 2003 (Feigenbaum, 2003). "But the importance of these methods pales in comparison with the importance of the body of domain knowledge—the artifact's knowledge base." He illustrates the point: "As I write, I can look out my window and see the Stanford Mathematics Department on my left and the Stanford Medical School on my right. In the Math building are some of the most powerful reasoners in the world, at least one of whom has won the Fields Medal. If I were, at this moment, to have a threatening medical event, I would ask my CS colleagues to rush me to the Medical School, not the Math Department." The Medical School is home not to elaborate reasoning procedures, but to scruffy knowledge, applied with fairly simple reasoning tools, which would save the day. As for those elaborate reasoning tools, he added, we have "an overabundance of such methods" yet to be tested properly or integrated into AI systems.

A determinedly Scruffy enterprise is Cyc (short for Encyclopedia), a 20-year effort directed by Doug Lenat, who calls it "the human memome project." Cyc was originally intended to be a world encyclopedia of common sense, real-world knowledge, and indeed, by now it carries in its knowledge base several million assertions, of the form:"Birds fly"; "Cats do not fly"; "Dead birds do not fly"; and "A small number of birds do not fly: They are ostriches, emus, and ..." Common-sense statements like these are found in its core knowledge base, and linked to that core knowledge base are highly specialized knowledge bases for applications that the company has developed for its various patrons. Cyc is a stand-alone firm, living on contracts from DARPA, other military agencies, and corporations that use its knowledge base for proprietary purposes.

OpenCyc, however, is publicly available, and consists of a growing subset of Cyc's knowledge base to be used by anyone (though it will always lag the larger private knowledge base by years). Lenat claims Cyc now knows enough to help with the knowledge-entry process, and there are plans to allow the public to contribute to that knowledge. Public contributions must first be logically consistent with what Cyc already knows, or Cyc will push those contributions aside. Those that pass that test will then be vetted by a human committee. No one knows what rough beast will slouch out of Cyc's Austin, Texas headquarters one of these days. Even Lenat himself won't predict what happens when Cyc reaches or surpasses the 100 million things a typical human knows about the world, a point he expects to cross in about five years (Anthes, 2002).

Ah, But Creativity

At the turn of the millennium, Bruce Buchanan, now University Professor Emeritus at the University of Pittsburgh, assumed the presidency of the AAAI, and took the occasion to examine another of the last human redoubts: creativity. Trained as a philosopher, Buchanan began by citing the opinion of the contemporary *Encyclopedia of the Philosophy of Mind*, that if creativity is a human process that cannot be described mechanistically, then human minds cannot be symbol-manipulation machines—a direct challenge to a fundamental working assumption of AI.

On the contrary, Buchanan said. From an experimentalist's point of view, AI is the perfect medium for understanding creativity because implementing ideas in computer programs gives us the means to test these ideas. No rigorous definition of creativity exists, either in common usage or in psychology. In fact, a strong school of thought in psychology says that Freud was right: creativity cannot finally be understood.

But AI, said Buchanan, has always been somewhat more optimistic about the problem. The field has been bold about examining and perhaps trying to define creativity since the late 1950s.

Merely breaking the rules, or having great control of much knowledge, has been no guarantee of creativity (in fact, the latter sometimes impedes fresh thinking—a strongly held tenet of DARPA, which insists on relatively brief tenures of its scientific program managers for exactly that reason), though an acquired set of skills and knowledge must let us see the "creative" work not as mere accident, but a departure from a solid base (Buchanan, 2001).

The results of successive studies had eventually allowed psychologist Margaret Boden to propose a working definition of creativity, that it involves generating ideas that are both novel and valuable. "Novel" and "valuable" are elastic concepts, and can be defined narrowly (a few instances per century) or widely (the geniuses who founded our startup). In 1995, Schank and Cleary had argued that small acts of creativity, though they differ in scope, are not different in kind from the brilliant leaps of an Einstein.

~

Did creative AI programs exist? Yes, beginning with Newell, Shaw, and Simon's Logic Theorist, which had discovered a novel proof in logic, and spanning poetry, music, art, historical science, chemistry, and mathematics. Buchanan's discussion of Harold Cohen's program AARON (which I have also written about (McCorduck, 1990)) was particularly interesting, for it raised further questions about the nature of creativity. AARON, the result of nearly four decades of work by Cohen (who started his career not as a programmer, but as a painter, with a significant reputation in the art world), autonomously produces paintings that are both novel and valuable every time it puts brush to paper. But as Cohen points out, even though each is unique (thanks to the combinatorial space AARON moves in), not one of them is based on prior experience from previous drawings or even a previous history, in a way that Cohen believes creative visual artists are influenced.[29]

---

[29] That was then. In September 2003, Cohen admitted that he was starting on the archival memory problem for AARON. "Don't mean to suggest I know what I'm doing and I'm a bit surprised to find myself doing it at all. (Of course, it may not last!) I thought I'd finally whacked the autonomy/creativity problem by the simplest of strategies: ignoring it and

On the other hand, AARON has taught its creator a few things, especially since Cohen began to tackle the problem of coloring in the early 1990s. I've written that Cohen is a gifted colorist, but in a different way, so is AARON (Cohen, 2003a). I'll say more about this later.

Other creative programs Buchanan mentioned included Doug Lenat's AM, that had discovered concepts and conjectures in number theory, and his EURISKO, that had discovered heuristics for playing and winning games; Pat Langley and his group (that included Herb Simon) had rediscovered important findings in the history of chemistry and physics.

Buchanan himself had participated in developing one of the most celebrated of scientific discovery programs, Meta-DENDRAL, which produced the first scientific results new and useful enough to be published in the refereed molecular biology literature. Meta-DENDRAL was so successful that, along with MYCIN and MOLGEN, it laid the foundation for the enduring collaboration between AI and molecular biology described earlier.

What these programs had in common was first, the assumption that creativity is another facet of cognitive activity that *can* be explained. Then, they fell into four classes: (1) combinatorial, (2) heuristic or guided search, (3) transformational, and (4) layered search, which he examined in some detail.

Research (and experience) had shown that combinatorial search (1) can be made more effective using heuristics, (2) (indeed, Newell, Shaw, and Simon had suggested that heuristic search might well be sufficient for explaining creative behavior) but such search could also be guided by other methods, including the neo-Darwinian approaches of genetic algorithms, or (3), the addition of analogies (adapting old patterns to new situations) and other heuristics for knowing when to keep alive seemingly useless hypotheses.

---

recasting AARON as my assistant-with-no-pretensions. And I've certainly been enjoying a remarkably productive run as a result: my studio's full of new work .... But some things just don't go away. The rationale is simply that new capabilities always (seem to) spawn new ideas about what to do with them and a few new ideas never hurt. And anyway, it's clear that further growth of autonomy isn't possible without an archival memory. And anyway, it's about bloody time: I've been yakking about it for years."

Layered search, (4), perhaps more subtle than the first three, refers to higher levels of search, ways of modifying the generator of possible solutions so that the framework within which search is conducted itself changes, an idea originally implied in John McCarthy's early paper about the advice-taker, "Programs with Common Sense" (McCarthy, 1958) and demonstrated in Randy Davis's 1976 PhD thesis (Davis and Buchanan, 1977). Arthur Samuel had also used layered search in his checker-playing program (a *tour de force*, Buchanan commented, which no one would come close to for at least 25 years).

Buchanan, his colleagues, and other AI and psychology researchers had devoted significant effort to define what metalevel reasoning meant to the improved performance of a problem solver or a learning program. Gradually, domain-independent criteria had emerged for say, what made a conjecture in science interesting. "For example, singularities and exceptions are interesting, and attributes that have a great deal of explanatory power are also interesting."

What, then, had been learned about creating creative programs? Knowledge, skill, and prior experience counted significantly; the ability to modify the ontology (the logical structure for organizing knowledge), the vocabulary, and the criteria that are used; motivation, persistence, and time on task; the ability to learn from prior experience and adapt old solutions to new problems; and finally, the ability to define new problems for oneself. Some of these issues were understood better than others, but Buchanan was optimistic that more creative programs were on the way, especially given the understanding AI researchers had achieved over the last 50 years in models of creativity that stressed heuristic search. For one thing, computing power that exceeded earlier amounts by orders of magnitude was now available.

"It is not unthinkable to solve the same problem in thousands of different ways to find the best framework and assumptions empirically even when we do not know enough to choose the right ontology, evaluation criteria, methods and so on at the onset. To achieve more creativity in problem solving, however, it is also important that we can reason analogically and carry on a search at the metalevel." Today's programs, however, could not accumulate experience, and thus, could not reason about it; they worked within fixed frameworks, including fixed assumptions, methods, and criteria of success; and they lacked the means to transfer concepts and methods from one program to another. Let them accumulate experience, have the capacity to reflect on a problem so as to be able to shift its representation (which might change a hard problem into an easy one, for example) or use representations that other programs had already used successfully, or introduce randomness (like the mutation operator in genetic algorithms) though guided by a semantic net; let them search through variations of conceptual frameworks to invent a new framework for a problem. Finally, they must be able to transfer what is known in one problem domain to another one, by sharing ontologies, by using analogy engines, by importing concepts from an old domain, and by modifying previously successful methods.

"Creativity is not a mystery," Buchanan concluded. "It does not require any noncomputable elements collectively called intuition or gestalt by some philosophers. It does require persistence, background knowledge, programming skill, and considerable experimentation, that is, creativity, on the part of the researchers." Might such programs help us with problems that today seem impossibly hard? What will they tell us about human creativity? Finally, he asked, will we be able to put all this effort to use in ways that alleviate human suffering and safeguard the planet?

A Cry to Halt

Buchanan's last questions were apposite to a new public controversy that was to envelop artificial intelligence at almost the same time. For no sooner did we breathe easy about the potential cultural meltdown caused by Y2K than Bill Joy, the cofounder and then chief scientist of Sun Microsystems, one of the largest and most successful computer firms, raised his own alarms in the April 2000 issue of *Wired* (Joy, 2000). A prepublication version of Ray Kurzweil's *The Age of Spiritual Machines* had crystallized a growing unease he'd felt for a while (Kurzweil, 1999).

Joy began by citing a dystopian scenario that appears in Kurzweil's book. Suppose intelligent machines are developed, and thus all work is done by vast, highly organized systems of machines without the necessity for human effort. As the machines improve, more and more decisions might be left to them because they can cope with complexity better than humans can (they're more intelligent?); then, a point comes when human oversight is effectively useless. But autonomous machines are unpredictable, and the human race would be at their mercy. Or, suppose a small elite group of humans manages to retain control over the machines. Do the nonelite disappear, or are they reduced to being domestic animals?

Joy then revealed that the scenarist was not Kurzweil, but Theodore Kaczynski, the Unabomber. That Kaczynski is a murderer does not dismiss his arguments, Joy said, for as he began showing the scenario to friends without revealing its author, they too agreed that there was something to be concerned about. Meanwhile, he consulted the web site of Hans Moravec, now one of the leaders in robotics research at Carnegie Mellon. Here, Joy found further disturbing arguments.

Moravec claimed that in evolutionary history, biological species almost never survive encounters with superior competitors. Thus, left alone, superior robots would squeeze humans out of existence. But they would not be left alone. "Judiciously applied, government coercion could support human populations in high style on the fruits of robot labor, perhaps for a long while." But not forever. Moravec's view was that the robots would eventually succeed us.[30]

Joy's friend Danny Hillis, not only a computer architect, but also a respected futurist, agreed that the process of human-machine melding would probably take place, and he was at peace with the idea and its risks. Hillis had already told Kurzweil that he was as fond of his body as anyone, but if offered the chance to live to be 200 years old with a body of silicon, he'd take it.

---

[30] Compare this to Fredkin's view in Chapter 11, where he agrees smart machines will replace us as the smartest beings on the planet, but that will not be our demise. On the contrary, we'll occupy a limited niche, burdens will be lifted from our shoulders, and we might actually be a lot happier for it.

Joy got to the nub of his argument. "Robots, engineered organisms, and nanobots share a dangerous amplifying factor: they can self-replicate. A bomb is blown up only once—but one bot can become many, and quickly get out of control."[31] Joy was well aware of the advantages these technologies offered: cures for intractable diseases, a significant extension of human life, a lifting of heavy labor. But "the 21st-century technologies—genetics, nanotechnology and robotics—are so powerful that they can spawn whole new classes of accidents and abuses. Most dangerously for the first time, these accidents and abuses are widely within the reach of individuals or small groups. They will not require large facilities of rare raw materials. Knowledge alone will enable the use of them."

"I think it is no exaggeration," Joy added, "to say we are on the cusp of the further perfection of extreme evil, an evil whose possibility spreads well beyond that which weapons of mass destruction bequeathed to the nation-states, on to a surprising and terrible empowerment of extreme individuals." (These words seemed all the more chilling a few months later, after al-quaeda's attack on the World Trade Center in New York City and the Pentagon in Washington, DC.)

Joy was at pains to stress that he is not a Luddite, that all his life he'd believed in science and technology as steps toward material progress. He had not been disappointed. But now came a new idea, that he might be working to create tools that would eventually lead to the replacement of our species. This new idea made him very uncomfortable; remedies seemed elusive. Instead, he argued, let us relinquish this research, limiting the development of technologies that are too dangerous, limiting our pursuit of certain kinds of knowledge. If we hadn't reached a point of no return in 2000, that point was rapidly approaching.

Joy's deeply felt alarm evoked many responses, most of them respectful, most of them agreeing that vigorous debate was essential (though some protested correctly that the debate had already

---

[31] Like Internet viruses, worms, and other unwelcome electronic intruders.

been underway for a while, as readers of the first edition of *Machines Who Think* can see for themselves). But nearly none of them agreed we should relinquish our research. (Which *we*? somebody inquired. Bill Joy and his friends? Citizens of India and China, who might be persuaded to give up more and cheaper material progress, based on the possible risks?)

Brain scientist William Calvin reminded us that surprises are part of the human condition. Our challenge is to manage rare but high-risk situations. Suppose, he said, we learn to make humans think slightly faster about slightly more, by putting in place a more scientific education based on how the brain *really* learns. "All these mental improvements are just as capable of destabilizing the world as more familiar forms of technology. We have to worry about it for all the same reasons why we worry about the information technology have-nots." Yet at the same time, thinking better and faster might have vital advantages in a crisis, such as a large, fast climate change (Calvin, 2000).

John Seeley Brown, chief scientist of the Xerox Company and director of the Xerox Palo Alto Research Center, and Paul Duguid, a specialist in social and cultural studies in education at UC Berkeley, and also associated with PARC, responded directly to Joy with a chapter in AAAS Science and Technology Policy Yearbook (Brown and Duguid, 2002).

Technology and social forces not only cannot be divided from each other, they shape each other, the authors said. Runaway technology seldom happens. Thus, as each of the technologies Joy worries about finally becomes potent enough to come close to wreaking the damage he fears, social attitudes will also have evolved. An example is atomic energy: Perhaps an early and unwarranted social reaction, which led to the closure of that field, has deprived the world of cheap energy it might have had, and intensified the carbon effect on the environment. On artificial intelligence specifically, they wrote: "Worries about robotics appear premature, as well." Although Xerox has worked on self-aware, reconfigurable "polybots," pushing the way toward "morphing robots" that can move and change shape, these are far from human and cannot learn in any significant way. "Indeed, the thing that handicaps robots most is their lack of a social existence. For it is our social existence as humans that shapes how we speak, learn, think and develop common sense ... .These critical social mechanisms allow society to shape its future. It is through planned, collective action that society forestalls expected consequences (such as Y2K) and responds to unexpected events (such as epidemics)."

Ray Kurzweil, a gifted technologist and an indefatigable public champion of artificial intelligence, didn't deny that technology had peril as well as promise. At a panel discussion at the Washington National Cathedral (Kurzweil, 2001), that included Joy and others, he asked his audience to imagine being alive 200 years ago, and confronted with the possibility, the inevitability, of weapons capable of destroying all mammalian life on Earth, and further, with predictions of the actual suffering and deaths that would lie ahead in two great 20th-century wars. Offered the chance to relinquish such technologies, this civilization probably would have, gladly. But their sacrifice might have exacted an even greater price. "For one thing, most of us here wouldn't be alive," Kurzweil said, since life expectancies would have remained as brief as they were then. The vast majority of humanity would have continued living "lives that were labor-intensive, poverty-stricken, disease-ridden, and disaster-prone." Though these and other technologies could and probably will be applied in destructive ways, our defensive technologies and protective measures will evolve along with the offensive potentials. "If we take the future dangers such as Bill and others described, and imagine them foisted on today's unprepared world, then it does sound like we're doomed. But that's not the delicate balance we're facing. The defense will evolve along with the offense. And I don't agree with Bill that defense is necessarily weaker than offense. The reality is more complex."

Nevertheless, three years later, Joy's manifesto was still provoking responses. Freeman Dyson, on the occasion of reviewing a Michael Crichton thriller, *Prey*, asked: "What is the appropriate response to dangers that are hypothetical and poorly understood?" Dyson, a brilliant physicist who came to wider public attention when he wrote about ordinary citizens taking on the dangers of nuclear proliferation, agreed that the dangers Joy described are real, but he disagreed with some details of his argument, and strongly disagreed with his remedies. He suggested an analogy between the 17th-century fear of moral contagion by soul-corrupting books, and the fear of physical contagion by pathogenic microbes (or, by extension, artificial intelligences that might replace humans), citing John Milton's *Areopagitica*. "In both cases, the fear was neither groundless nor unreasonable. John Milton argued that the risks must nevertheless be accepted, for regulating 'things, uncertainly and yet equally working to good and to evil' is difficult, perhaps impossible before the fact" (Dyson, 2003).

John McCarthy put it succinctly: "The main danger is of people using AI to take unfair advantage of other people. However, we won't know enough to regulate it until we see what it actually looks like" (McCarthy, 2003).

Physicist Richard Feynman, himself no stranger to technologies with ambiguous outcomes, used to like to tell a story of visiting a Buddhist temple, where he met a man who said: "I am going to tell you something you will never forget. To every man is given the key to the gates of heaven. The same key opens the gates of hell." And so it is with science, Feynman went on. It's the key to the gates of heaven and the gates of hell, and we have no instructions as to which gate. "Shall we throw away the key and never have a way to enter the gates of heaven? Or shall we struggle with the problem of which is the best way to use the key? That is, of course, a very serious question, but we cannot deny the value of the key to the gates of heaven" (Feynman, 1998).

AI and the Sacred

Feynman's puzzle is an apt introduction to the work of people who, at the same time Bill Joy and others were expressing their alarm, were asking whether a theological approach to artificial intelligence is appropriate. For several years, Anne Foerst, trained as a theologian and a practicing Lutheran minister, has been in residence at the MIT Laboratory for Computer Science. She describes herself as "theological advisor" to some of its more famous robots, but she has also run a program where well-known computer scientists lectured about the divine and its relationship to their work. She describes herself as "engaged in a kind of dialogue to integrate the world views of AI and theology," attempting to answer, among other difficult questions, what constitutes dignity between human and robot, and where is the threshold when you cannot just switch a robot off? How will robots react to us when they realize we have endowed them with human limitations?

In her book, *In Our Image*, Noreen L. Herzfeld, of St. John's University in Collegeville, Minnesota, draws parallels between the human in the world of the computer, and God in the human world. Any artificial intelligence that does not exhibit both what separates the human from the animal, and what is necessary rather than contingent to our nature, she argues, would likely be judged insufficient, not a true image of ourselves, and hence not fully intelligent. Hopefully, we'll have a better idea of just what's necessary and what's only contingent to human nature by the time that debate is more than hypothetical.

Theological questions may be, like cries to stop, premature. It gives me pause, at the very least, to think about applying the precepts of Paul Tillich to artificial intelligence. In this book's first edition, I suggested that our attitudes toward intelligent machines have historical, if not theological, roots, which predispose us to welcome them (the Hellenic view) or fear them (the Hebraic point of view). I might have added the Shintoist view, which ascribes life, or sentience, to everything on the planet, a view that has predisposed the Japanese, at least, to welcome robots into their lives. But these inquiries, now underway in university bioethics courses and in theological seminaries, lay out possible futures for us to examine, and, perhaps more important, give us fresh insight into our present, with the possibility of changing our views and our circumstances. People like Bill Joy hope that change consists of stepping back; people like Ray Kurzweil (and many others, myself included) hope that we'll at least go about artificial intelligence research mindfully. The entire project of science, unlike dogma, is about changing views and circumstances, so the effort shouldn't be alien.

Let me repeat what can be found in Chapter 14 of the first edition: "And wherefore was it glorious" Victor Frankenstein once cried out to the frightened. "Not because the way was smooth and placid as a southern sea, but because at every new incident your fortitude was to be called forth and your courage exhibited, because danger and death surrounded it, and these you were to brave, and overcome."

Ah. We nearly forgot.

What Lies Ahead for Machines Who Think?

In January 2003, to celebrate its 50th anniversary, the *Journal of the Association for Computing Machinery* asked living winners of the Turing Prize, its most prestigious award to computer scientists across the spectrum of specialties, to suggest the challenges for computing in the next 50 years. Though only three of the contributors had spent their careers and earned their Turing Awards for work in artificial intelligence, it's striking that many of the contributors from other specialties proposed AI challenges.

For example, Frederick P. Brooks of the University of North Carolina at Chapel Hill, a software engineer famous for his book, *The Mythical Man-Month* (Brooks, 1996), about the difficulties of building large software systems, suggested three problems: the quantification of information embodied in structure; a method to make software engineering as predictable a discipline as civil or electrical engineering; and the transformation of user interface design from art into engineering discipline. At least two of those, the first and the third, are partially, or even mainly, AI problems.

In the same 50th anniversary volume, Jim Gray of Microsoft Research, San Francisco, best known for computer architecture, suggested that, among other problems to be addressed by computer scientists in the next 50 years, was for machines to pass the Turing Test. Despite past failures, he was optimistic because, "I am persuaded by the argument that we are nearing parity with the storage and computational power of simple brains." But something further, perhaps very fundamental, is missing. "We have been handed a puzzle: genomes and brains work. They use much more compact programming languages than we do (their programs are a lot smaller than the ones we are familiar with). Understanding the answer to these puzzles is a wonderful long-term research goal." He also proposed a "personal Memex," a box that records everything you see, hear, or read (Memex is the term Vannevar Bush once proposed for a giant accessible library, and the idea emerged, a few months after Gray's paper was published, as a call for research proposals from DARPA under the term "LifeLog"). A world Memex should follow, a device that analyzes a large amount of material and presents it to you in a convenient way (its early version is perhaps the Semantic Web, now under development). We might be able to record everything (and Gray gives plausible arguments why) in a way that would lead to teleobserving (a virtual environment on demand that allows the observer to experience the event as well as actually being there), and then to telepresence, allowing the telepresent person to interact physically with the world via a robot. Gray had other ideas for computer scientists to put their minds to in the coming 50 years, but these I've described were clearly in the center of artificial intelligence research.

Butler Lampson, a distinguished computer architect, who has contributed to the design of local area networks, raster printers, page description languages, operating systems, programming languages, and more besides, now at Microsoft Research in Redmond, Washington, said he thought getting computers to understand was one of the great challenges for the field of computer science (Lampson, 2003). He had concrete proposals: cars that don't kill people (cars that understand roads), and automatic programming, writing a program from its specifications as well as a team of human programmers could do it.

From Harvard's Leslie Valiant, a specialist in computational complexity, computational learning, neural computation, and large computer systems, came three proposed challenges, two of which were "characterizing a semantics for cognitive computation" and "characterizing cortical computation."

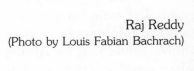

Raj Reddy
(Photo by Louis Fabian Bachrach)

Since John McCarthy, an emeritus professor at Stanford, had made multiple contributions to computer science in several specialties, his suggestions spanned those specialties (McCarthy, 2003). AI's goal, he said, should be human-level AI, that is, computer programs with at least the intellectual capabilities of humans, though he was doubtful that it would arrive by 2050—"I'll guess 0.5 probability in the next 49 years, but a 0.25 probability that 49 years from now, the problems will be just as confusing as they are today."

Raj Reddy of Carnegie Mellon, well known for his work in artificial intelligence, proposed three open problems toward the goal of achieving human-level AI: first, a computer that could read a chapter in a book and answer the question at the end of the chapter; second, remote repair, a machine that could repair a mobile robot and successfully demonstrate the capability by repairing one on Mars (or with appropriate simulated time-delay on earth); and third, an encyclopedia on demand, the synthesizing of information from various sources, summaries in a convenient size, and a finished summary in natural and intuitive language.

Finally, Edward Feigenbaum, who, like Reddy, has spent his professional career in artificial intelligence, proposed AI challenges, too. We've learned much from partially intelligent artifacts, he said, in natural language processing, computer vision, and reasoning tasks (on very hard problems, it's important to add, such as medicine, the physical sciences and engineering, in military applications, and the analysis and control of many business and manufacturing processes). While some of these systems achieve world-class performance, none of them would pass the Turing Test because they lack breadth of behavior and flexibility in interacting with humans.

What if the Turing Test were revised to test the quality (the complexity, the depth) of reasoning that distinguishes the Einstein from the rest of us? This game, called the Feigenbaum Test, requires two players and a judge. One player is a computational intelligence, and one player is a human, chosen from among the elite practitioners, a member of the National Academy, in each of three preselected fields of natural science, engineering, or medicine. (The number could be larger, but for this challenge not greater than ten.)

In each round of the game, the performance of the two players (elite scientist and computer) is judged by another Academy member whose specialty is that particular domain of discourse—for example, an astrophysicist would judge astrophysics behavior, or a molecular biologist would judge molecular biology behavior. Of course, the identity of the players is hidden from the judge, as it is in the Turing Test. The judge must pose problems, ask questions, and ask for explanations, theories, and so on—as one might do with a colleague. Can the human judge choose, at better than chance level, which is his National Academy colleague and which is the computer?

Let the game be played several times to enhance the statistics of the test, using different pairs of Academy members from the selected domain, all of them— judges, humans, and computers alike—using the jargon of the field. The challenge will be considered met if the computational intelligence wins one out of three disciplinary contests, that is, one of the three judges is not able to choose reliably between human and computer performer.

This is a formidable grand challenge, he continued, but it's far from the extraordinary grand challenge of the ultraintelligent computer. Paradoxically, the UIC would be easily distinguished from the elite human performer because it would be offering inductions, problem solutions, and theories that were not yet reached by any human, yet were plausible, rigorous upon explanation, and either correct or interesting enough to subject to experimental test.

Building a knowledge base has been an AI bottleneck for two decades. Here, Feigenbaum proposes a challenge that encompasses both machine learning and educational strategy: first, manually encode a novice-level view of a domain (say, an introductory textbook). Then, write the software for a system that will read the next level text in the field, augmenting as it reads the kernel novice-level view it already has. Queries for clarification should be answered, and occasional direct intervention into the symbolic structures may be needed to introduce a missed concept or correct a misunderstanding, but these must not exceed ten percent of all the symbolic structures. Then, the computational intelligence must continue to educate itself, and "keep up with the literature," so that it can participate in subsequent Feigenbaum Tests.

Since the World Wide Web is a mirror of our culture, and, for knowledge engineers trying to build a broadly knowledgeable computational intelligence, the most tempting data base of all, figure out how to turn this variegated lot of data into machine-useful knowledge. This task is already underway, he pointed out, supported by both the US government and the European Union, with help from the World Wide Consortium under the label of the Semantic Web.

He concluded: "I hold no professional belief more strongly than this. I call computational intelligence the manifest destiny of computer science."

⁓

Missing from the contributions was an essay by Turing Award winner Marvin Minsky, perhaps because he was at work on a new book which, while not finished, is available for browsing and comment on his web site (Minsky, 2003). Tentatively entitled *The Emotion Machine*, it's a sequel to *The Society of Mind*, in that it proposes that minds are not single things, but networks and layers of processes, an understanding we have thanks to our experience with the behavior of computers. As Minsky puts it, unlike physics, which seeks simple answers to complex questions, the science of mind (or at least his proposed version of it) seeks complex answers to questions about the mind that seem simple. He lays out the role of emotions, and their contributions to our intelligent behavior, traditionally considered separate from emotions. Emotions, he argues, are a set of processes for learning and coping, and belong just as surely in a theory of intelligent behavior as logic and rationality do. The "narrative intelligence" point of view was deeply influenced by Minsky's work when it began, and it may be that he is returning the favor.[32]

Certainly some of the most successful simulators, in particular those used for military training, feature animated agents who exhibit a variety of behaviors. Agents display both intelligence and emotions, vital for training young soldiers who will have to deal, human-to-human, with such situations in combat. The same can be said of the most sophisticated games, which use proprietary technology, and whose agents can exhibit intelligence, cooperative behavior, or the most wicked sort of villainy as they engage human players.

Can Revolution Be Fomented?

A few years ago, DARPA began letting contracts for what has become known as the Semantic Web, a next-generation World Wide

---

[32] Minsky also stirred up some controversy in the summer of 2003 by declaring in various venues that AI was "brain-dead" and had been for the past 30 years. It's a strong charge. Whether it refers to the incremental rather than revolutionary progress made in AI in the past few decades (surely vexing for a former revolutionary himself), to the hegemony of the Neats, or even to his quarrel with the "reactive" robots that have dominated MIT's robotics research for nearly 20 years, is anybody's guess. But as even Marvin Minsky might concede, MIT is not the world, and incremental progress is not brain-death.

Web "in which information is given well-defined meaning, better enabling computers and people to work in cooperation" (Bernes-Lee et al., 2001). The Semantic Web will lead to more precise and deeper searches, allowing your personal agent to find information (in any language, in any form) and eventually to break out into the physical world, scheduling appointments for you, and allowing you to automate and coordinate the disparate appliances in your house with minimum effort on your part. Though your online query might be vague, your agent will quiz you until it knows what you mean, and information on web sites, however unstructured, will be queried according to the *meaning* of content on that web site, not only in text, but in images, music, or other media. Eventually, researchers foresee that this Web, if properly designed, can assist the evolution of human knowledge as a whole, as agents learn to exchange meaningful information.

"Adding logic to the Web—the means to use rules to make inferences, choose courses of action and answer questions—is the task before the Semantic Web Community at the moment," write Tim Berners-Lee and his colleagues. Already in place are XML, the eXtensible Markup Language, which lets everybody create their own tags, hidden labels that annotate the site, or add arbitrary structure to the documents; and RDF, the Resource Description Framework, which expresses and encodes meaning. The problems are challenging, because the Semantic Web requires an agent to understand and even construct an ontology (a logical structure that formally defines the relationships between terms so that knowledge can be organized) for each query. Moreover, the new structures must work not only for sites written in the new language, but must be able to go back over legacy sites, written in the language of the first-generation web, and incorporate those in any search, too. The project is jointly supported by DARPA and the European Union, in association with the World Wide Web Consortium. It's a huge and enormously difficult project, and whether present techniques in, say, natural language processing, or logical structures, can be appropriately extended, or whether the task will require revolutionary science, remains to be seen.

In 2003, DARPA announced three new programs it hopes to implement. First was the "LifeLog," a lifetime-long electronic record that would capture, store, and make accessible the flow of one person's experience in, and interactions with, the world. Such a log would act as a dynamic and instantly accessible journal, capturing events, states, and relationships from a few seconds earlier or the distant past. It would be, in short, an auxiliary memory.

Its first stage would be a single individual's stand-alone system, to serve "as a powerful multi-media diary and scrapbook" that allows the user to search for a specific thread ("When did I first hear about this? What did I think then?") with as much detail as desired, including imagery, audio, or video replay of the event. It could eventually support personal, medical, financial, and other kinds of assistants and serve as a teaching or training tool. As more people acquired LifeLogs, individual LifeLogs could engage in collaborations, including medical research (the early detection of an epidemic, the long-term exposures and practices that lead to diseases), but this is for later. DARPA is sensitive to the privacy issues such an auxiliary memory raises: the specifications address these directly.

A second program solicits proposals for a new class of cognitive systems that can reason in a variety of ways, using substantial amounts of appropriately represented knowledge. These cognitive systems can learn from experiences so that performance improves as knowledge and experience accumulate; can explain themselves and accept direction; can be aware of their own behavior and reflect on their own capabilities; and can respond robustly to surprises. One motive for this program is the growing complexity of systems, "dauntingly difficult to debug and test" that "regularly fail in practice and are increasingly vulnerable to attack." They should be not only faster and smaller, but smarter, with awareness of their goals (and thus the ability to help extend or debug themselves) and have intelligent user interfaces that adapt to their users, rather than the other way around.

A third program solicits proposals for a new program on real-world reasoning. Its objective is to explore and develop founda-

Ron Brachman
(Courtesy of Defense Advanced
Research Projects Agency)

tions, technology, and tools to allow effective, practical automated reasoning of the scale and complexity required for computers to perform complex tasks in the real world that require intelligence. Effective real-world machine reasoning requires inference in environments that are far more complex in scale and scope than those tackled by current machine reasoning methods, with vast amounts of knowledge and information, often concerning dynamic and intentional phenomena. In addition, beliefs about the environment are often uncertain and involve plausible, but not provable assumptions. Thus, this program aims to expand the breadth of machine reasoning, and to combine multiple reasoning methods—deductive reasoning, reasoning by analogy, strategic reasoning in multiplayer or multiagent contexts, and so forth—so that their different advantages can be exploited.

Though these major projects overlap somewhat, one thing common to them is that they're unlikely to be achieved by incremental methods. AI is ready for new techniques, new approaches, new ideas. DARPA, frankly, is hoping to foment a new revolution.

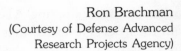

When I spoke to Ron Brachman, who heads DARPA's Information Processing and Technology Office, a chair once occupied by

J. C. R. Licklider, Bob Kahn, and other legendary visionaries, he seemed suitably awed by his predecessors, but full of hope, and certainly great enthusiasm, that his tenure might make a difference, too. Brachman spent many years as an AI researcher, first at Bolt Beranek and Newman, and later at Bell Labs, and AT&T Labs, and he's well acquainted with the field. He agreed that for the most part, AI research had recently been "normal," that is, good science, sound science, done step by step, implementation by implementation.

IPTO's proposals call for "revolutionary" science. "Can you foment revolution?" he asked, repeating my question to him. "I don't know. It's worth trying." Worth trying because "everybody—even the people who claim to distrust AI, or don't like it, or think it can't be done—needs it." Then he grinned. "We are actually casting our net a little more broadly, and a little bit differently than has been the case with traditional AI. We want to bring into the picture more computer science, along with relevant work on neuroscience and psychology. We are calling our area of interest 'Cognitive Systems' – the folks that never liked 'AI' seem to be able to deal with that." To borrow an idea from colleagues at NASA, take deep space explorations, he went on. A probe travels to a moon of Jupiter, and we have no idea what's beneath the surface—that's why we're probing it. Will it be ice? Will it be liquid? What viscosity? Filled with floating objects? The answers may tell us something deeply significant about the origins of our solar system, perhaps even the origins of the universe. The deep-space communication time lag is far too great for the robot probe to get its instructions from Earth in real time: It must figure out for itself what to do, depending on the conditions it finds. It must be able to respond in a reasonable way, no matter what it encounters, even if it runs into things no one has ever anticipated.

In science, in the military, in everyday applications, those autonomous decisions must be made, made quickly, and made well. Not perfectly, but well enough. That led Brachman to muse about the tension among generality, versatility, and expertise in intelligent behavior. Some of the best work in AI has been in *expert* systems (narrow, but often world-class, performances in a given domain). Can we create a system that trades its high expertise—that performs only modestly—for great generality? A system that initiates appropriate problem solving on its own? In other words, a human-like intelligence? If so, how will we know? What task can we devise that will measure such a thing? If it's too narrow, we'll be able to engineer the problem; if it's too broad, we won't succeed anytime soon. Certainly as vague as it now is, it would require a third-party evaluation.

Will the problems that 20 years earlier beset the Strategic Computing Initiative also beset this new initiative? Perhaps, he conceded. DARPA, high-risk, high-payoff, is committed to rotating scientific management out after four or so years, do or die. Managers know they have to make their mark in a brief time, so they work hard and act boldly. They aren't interested in projects other funding agencies could handle—they want to pursue the difficult, the DARPA-hard projects—and they aren't hampered by memories of things that didn't work in the past; after all, they might work now. In short, DARPA is about as distant from the usual government bureaucracy as it can be.

"A hard problem," I said.

"A hard problem—but we've got some great proposals already! People are out there, the ideas are out there. And not just in the US." Brachman and I were speaking before the day's events got underway at the IJCAI 2003 meeting in Acapulco, and across the room from us, German AI researchers were in conversation with UK researchers, Japanese AI researchers had brought families to enjoy a small and exotic vacation in Mexico, and you could hear Italian AI researchers laughing together. A large group of young Mexican researchers had taken advantage of the proximity of this international meeting and roamed the halls, having come to see people and exchange ideas. "Maybe my biggest problem is managing human expectations," Brachman said.

I could guess what he meant. If they want it at all, people want AI yesterday. Results have been so *disappointing*, they pout, and I must wonder, compared to what? The 2,500 years of research in physics? The hundreds of years of research in biology? Medicine, which is only edging toward science after millennia as an art? Cosmology? Over dinner recently, a distinguished brain researcher defined for me "just where AI went wrong." I did not make myself disagreeable by launching into a lecture about levels of scientific modeling, nor did I mock the pre-Vesalian state of brain research and cognitive psychology when AI began in the mid-1950s (rat mazes, stimulus-response, a few poor souls with split hemispheres who could be shown and sometimes respond to flashcards). After all, now that new computer-based technologies have permitted brain research to become more like a real science, it might be on the verge of making some substantial contributions to AI, including how to avoid the shortcomings of natural brains. (If the most we ultimately want from AI is human-level intelligence, why have AI at all? The planet is not short of humans.)

Both AI's loudest public champions and its loudest public enemies raise expectations or fears that are, to put it generously, premature. (Remember me, standing at the urban rescue demo, looking nervously over my shoulder for an imminent hurricane and hoping if it came, they'd send in the K-9 corps?) Neither the field of dreams nor the field of nightmares portrayed, AI is science. Its earliest results have already had an impact; as it gets better, it will have a bigger impact. But isn't magic. It's, well, clever programming. Very clever. Three cheers for the programmers.

Though governments from Japan to Europe to Latin America (and perhaps China; nobody knows for sure) are putting bets on AI research, not everyone is waiting for their government. For example, after the catastrophe of September 11, 2001 in the United States, the government itself turned for help to the gaming industry, which had already developed sophisticated proprietary data-mining programs to identify undesirable clients, or hidden collusion between clients and casino employees. Data mining could suggest, even uncover, groups of conspirators who would otherwise go undetected.

Unfortunately, however useful, even essential, this kind of data mining might be to national security, the issue was handled with teeth-grinding clumsiness by the administration, which put polarizing political figures on the project's public face (Poindexter, 2003). Civil rights and privacy lobbyists soon changed the terms of the argument: instead of a reasoned debate about balance, and a consensus about fair tradeoffs between privacy and security, many American citizens began to think that their own government was as wicked as the 19 terrorists who'd attacked. Journalists confused DARPA's data mining program, "Terrorism/Total Information Awareness," with the LifeLog proposal, and nearly killed it. (The European Union, whose privacy rules are far stricter than those of the US, is nevertheless pursuing a similar program of data mining to head off terrorists.)

Nobody believes that data mining is free of ethical issues. Ordinary citizens need protection from privacy invasion, and from government interference with peaceful dissent. But Bill Joy was right to be alarmed when, even before the 9/11 attack, he saw that a small group of enemies could wreak damage far beyond their numbers. The issue needs thoughtful and open discussion, not predictable ideological posturing.

These days, AI is nearly ubiquitous from the most esoteric scientific applications to the most mundane, everyday stuff. Imagine a billion conversations per second, your job to pick out and study the interesting ones. Unfortunately, you know that of those billion conversations per second, only one in ten million will be interesting to you. In addition, each of these conversations is long and complex, making the interpretation of each one a difficult and subtle task. Identifying and interpreting those interesting events, plus presenting the evidence in ways scientists can study, is the task of a smart program used at the ATLAS experiment, soon to be in progress at the Large Hadron Collider at CERN, the international physics research institute in Geneva, Switzerland. ATLAS, involving 34 countries and 2,000 physicists, is dedicated to exploring the fundamental nature of matter.

In ATLAS, most collisions between an individual quark or gluon from one proton, and a quark or gluon from another proton, are "ordinary," but a tiny fraction—one in ten million—is "interesting." The quality of "interesting" is extremely subtle, based on the analysis of many variables, and could be the sign of a new particle, or be confirming evidence of certain theories, including the Standard Model. The smart program must pick through the billion events per second, and find and interpret the "interesting" ones. The software to do this has been developed over many years and has been utilized in a number of high-energy physics experiments. It is based on aspects of artificial intelligence, including distributed processing and neural networks, as well as very sophisticated pattern recognition and imaging techniques.

Across the Atlantic from CERN, in Rockville, Maryland, the National Association of Security Dealers monitors not only trading among its dealers, but news stories and other activity, looking for signs of insider trading or fraud. Its programs use a variety of AI techniques, including natural language text mining, statistical regression, rule-based inference, uncertainty, and fuzzy matching. Prediction Company, in Santa Fe, New Mexico, which trades second by second on the international currency and other markets, uses a variety of AI techniques to model those markets in real time and make trading decisions. More modest and mundane, train and airline crews are routinely scheduled using AI techniques. My morning paper tells me a patent has just been awarded for a robot that will perform hair implants on bald human heads, exacting but tedious.

Some of the most interesting examples of applied artificial intelligence can be found in the virtual players in certain kinds of electronic games. Some are opponents, and can outthink you; some are allies, and collaborate with you. Firms small and large continue to deploy AI techniques, sometimes disguising them (Red Pepper Software, which adapted and licensed the intricate scheduling software first built by the AI group at NASA-Ames for the Space Shuttle, calls their product scheduling software; the Triology firm, having adapted the R1 computer configuration expert system first developed at Digital Equipment Corporation, calls it simply "the Sales

Builder"). AI has simply become background technology in a complicated and complex world. Good. Let artificial intelligence, cognitive intelligence, computational intelligence, whatever its label, slip into our lives ever further. In tasks that are beyond human wit, too complex, too fast, too detailed, AI steps in and allows us to do them anyway. We are building in our own image, but better, we hope. If not, we'll find out soon enough.

The Singularity

How soon will we find out? Two quite detailed scenarios have emerged here, one the Moravec/Kurzweil scenario, which we might call the "Out to Pasture in the Elysian Fields," that foresees machines as intelligent as humans, maybe more so, in 50 years and on the whole, a good thing. This leads to questions both Moravec and Kurzweil, to their credit, raise about whether those machines will take over for us (or from us), the basis of the second scenario, Bill Joy's quite opposite and dark vision, which posits the same improvement in machine intelligence, but with a horrifying outcome, the "NanoGenRoboNightmare." Some believers in the Elysian Fields scenario have been arguing about "the singularity," borrowed from science fiction writer Vernor Vinge, the moment AI becomes powerful and ubiquitous enough so that all of the rules change and there's no going back. The only argument is over whether the singularity will come soon, or has already arrived.

I don't consider either of these scenarios above implausible. But let me suggest other considerations. Both scenarios above are based on a projection of 40 years' growth in computing power, the marvelous Moore's Law, extrapolated 50 years out. But nobody believes that Moore's Law (which is an observation, not a law) is going to hold for more than another 10 or 15 years using conventional silicon technology. Government agencies and a few wealthy firms are putting significant effort and resources right now into novel computers—quantum computers, biological computers, molecular computers.

Suppose we reach the physical limits of conventional computing in 10 or 15 years, and there's nothing there to take its place? Quantum computers turn out not to scale, or biological computers are not general purpose. Suppose we fail to find that clever natural programming language that human intelligence uses, as Jim Gray puts it, that's so much more economical than the artificial ones we use now (Gray, 2003)? Or suppose instead that all these work, but we still can't widen the knowledge-acquisition bottleneck? Then the scenario looks more like John McCarthy's 0.25 probability "that 49 years from now, the problems will be just as confusing as they are today" (McCarthy, 2003).

The required research is a very expensive undertaking. Few firms have the money or the motivation to carry on this high-risk research, development, and engineering. Our governments, with more resources and more willingness to take research risks for the common good, have usually supported the difficult, early basic research for us, work that was then exploited by private enterprise. The US government (and the now nearly defunct large industrial laboratories) led the way for most of these past 50 years; the Japanese government's Fifth Generation (a part of the program that didn't necessarily involve AI) laid the foundation for that country's present first rank in supercomputing—though a first-place ranking always depends on how you count, and changes quickly. In short, high-risk research is a government game.

As it happens, at the moment and for the foreseeable future, both these governments face severe revenue shortages and pressing demands on such revenues as they do have. Can they keep support up at the level needed to achieve these intelligence goals, even if we're smart enough to be able to do the job in principle? Or suppose an ideologue and his cronies capture the US administration. "That phrase, computational intelligence, doesn't fool me," he'll say. "What you're doing is specifically prohibited in the scripture." Then AI research would face the prohibitions that stem-cell scientists faced with the George W. Bush administration. In that case, for better or for worse, we can be sure that the Indian or Chinese governments would quickly step in to become AI's patrons, which leads to a group of other scenarios, left as exercises for the reader. If our governments won't do it, would we be willing to turn the task over to a for-profit firm, or a consortium of them, even if it promised to share its findings? That particular little scenario runs Joyesque chills up and down my spine, believe me.

Or suppose everything works, but only sort of. We have more computational power, we learn to program it economically, and we have some fine ultraintelligent computers, as Feigenbaum calls them, doing superb work in fields where it has been important, or economical, enough to force-feed knowledge into a knowledge base. The UICs will be "offering inductions, problem solutions and theories that were not yet reached by any human, yet were plausible, rigorous upon explanation, and either correct or interesting enough to subject to experimental test," in their particular field of expertise (Feigenbaum, 2003). These would be exceptionally valuable to the human race, but not quite the dazzling moment of the singularity, where we back out humbly from history with our hats in hand. No use even mentioning the wild cards, where some mad genius in a garage or a cave in the Middle East—oh well, you know those. I only mean to say that, as plausible as "Out to Pasture in the Elysian Fields" or the "NanoGenRoboNightmare" are, any of these others, or a combination of them, could happen in the next 50 years.

We Summarize

What has AI taught us in the last 25 years?

The examples of molecular biology and high-energy physics are just that—examples—of the myriad ways AI has not only insinuated itself into the everyday practice of science, and taught us new things we didn't know, or confirmed things we only guessed. The same thing has happened in other fields as well.

But we've also learned what we couldn't know except with the help of artificial intelligences. Consider what happens when a programmer goes head to head with his or her own program. Harold Cohen, whose work on the art-making program, AARON (previously mentioned), serves as an illuminating example. He'd spent many years making his own knowledge of art explicit enough to cause AARON to draw interestingly, and then to begin to handle color. To make his program do something, Cohen has asked how he himself did it, then turned the answer into executable computer code. But with color, he says, "it not only hadn't worked, it couldn't work. We (humans, machines …) are all hardware bound." The human colorist is constrained by the human visual system, which is splendid, but has no ability to form a stable internal representation of a complex set of color relationships, whereas a computer has an almost limitless capacity to do so. Though Cohen has always been considered a gifted colorist, he says that whatever gift he has "would never have been objectified as it has [appearing in the work of AARON] unless I'd found a way of representing it as a problem to be solved in terms appropriate to a computer program" (Cohen, 2003b). The colors of the lush new AARON paintings are not only beautiful, but enthrall me: how did it *do* that?

AI has taught us more about human intelligence, that it is not only the function of brain activity (search, prune, recombine, all the techniques mentioned here), but that general intelligence exists by being embedded in a milieu, an environment, cultural, social, and physical, that contains not only other human beings (and now partially intelligent machines), but also the accumulated human knowledge and artifacts of the past, as well as nature itself. We also now know more certainly that expertise (perhaps taking ten years' study and practice) is essential to expert behavior, but we also know that assiduous search can compensate for lack of knowledge, and knowledge can compensate for lack of search, and the two can, in some cases, be traded off. These findings are the result of empirical experiments, not inspired guesses, and are safely inside the boundaries of the scientific method.

More tentatively, a number of researchers propose that, based on their research, creativity, even genius, is like the rest of us, only more so. I keep an open mind and welcome proof on that one.

I claimed in the first edition of this book, that AI would probably make us see ourselves differently, and though that's beginning, it hasn't happened as fast as I'd thought. A report on a recent bioethics symposium at Yale has people suddenly concerned that AIs, smarter than we are, might treat us as badly as we've traditionally treated the less intelligent or less powerful creatures on the planet, and oh boy! Maybe we'd better reform to set a good example for our coming superiors. Well, whatever makes us behave ourselves better is all to the good, I think, though this isn't quite what I had in mind.

It may be that the manifest destiny of computer science is computational intelligence, as Ed Feigenbaum says, but I'd go further.

Intelligence means literally "to choose among what has been gathered." This etymology casts the history of artificial intelligence into a vastly larger landscape, revealing it to be not some hubristic overreaching (though it has sometimes seemed so), but instead another natural stage in the flow of that most enduring, even noble, of human urges: the passion to gather, organize, and share knowledge so that we all benefit. If competition typifies human behavior (and it does) then alongside it, possibly in equal measure, is cooperation. From the beginning, it seems, we've gathered and organized information, knowledge, and techniques largely to share all these with our fellow humans, to amplify their knowledge as well as our own. (We aim to serve; but we also aim to astound, delight, profit from, make envious, and so forth. Humans entertain a host of motives simultaneously.)

This enduring urge to collect, organize, and share knowledge has been shaped by the moment's technology. For example, we only know about the earliest libraries and encyclopedias because their contents were recorded on clay tablets and thus preserved for us. Surely elsewhere and surely earlier, oral repositories of knowledge were likewise collected. However different from us those distant cultures and societies were, we must nod in recognition at the evidence of gathering, organizing and preserving knowledge, the better for others to know and think.

Now we can re-read history. The slow-moving but persistent river of effort to collect, organize and disseminate knowledge enters new terrain thanks to the computer. It arrives first as unfinished dreams—Vannevar Bush's Memex, Ted Nelson's Xanadu. Then comes the Internet, where scientists learn they can collaborate quickly and effectively, discovering and exchanging knowledge over great distances. Whole scientific fields, such as biology, the earth sciences, physics, and astrophysics, are transformed.

Meanwhile, here comes the once separate stream that represents the mechanization of thought, its current charged by a better, faster version of those human qualities that have always served us best: our reason, our collective knowledge, and our active collaboration with each other. This stream joins the great river and transforms it: the World Wide Web, daughter system of the Internet, will evolve into the Semantic Web, a repository and amplifier of intelligence for the entire planet.

Yes, opportunities for mischief, even catastrophe, lurk. Vigilance will help, but is no warranty. Yet to back off from all this is to back off from one of our noblest, most enduring and significant human projects. I don't think we should. I don't think we can.

Up the hill from where I write this, a slogan floats around the Santa Fe Institute: "The world is not only more complex than you imagine; it's more complex than you *can* imagine." Undoubtedly so, and without help, we might as well resign ourselves to brief lives of exceeding nastiness and brutality. But the slogan doesn't say we can never imagine the world's complexity. Look at Cohen, discovering news about color from AARON because the program can do things even beyond the marvelous visual system that nature endowed Cohen with. With help—yes, the help of our own artifacts—we might at last be able to imagine new levels of complexity. In that case, our chances for survival, for even more than mere survival, look brighter.

AI is a godlike enterprise, I said in the first edition, and I stand by that. To repeat one of Stewart Brand's aphorisms, we are as the gods, and we may as well get good at it. My own faith, that our smart machines are going to help us do that, can rest on much more evidence than when I first declared it—more, but not yet altogether conclusive. I confessed then that I was of a Hellenic turn of mind, and I still am. Another pause here, before I have to call forth fortitude and exhibit courage: I pause to cheer on the people who are actually bringing this about, a heroic enterprise, I think. And, of course, I pause to savor the thrill of sharing in something awesome.

Time Line:
The Evolution of Intelligence

This time line is different from one that might have been constructed when *Machines Who Think* was first published in 1979. Most salient, it includes the first great collections of static as well as dynamic knowledge, and both individual and collective thinking. It shows how early and how often humans were impelled to collect, organize, and disseminate what was known, and how deep the roots are of the impulse to mechanize thought. In the 20th and early 21st centuries, all these elements are converging, thanks to the development of the computer and telecommunications. Automated intelligence will be crucial to the success of that convergence.

Before the Common Era

40,000	Symbolic communication among humans, probably rudimentary until now, flowers in language, tool making and art; human social organizations grow more complex
Twenty-First century	Earliest known "library" (on clay tablets) in Babylonia
1270	Assyrian scholars compile early encyclopedia
Seventh century	World's first great library founded at Ninevah (Babylonia) with evidence of a well-developed cataloguing system
Sixth century	Composed earlier in oral form, Homer's poem *The Iliad* is codified, introducing into written literature assorted automata from the workshops of the Greek god Hephaestos

Sixth – fifth centuries	Hebrew Torah is canonized, including the second of the Ten Commandments, the prohibition against making graven images
Fifth century	Aristotle lays out the epistemological basis in the West for the division of knowledge into categories, with theory the most important and art the least important; he also introduces syllogistic logic, the first formal deductive reasoning system
	The *Erh Ya*, a collaboratively researched book of general knowledge, is compiled in China
c. 350	Speusippus (407-339), nephew of Plato, tries to collect all human knowledge into one volume
c. 330	Ptolemy I establishes the famous library in Alexandria, Egypt, which at its peak has 400,000 – 700,000 scrolls, the greatest collection of knowledge in the ancient world
Third century	Sporadic trade of the previous 3,000 years now grows markedly along the Silk Route, providing indirect but regular contact between China and Europe
239	20,000 word volume of collected knowledge from many and far-flung Chinese scholars, commissioned by Pu-Wei, minister of the State of Ch'in, is completed

Common Era

c. 39	Emperor Augustus of Rome establishes the public library envisioned by his predecessor, Julius Caesar
77	Pliny the Elder completes the 37-volume *Historia Naturalis*, a classified anthology of information

Late First century	Heron of Alexander builds fabled automata and other mechanical marvels
c. 600	*Ch'u usueh chi* (*A Manual for First Steps in Learning*), an encyclopedia of knowledge, is compiled by the Chinese government to help candidates prepare for civil service exams
622	Isidore of Seville produces an encyclopedia of arts and sciences
Ninth century	Caliph al Ma'mun founds the Bakyt al-Hikmah (House of Wisdom) in Baghdad, where Greek texts are collected and translated into Arabic
c. 825	Persian mathematician and scholar al-Khwarizmi (hence *algorithm*) compiles an encyclopedia beginning with jurisprudence and scholastic philosophy, concluding with practical topics of medicine, mathematics, and mechanics (labeled as "foreign knowledge")
Fifteenth century	Growth of European libraries as centers of knowledge, thanks to the printing press; Pope Sylvester II, Bishop Grosseteste, Roger Bacon, and Albertus Magnus said to have "brazen heads," simultaneously sources and proof of their owners' wisdom; Ramon Llull, Catalonian mystic and theologian, invents his "Ars Magna," a machine for discerning truth by "bringing reason to bear on all things," and based on the Arabic *zairja* he had seen
1403-1407	Ming Dynasty scholars compile an encyclopedia of 11,095 books, the *Yongle Canon*, which remains in manuscript form owing to its great length, until it is republished digitally in 2002

Fifteenth – sixteenth centuries	Mechanical clocks, the first modern measuring machines, appear in European towns; Paracelsus provides a recipe for homunculus, an intelligent "little man"
1548	Swiss Conrad Gessner attempts to compile all current knowledge in 21 volumes, but dies after volume 19
1580	"The Golem," said to be created by Rabbi Judah ben Loew in Prague
Seventeenth century	Mechanical automata appear on European clocks and are also produced to work alone as amusements
1642	Blaise Pascal invents a mechanical calculator, the Pascaline
1664	*Treatise on Man*, by René Descartes, is published posthumously and codifies the mind/body problem
1673	Gottfried Wilhem Leibniz invents the Step Reckoner, an improved mechanical calculator, and envisions a universal calculus of reasoning to decide arguments mechanically
Eighteenth century	Philosophers (Leibniz, Spinoza, Hobbes, Locke, Kant, and Hume) and scientists (La Mettrie, and Hartley) try to formulate laws of thought
1728	Ephraim Chambers publishes the *Cyclopaedia* in London, a compilation of knowledge that is the first to use extensive cross-references and to publish supplements
1738	Jacques de Vaucanson presents his mechanical duck to the European public

1751 - 1772	17-volume *Encylopédie, ou Dictionnaire Raisonné des Sciences, des Arts et des Métiers,* by Denis Diderot and Denis Jean d'Alembert, published in France, which postulates that everything can be reasoned
1768	First volumes of *Encyclopedia Britannica* published in Scotland
Late 1700s	Napoleon envisions the Bibliothéque Nationale, which becomes the model for national libraries
Nineteenth century	Literary artificial intelligences proliferate, such as Hoffman's *The Sandman,* Goethe's *Faust* (part II), and Mary Shelley's *Frankenstein;* the beginning of empirical psychology
1800	US President John Adams establishes the Library of Congress (President Thomas Jefferson will enlarge it to include a wide range of books in many different languages)
1822	Charles Babbage begins but never finishes the Difference Engine
1843	Ada, Countless Lovelace, publishes her account of Charles Babbage's Analytical Engine
1844	Samuel F. B. Morse sends the message "What hath God wrought?" through his invention, the telegraph, from Baltimore to Washington, DC
1854	George Boole publishes *An Investigation of the Laws of Thought;* von Kempelen's fraudulent chess-playing machine perishes in a fire
1876	Alexander Graham Bell demonstrates the telephone at the World's Fair
1887	Montgomery Ward issues a 500-page catalog of general merchandise to be sold by mail

1890	Herman Hollerith conducts the US census using machines that encode information on punch cards
1896	G. Marconi begins developing the wireless telegraph, or radio
1906	The first edition of the Yellow Pages appears
1914	A. Torres y Quevedo builds electromechanical machines for chess endgames
1923	"Robot" introduced into English in a London production of Karel Capek's play, *R.U.R.*, *Rossum's Universal Robots*
1937	Alan Turing proposes an abstract universal computing machine
1938	Developing his Z1 computer in Berlin, Konrad Zuse recognizes that the technology will eventually become an artificial brain
1940	Automatic decryption of German intelligence messages undertaken by Turing and others at Bletchley Park, England
1941	McCulloch and Pitts publish "A Logical Calculus of the Ideas Immanent in Nervous Activity;" Rosenblueth, Wiener, and Bigelow publish "Behavior, Purpose and Teleology," introducing the term cybernetics
1944	ENIAC (Electronic Numerator, Integrator and Computer) developed by Eckert and Mauchly, comes online at the University of Pennsylvania
1945	Vannevar Bush proposes the Memex as an automatic means to store and distribute information

1946	Turing writes a pioneering, but unpublished paper, "Intelligent Machinery"
1947	Norbert Wiener publishes *Cybernetics*; Grey Walter builds his electromechanical "turtle"
1949	Mark I, the first stored-program computer, comes online at Manchester University; Turing and his colleagues attempt to program it to play chess
1950	Turing publishes "Computing Machinery and Intelligence," proposing the Turing Test; Isaac Asimov offers his "Three Rules of Robotics" in *I, Robot;* Shannon publishes a detailed analysis of chess-playing as search
1951	IAS machine, proposed by John von Neumann in 1945, comes online at the Institute for Advanced Study, Princeton
1952	Arthur Samuel begins work on a checkers-playing machine that learns and eventually competes with human champions
1956	Dartmouth Conference, where John McCarthy proposes the term "artificial intelligence" and Newell, Shaw, and Simon demonstrate the first working AI program, the Logic Theorist
1957	Newell, Shaw, and Simon demonstrate the General Problem Solver; McCarthy proposes the Advice-Taker; US President Dwight Eisenhower approves funding for the Defense Department's Advanced Research Projects Agency (ARPA)
1957	McCarthy invents LISP; H. Gelernter and N. Rochester produce a geometry theorem prover with a semantic component

1959

Jack Kilby and Robert Noyce independently apply for US patents for an integrated circuit, which leads to the technological improvements and increasing economies described by Moore's Law

1961

1960s John Kemeny develops the first time-shared system at Dartmouth College where all undergraduates are required to be "computer literate"

1961

T. Evans's ANALOGY program solves the same analogy problems that appear on IQ tests

1962

J. C. R. Licklider envisions in a series of memos, what will eventually become the Internet, a worldwide medium for collaboration, information dissemination, broadcasting and interaction between individuals, regardless of geographic location. DARPA begins and sustains its support, and a number of scientists, especially L. Kleinrock, L. Roberts, P. Baran, R. Kahn, and V. Cerf, make crucial breakthroughs in the next few years. J. Slagle's SAINT program solves calculus problems at the college freshman level

1964

D. Bobrow's STUDENT program understands natural language well enough to solve algebra word problems

1965

Lotfi Zadeh invents fuzzy logic; Ted Nelson begins but never completes his Xanadu hypertext system; publishes his first papers about hypertext; D. Engelbart develops the computer mouse as a way of implementing his NLS (oN Line System) hypertext and collaborative workspace; J. Weizenbaum's ELIZA interactively mimics a psychotherapist

1966

R. Quillian's PhD dissertation demonstrates the power of semantic nets

1967	*A scientific turning point in AI, where knowledge is seen to be as important as reasoning in intelligent behavior.* DENDRAL, the first successful knowledge-based program for scientific reasoning; MACSYMA, the first successful knowledge-based program in mathematics; MacHack, a knowledge-based chess-playing program, achieves a class C rating in tournament play; first version of LOGO, an interactive learning environment, appears
1968	Intelsat, first of a series of new communications satellites, is launched
1969	ARPANET, the precursor to the Internet, is established; Shakey, a mobile "intelligent" robot, roams SRI's halls
1970	M. Hart begins Project Gutenberg to distribute free electronic versions of texts in the public domain
1971	H. Cohen first demonstrates AARON, an autonomous art-making program; T. Hoff invents the microprocessor
1974	First expert system, T. Shortliffe's MYCIN program demonstrates the power of rule-based systems for knowledge representation and inference in medical diagnosis and therapy; first personal computer, the Altair, goes on sale in kit form for $400
1975	Minsky proposes frames as a representation to integrate different sources of knowledge; MetaDendral produces the first scientific discoveries by a computer to be published in a refereed journal; D. Farmer and friends deploy the first wearable computer (big-toe operated inside their shoes) to beat roulette dealers in Las Vegas

Late 1970s — Stanford's SUMEX-AIM Lab demonstrates the power of the ARPANET for scientific collaboration

1978 — Simon wins the Nobel Prize in Economics for his theory of bounded rationality, a keystone of AI (and human behavior) known as "satisficing;" Moravec's cart is the first computer-controlled autonomous vehicle

1981 — Commercialization of AI begins; the Japanese announce the Fifth Generation project with significant AI goals

1982 — Newell et al., create SOAR, an architecture for general intelligence; US embarks on the Strategic Computing Project to achieve AI goals

1985 — R. Brooks demonstrates "Allen," the first of his autonomous reactive robots, to be followed by an explosion of this species

1987 — ARPANET becomes a civilian entity, NSFNET; Minsky publishes *Society of Mind*

1988 — Berners-Lee begins work on the World Wide Web at CERN in Geneva

Late 1980s — The AI Winter

Early 1990s — *Another turning point in AI: intelligent behavior is recognized to be collaborative as well as single-agent*

1990 — Human genome project begins

1993 — M. Andreessen at the National Center for Supercomputer Applications (University of Illinois) releases the Mosaic web browser

1997	Deep Blue defeats Garry Kasparov, the world's chess champion, ending the single-agent, single-task model of intelligence as a significant AI goal; first official Robo-Cup soccer match, the new paradigm
1998	Open Directory Project (ODP) begins, intended to edit and catalog the World Wide Web
2000	Robot pets, smart toys, become commercially available; C. Breazeal creates Kismet, a robot that exhibits emotions
2001	Working draft of the Human Genome Project is published in *Science;* Berners Lee et al., begin work on the Semantic Web, an international effort to bring about the global exchange of commercial, scientific, and cultural data on the World Wide Web using logic, inference, action, and offering answers to questions; the Wikipedia, an online, multilingual, free content encyclopedia is begun, to which anyone can contribute or edit
2002	Bibliotheca Alexandrina opens in Alexandria, Egypt, aiming eventually to be a digital equivalent of the original Alexandria Library
2003	DARPA initiates three major AI projects: the "LifeLog," new reasoning cognitive systems, and new real-world reasoning systems; his and hers multifunction robots are offered in the NeimanMarcus Christmas catalog for $400,000 (by coincidence, the same sum that John von Neumann requested in 1945 to build his IAS machine)

Bibliography

Amarel, S.; Brown, J. S.; Buchanan, B.; Hart, P.; Kulikowski, C.; Martin, W.; and Pople, H. 1977. "Report of a Panel on Applications of Artificial Intelligence." *Proceedings of the Fifth International Joint Conference on Artificial Intelligence.*

Anthes, Gary H. 2002. "Computerizing Common Sense." *Computer World,* April 8. Full text available from the World Wide Web (http://www.cyc.com/cyc/company/news/computerizing).

Appleman, P. 1970. *Darwin: Texts, Backgrounds, Contemporary Opinion, Critical Essays.* New York: Norton.

Ashby, W. R. 1952. Design for a Brain. New York: Wiley. Also appeared as "Design for a Brain," *Electronic Engineering,* 20 (1948), 379-383. These references are from the corrected 1954 edition of the book.

Asimov, I. 1950. *I, Robot.* New York: Gnome Press.

Balch, T. and Yanco, H. A. 2002. In "Ten Years of the AAAI Mobile Robot Competition and Exhibition." *AI Magazine,* Spring.

Bar-Hillel, Yehoshua. 1964. *Language and Information.* Reading, Mass.: Addison-Wesley.

Bernal, J. D. 1974. *Science in History.* Vol. 2. Cambridge, Mass: MIT Press.

Berners-Lee, T., Hendler, J., and Lassila, O. 2001. "The Semantic Web." *Scientific American,* May.

Block, C. 1925. *The Golem: Legends of the Ghetto of Prague.* New York: Rudolf Steiner Press.

Bobrow, D. G. 1991. "Dimensions of Interaction," *AI Magazine,* Fall.

Boden, M. 1977. *Artificial Intelligence and Natural Man.* New York: Basic Books.

Boring, E.G. 1946. "Mind and Mechanism." *American Journal of Psychology,* 2, April.

535

Boswell, James. 1968. *The Life of Samuel Johnson*, Frank Brady, ed. New York: New American Library.

Bowden, B. V. 1953. *Faster than Thought*. London: Sir Isaac Pitman.

Brooks, F. P. 1996. *The Mythical Man-Month*. Reading, Mass.: Addison-Wesley (anniversary edition).

Brooks, R. R. 2002. *Flesh and Machines*. New York: Pantheon Books.

Brooks, R. A. and Flynn, A. M. 1989. "Fast, Cheap, and Out of Control: A Robot Invasion of the Solar System." *Journal of the British Interplanetary Society, 42(10)*.

Brown, J. S. and Duguid, P. 2001. "A Response to Bill Joy and the Doom-and Gloom Technofuturists." In *AAAS Science and Technology Policy Yearbook*, edited by A. H. Teich, S.D. Nelson, C. McEnaney, and S.J. Lita, pp. 77-83. American Association for the Advancement of Science. Available from theWorld Wide Web (http:/www.aaas.org/spp/rd/ch4.pdf).

Bruner, J. 2002. "The Narrative Construction of Reality." In *Narrative Intelligence*, Mateas, M. and Sengers, P. eds. Amsterdam and Philadelphia: John Benjamins Publishing Company.

Cadwallader-Cohen, J.; Zysiczk, W. W.; and Donelly, R. R. 1961. "The Chaostron: An Important Advance in Learning Machines," *Journal of Irreproducible Results*, 10, 30.

Calvin, W. H. 2000. "Brains and the World of 2025." Potomac Institute for Policy Studies Conference, Washington D.C., June 27.

Chapuis, A., and Droz, E. 1949. *Les Automates*. Neuchatel: Editions du Griffon.

Chessbase News. 2003. Available from World Wide Web. (http://www.chessbase.com/newsdetail.asp?newsid=782), February 8.

Cohen, J. 1966. *Human Robots in Myth and Science*. London: Allen and Unwin.

Cohen, H. 2003a. Personal communication.

Cohen, H. 2003b. Personal communication.

Colby, K. M. 1976. "On the Morality of Computers Providing Psychotherapy." Memo Al HMF-6, Department of Psychiatry, University of California at Los Angeles.

Davis, M. and Travers, M. 2002. "A Brief Overview of the Narrative Intelligence Reading Group." *Narrative Intelligence*, M. Mateas and P. Sengers, eds. Amsterdam and Philadelphia: John Benjamins Publishing Company.

Davis, R. 1998. "What Are Intelligence? And Why?" *AI Magazine*, Spring.

Davis, R. and Buchanan, B. G. 1977. "Meta-Level Knowledge: Overview and Applications." *Proceedings of the Fifth International Joint Conference on Artificial Intelligence*. Menlo Park, Calif.: International Joint Conferences on Artificial Intelligence.

Dinneen, G. P. 1955. "Programming Pattern Recognition." *Proceedings of the 1955 Western Joint Computer Conference*, IRE, March.

Dreifus, C. 2003. "A Passion to Build a Better Robot, One with Social Skills and a Smile." *New York Times*, June 10.

Dreyfus, H. 1972. *What Computers Can't Do; A Critique of Artificial Reason*. New York: Harper & Row.

Dyson, F. 2003. The Future Needs Us!" *The New York Review of Books*, vol. 50, no. 2, February 13.

(unsigned editorial; leader). 2003. "Computers and Chess, Not So Smart." *The Economist*, February 1.

Evans, T. G. 1968. "A program for the solution of geometric-analogy intelligence test questions." In *Semantic Information Processing*, Minsky, M. ed., Cambridge, Mass.: MIT Press.

Evert, W. 1974. "Frankenstein." Four talks delivered on WQED-FM, Pittsburgh, Penn.

Feigenbaum, E. A. 1977. "The Art of Artificial Intelligence: Themes and Case Studies of Knowledge Engineering." *Proceedings of the Fifth International Joint Conference on Artificial Intelligence*.

Feigenbaum, E. A. 2003. "Some Challenges and Grand Challenges for Computational Intelligence. *Journal of the ACM*, January.

Feigenbaum, E. A.; Buchanan, B. G.; and Lederberg, J. 1971. "On Generality and Problem Solving: A Case Study Using the DENDRAL Program." In *Machine Intelligence*, 6. Edinburgh: Edinburgh University Press.

Feigenbaum, E. A., and Feldman, J. 1963. *Computers and Thought*. New York: McGraw-Hill.

Feigenbaum, E. and McCorduck, P. 1983. *The Fifth Generation: Artificial Intelligence and Japan's Computer Challenge to the World*. Reading, Mass.: Addison-Wesley.

Feigenbaum, E., McCorduck, P. and Nii, H. P. 1988. *The Rise of the Expert Company*. New York: Times Books.

Feynman, R. 1998. *The Meaning of it All: Thoughts of a Citizen-Activist*. Reading, Mass.: Perseus Books.

Fong, T., Nourbakhsh, I., and Dautenhahn, K. 2003. "A Survey of Socially Interactive Robots." *Robotics and Autonomous Systems, 42.*

Gardner, H. 1983. *Frames of Mind: The Theory of Multiple Intelligences.* New York: Basic Books.

Gardner, M. 1958. *Logic Machines and Diagrams.* New York: McGraw-Hill.

Garreau, J. *Our Next Humans: The Future of Human Nature.* To appear.

Goldstein, I., and Papert, S. 1976. "Artificial Intelligence, Language, and the Study of Knowledge." AI Memo 337, MIT AI Laboratory.

Goldstine, H. 1972. *The Computer from Pascal to von Neumann.* Princeton, N. J.: Princeton University Press.

Gray, J. 2003. "What Next? A Dozen Information-Technology Research Goals." *Journal of the ACM,* January.

Greenberger, M. 1962. *Management and the Computer of the Future.* New York: MIT Press and Wiley.

Grosz, B. J. 1996. "Collaborative Systems." *AI Magazine,* Summer.

Hiller, L. A., and Isaacson, L. M. 1959. *Experimental Music.* New York: McGraw-Hill.

Hillis, W. D. 1998. *The Pattern on the Stone.* New York: Basic Books.

Hofstadter, D. R. and Dennett, D. C. 1981. *The Mind's Eye: Fantasies and Reflections on Self and Soul.* New York: Basic Books.

Hunter, L. 2003. "AI & Molecular Biology: A Growing Success Story." The Robert S. Engelmore Memorial Lecture, IJCAI, Acapulco, Mexico, August 14.

Jonas. G. 1976. "A Reporter at Large (Stuttering)." *The New Yorker, 52,* November 15.

Joy, B. 2000. "Why the Future Doesn't Need Us." *Wired,* April.

Kasparov, G. 2003. "Man vs. Machine: Deep Junior Makes the Fight Worth It." *The Wall Street Journal,* February 11.

Koberg, D., and Bagnall, J. 1972. Revised 1973, 1974. *The Universal Traveler.* Los Altos, Calif.: William Kaufmann.

Kurzweil, R. 1999. *The Age of Spiritual Machines.* New York: Viking Penguin.

Kurzweil, R. 2001. "Are We Becoming an Endangered Species? Technology and Ethics in the 21st Century." A panel discussion at the Washington National Cathedral, Available from the World Wide Web (http://www.kurzweilai.net/articles/art0358.html).

Labat, R. 1963. *History of Science.* Vol. 1. New York: Basic Books.

Laird, J. E. and P. S. Rosenbloom, 1992. "Allen Newell, In Pursuit of Mind…" *AI Magazine*, Winter.

Lampson, B. 2003. "Getting Computers to Understand." *Journal of the ACM*, January.

Lattimore, R. 1951. *Homer's Iliad*. Chicago: University of Chicago Press.

Lederberg, J. 1976. "Review of J. Weizenbaum's Computer Power and Human Reason." In *Three Reviews of J. Weizenbaum's Computer Power and Human Reason*. Memo AIM-291, Stanford AI Laboratory, November.

Lesser, V. 2003. Private communication.

Lettvin, J. Y.; Maturana, H.; McCulloch, W. S.; and Pitts, W. 1959. "What the Frog's Eye Tells the Frog's Brain." *Proceedings of the IRE*, 47, November.

Lewis, C. I., and Langford, C. H. 1956. "History of Symbolic Logic." In *The World of Mathematics*, J. R. Newman, ed. New York: Simon and Schuster.

MacKay, D. 1951. "Mindlike Behavior in Artefacts." *British Journal for the Philosophy of Science*, 2, 6.

MacKay, D. 1955. "The Epistemological Problem for Automata." In Automata Studies, *Annals of Mathematical Studies*, 34, C. Shannon and J. McCarthy, eds. Princeton, N. I.: Princeton University Press.

MacNiece, L. 1951. *Goethe's Faust*. New York: Oxford University Press.

Marshall, J. 1889. *The Life and Letters of Mary Wollstonecraft Shelley*. 2 vols. London: Richard Bentley and Son.

McCarthy, J. 1958, 1968. "Programs with Common Sense." Presented at the Teddington Conference on the Mechanization of Thought Processes, December 1958; reprinted in *Semantic Information Processing*, ed. M. Minsky, Cambridge, Mass.: The MIT.Press, 1968.

McCarthy, J. 1972. "Mechanical Servants for Mankind." *1973 Britannica Yearbook of Science and the Future, Encyclopedia Britannica*. Chicago: William Benton.

McCarthy, J. 1976. "An Unreasonable Book." In *Three Reviews of J. Weizenbaum's Computer Power and Human Reason*. Memo AIM-291, Stanford AI Laboratory, November.

McCarthy, J. 1990. "Review of *The Emperor's New Mind* by Roger Penrose," *Bulletin of the American Mathematical Society*, 23(2) October.

McCarthy, J. 2003. "Problems and Projections in CS for the Next 49 Years." *Journal of the ACM*, January.

McCarthy, J., and Hayes, P. J. 1969. "Some Philosophical Problems from the Standpoint of Artificial Intelligence." In *Machine Intelligence*, 4, D. Michie, ed. New York: Elsevier.

McCorduck, P. 1979. *Machines Who Think*. San Francisco: W. H. Freeman. "Language, Scenes, Symbols and Understanding."

McCorduck, P. 1985. *The Universal Machine*. New York: McGraw-Hill.

McCorduck, P. 1990. *AARON's Code*. New York: W. H. Freeman.

McCracken, D. 1976. "Review, *Computer Power and Human Reason* by Joseph Weizenbaum." *Datamation*, April.

McCulloch, W. 1965. *Embodiments of Mind*. Cambridge, Mass.: MIT Press.

Medawar, P. B. 1977. "Unnatural Science." *New York Review of Books XXIV*, 1, February.

Meltzer, B. 1971. "Personal View: Bury the Old War-Horse! " *ACM SIGART Newsletter*, 27, April.

Minsky, M. 1956. "Heuristic Aspects of the Artificial Intelligence Problem." Group Report 34-55, ASTIA Document AD 236885 (MIT Hayden Library H-58), Lincoln Laboratories, MIT, Lexington, Mass.

Minsky, M. 1959. "Some Methods of Heuristic Programming and Artificial Intelligence." In *Proceedings of the Symposium on Mechanisation of Thought Processes*, D. V. Blake and A. M. Uttley, eds. 2 vols. London: H. M. Stationery Office.

Minsky, M. 1963. "Steps Toward Artificial Intelligence." In *Computers and Thought*, E. A. Feigenbaum and J. Feldman, eds. New York: McGraw-Hill.

Minsky, M. 1968. *Semantic Information Processing*. Cambridge, Mass.: MIT Press.

Minsky, M. L. 1985. *A Society of Mind*. New York: Simon and Schuster.

Minsky, M. 2003. "Marvin Minsky." Available from World Wide Web (http://web.media.mit.edu/~minsky).

Minsky. M., and Papert, S. 1968. *Perceptrons*. Cambridge, Mass.: MIT Press.

Minsky, M., and Papert, S. 1971. *Artificial Intelligence*, Project MAC Report.

Mitroff, I. 1974. *The Subjective Side of Science: A Philosophical Inquiry into the Psychology of the Apollo Moon Scientists*. New York: EIsevier.

Moravec, H. 1977. "Intelligent Machines: How to Get There From Here and What to Do Afterwards." Unpublished.

Moravec, H. 1988. *Mind Children: The Future of Robot and Human Intelligence*. Cambridge, Mass.: Harvard University Press.

Moravec, H. 1998. *Robot: Mere Machine to Transcendent Mind*. New York. Oxford University Press.

Moravec, H. 2000. "Robots, Re-Evolving Mind" Available from World Wide Web. (http://www.frc.ri.cmu.edu/~hpm/project/archive/robot.papers/2000/Cerebrum.html).

Morgenstern, O., and von Neumann, J. 1944. *Theory of Games and Economic Behavior*. Princeton, N.J.: Princeton University Press.

Morrison, P., and Morrison, E. 1961. *Charles Babbage and His Calculating Engines*. New York: Dover.

Moses, J. 1971. "Symbolic Integration—The Stormy Decade." *Communications of the ACM*, 14, August.

Murphy, G. 1972. *Historical Introduction to Modern Psychology*. New York: Harcourt Brace Jovanovich.

Newell, A. l973a. "Production Systems: Models of Control Structures." In *Visual Information Processing*, W. C. Chase, ed. New York: Academic Press.

Newell, A. 1973b. "Artificial Intelligence and the Concept of Mind." In *Computer Models of Thought and Language*, R. Schank and K. M. Colby, eds. San Francisco: W. H. Freeman and Company.

Newell, A. ed. 1973c. *Speech Understanding Systems: Final Report of a Study Group*. New York: Elsevier.

Newell, A. 1981. "The Knowledge Level." *AI Magazine*, Summer.

Newell, A. 1990. *Unified Theories of Cognition*. Cambridge, Mass.: Harvard University Press. Cited in Laird and Rosenbloom.

Newell, A. 1991. "Desires and Diversions." Distinguished Lecture, School of Computer Science, Carnegie Mellon University, December 4.

Newell, A.; Shaw, J. C.; and Simon, H. A. 1958. "Chess Playing Programs and the Problem of Complexity." *IBM Journal of Research and Development, 2*, 4. Reprinted in Feigenbaum and Feldman, 1963.

Newell, A., and Simon, H. A. 1956. "The Logic Theory Machine." *IRE Transactions on Information Theory*, September. Reprinted in Feigenbaum and Feldman, 1963.

Newell, A., and Simon, H. A. 1961. "Computer Simulation of Human Thinking." *Science*, 134, December.

Newell, A., and Simon, H. A. 1973. *Human Problem Solving*. Englewood Cliffs, N.J.: Prentice-Hall.

Newell, A., and Simon, H. A. 1976. "Computer Science as Empirical Inquiry: Symbols and Search." *Communications of the ACM*, 3, March.

Nilsson, N. J. 1974. "Artificial Intelligence." *Proceedings of the 1974 IFIP Congress*, Stockholm, Sweden. Also published as SRI Artificial Intelligence Center Technical Note 89.

Nilsson, N. J. 1983. "AI Prepares for 2001." *AI Magazine*, Winter.

Niwa, K., Sasaki, K., and Ihara, H. 1984. "An Experimental Comparison of Knowledge Representation Schemes." *AI Magazine*, Summer.

Papert, S. 1968. "The Artificial Intelligence of Hubert L. Dreyfus: A Budget of Fallacies." AI Memo 154, MIT AI Laboratory.

Papert, S. 1971. "Teaching Children Thinking." AI Memo 247, MIT AI Laboratory.

Papert, S. 1976. "Some Poetic and Social Criteria for Education Design." AI Memo 373, MIT AI Laboratory.

Penrose, R. 1989. *The Emperor's New Mind*. Oxford and New York: Oxford University Press.

Pentland, A. P. and Fischler, M. A. 1983. "A More Rational View of Logic, or Up Against the Wall, Logic Imperialists." *AI Magazine,* Winter.

Plank, R. 1965. "The Golem and the Robot." *Literature and Psychology*, 15, Winter.

Poindexter, J. M. 2003. "Finding the Face of Terror in Data." *New York Times* Op-Ed, September 10.

Pool, I., ed. 1977. *The Social Impact of the Telephone*. Cambridge, Mass.: MIT Press.

Pople, H. 1977. "The Formation of Composite Hypotheses in Diagnostic Problem Solving: An Exercise in Synthetic Reasoning." *Proceedings of the Fifth International Joint Conference on Artificial Intelligence.*

Price, D. J. 1963. *Little Science, Big Science*. New York: Columbia University Press.

Randell, B. 1973. *The Origins of Digital Computers: Selected Papers*. New York and Berlin: Springer-Verlag.

Raphael, B. 1970. "The Relevance of Robot Research to Artificial Intelligence." In R. B. Banerji and M. D. Mesarovic, *Theoretical Approaches to Non-Numerical Problem Solving*. Berlin and New York: Springer-Verlag.

Raphael, B. 1976. *The Thinking Computer: Mind Inside Matter*. San Francisco: W. H. Freeman and Company.

Reddy, R. 1988. "Foundations and Grand Challenges of Artificial Intelligence, *AI Magazine,* Winter.

Reid, C. 1970. *Hilbert*. New York: Springer-Verlag.

Roland, A. with Shipman, P. 2002. *Strategic Computing: DARPA and the Quest for Machine Intelligence, 1983-1993.* Cambridge, Mass.: MIT Press.

Rosen, S. 1959. *Doctor Paracelsus.* Boston: Little, Brown.

Sagan, C. 1973. *The Cosmic Connection.* New York: Doubleday.

Sagan, C. 1977. *The Dragons of Eden.* New York: Random House.

Searle, John. 1980, 1981. "Minds, Brains and Programs." Originally printed in *The Behavioral and Brain Sciences,* vol. 3, 1980. Cambridge, UK: Cambridge University Press, 1980, reprinted in D. R. Hofstadter and D. C. Dennett, *The Mind's Eye: Fantasies and Reflections on Self and Soul.* New York: Basic Books, 1981.

Selfridge, O. G. 1955. "Pattern Recognition and Modern Computers." *Proceedings of the 1955 Western Joint Computer Conference,* IRE, March.

Selfridge, O. G. 1959. "Pandemonium, a Paradigm for Learning." In *Proceedings of the Symposium on Mechanisation of Thought Processes.* 2 vols. D. V, Blake and A. M. Uttley, eds. London: H. M. Stationery Office.

Shannon, C. 1950. "A Chess-Playing Machine." *Scientific American,* 182, February.

Shannon, C. 1953. "Computers and Automata." *Proceedings of the IRE.*

Shannon, C., and McCarthy, J., eds. 1956. *Automata Studies. Annals of Mathematical Studies,* 34. Princeton, N.J.: Princeton University Press.

Simmons, R. F. 1965. Answering English questions by computer: a survey. *Communications of ACM,* 8, January.

Simmons, R. F. 1970. Natural language question answering systems: 1969. *Communications of ACM,* 13, January.

Simon, H. A. 1947. *Administrative Behavior.* New York; Macmillan.

Simon, H. A. 1967. "An Information-Processing Explanation of Some Perceptual Phenomena." *British Journal of Psychology,* 58, May.

Simon, H. A. 1969. *Sciences of the Artificial.* Cambridge, Mass.: MIT Press.

Simon, H. A. 1971. "Cognitive Control of Perceptual Processes." In G. V. Coelho and E. Rubinstein, eds., *Social Change and Human Behavior.* Rockville, Md.: National Institutes of Health.

Simon, H. A. 1972. "What Is Visual Imagery? An Information Processing Interpretation." In *Cognition in Learning and Memory,* L. W. Gregg, ed. New York: Wiley.

Simon, H. A. 1976. *Administrative Behavior.* 3rd ed. New York: The Free Press.

Simon, H. A. 1977. "Artificial Intelligence Systems that Understand." *Proceedings of the Fifth International Joint Conference on Artificial Intelligence.*

Simon, H. A. 1991. *Models of My Life.* New York: Basic Books.

Simon, H. A., and Barenfeld, M. 1969. "Information Processing Analysis of Perceptual Processes in Problem Solving." *Psychological Review*, 76, September.

Simon, H. A., and Siklóssy, L. 1972. *Representation and Meaning.* Englewood Cliffs, N.J.: Prentice-Hall.

Strachey, Lytton. 1948. *Literary Essays.* London: Chatto and Windus.

Sussman, G. J., and Stallman, R. M. 1975. "Heuristic Techniques in Computer Aided Circuit Analysis." *IEEE Transactions on Circuits and Systems CAS-22*, No. 11.

Sutherland, N. S. 1976. "The Electronic Oracle." *Times Literary Supplement*, July 30.

Taube, M. 1961. *Computers and Common Sense: The Myth of Thinking Machines.* New York: Columbia University Press.

Taylor, R. H. 1976. "Letter." *Datamation*, June.

Thomas, L. 1974. *The Lives of a Cell.* New York: Viking Press.

Turing, A. 1950. "Computing Machinery and Intelligence." In *Computers and Thought*, E. A. Feigenbaum and J. Feldman, eds. New York: McGraw-Hill, 1963.

Turing, A. 1969. "Intelligent Machinery." In *Machine Intelligence*, 5, B. Meltzer and D. Michie, eds. Edinburgh: Edinburgh University Press.

Turing, S. 1959. *Alan M. Turing.* Cambridge: W. Heifer and Sons.

Turney, C. 1972. *Byron's Daughter: A Biography of Elizabeth Medora Leigh.* New York: Charles Scribner's Sons.

Vartanian, A. 1960. *La Mettrie's L'Homme Machine.* Princeton, N.J.: Princeton University Press.

van Boehn, M. Undated. *Puppets and Automata.* New York: Dover Books.

Veloso, M. 2003. "Autonomous Robot Soccer Teams." *The Bridge*, The National Academy of Engineering, Spring.

von Neumann, J. 1951. "The General and Logical Theory of Automata." In *Cerebral Mechanisms in Behavior*, L. A. Jeffress, ed. New York: Wiley. Also reprinted in *Perspectives on the Computer Revolution*, Z. W. Pylyshyn, ed. New York: Prentice-Hall, 1970.

von Neumann, J. 1958. *The Computer and the Brain*. New Haven, Conn: Yale University Press.

Waltz, D. L. 1972. "Generating Semantic Descriptions from Drawings of Scenes with Shadows." AI Memo TR 271, MIT AI Laboratory.

Weizenbaum, J. 1972. "On the Impact of Computers on Society." *Science*, 176, May.

Weizenbaum, J. 1976. *Computer Power and Human Reason*. San Francisco: W. H. Freeman and Company.

Wiener, N. 1961. *Cybernetics*. 2nd ed. Cambridge, Mass.: MIT Press.

Winograd, T. 1973. "A Procedural Model of Language Understanding." In *Computer Models of Thought and Language*. R. Schank and K. M. Colby, eds. San Francisco: W. H. Freeman and Company.

Winston, P. H. 1975. *The Psychology of Computer Vision*. New York: McGraw-Hill.

Winston, P. H. 1977. *Artificial Intelligence*. Reading, Mass.: Addison-Wesley.

Winterbotham, F. W. 1974. *The Ultra Secret*. New York: Harper & Row.

Index

Machines Who Think

Designed by
Darren Wotherspoon
Composed in Adobe Garamond
with display lines in Souvenir LT

Printed and Manufactured by
Friesens, Altona, Manitoba

Cover Design by
Darren Wotherspoon

Pamela McCorduck
(© Jill Fineberg, 2004)

Pamela McCorduck is the author or coauthor of eight published books, two of them novels. Among her books are *Machines Who Think*, a history of artificial intelligence; *The Universal Machine*, a study of the worldwide impact of the computer; and *Aaron's Code*, an inquiry into the future of art and artificial intelligence. Her work has been translated into all the major European and Asian languages. A recent book, coauthored with Nancy Ramsey, is *The Futures of Women* (Addison-Wesley, 1996; Time-Warner paperback, 1997), containing four scenarios for women worldwide in the year 2015.

Ms. McCorduck has been an active member of PEN American Center, the author organization in New York City, serving on its executive board and as vice president for several years. In addition, she founded and chaired PEN's Readers and Writers Program, which sends authors and their books to newly literate adults all over the country. She has been a board member and treasurer of the New Mexico Committee of the National Museum of Women in the Arts, and currently serves on its advisory committee. Ms. McCorduck also works as a consultant, constructing future scenarios for firms in the transportation, financial, and high-tech sectors.

Educated at UC Berkeley and Columbia University, Ms. McCorduck divides her time between New York City and Santa Fe, and loves the visual and performing arts in both cities.